ANALYTICAL MICROBIOLOGY METHODS
METHODS
Chromatography and
Mass Spectrometry

ANALYTICAL MICROBIOLOGY METHODS
Chromatography and Mass Spectrometry

Edited by
Alvin Fox and Stephen L. Morgan
University of South Carolina
Columbia, South Carolina

Lennart Larsson and Göran Odham
University of Lund
Lund, Sweden

PLENUM PRESS • NEW YORK AND LONDON

Library of Congress Cataloging in Publication Data

Analytical microbiology methods: chromatography and mass spectrometry /
edited by Alvin Fox . . . [et al.].
    p.   cm.
   "Based on the proceedings of the First International Symposium on the Inter-
face between Microbiology and Analytical Chemistry, held June 3–7, 1987, at the
University of South Carolina, Columbia, South Carolina" – T.p. verso.
   Includes bibliographical references.
   ISBN 0-306-43536-5
   1. Chromatography – Congresses. 2. Mass spectrometry – Congresses. 3.
Diagnostic bacteriology – Technique – Congresses. I. Fox, Alvin. II. International
Symposium on the Interface between Microbiology and Analytical Chemistry
(1st:1987:University of South Carolina)
QR69.G27A53   1990   90-34252
576 – dc20   CIP

Based on the proceedings of the First International Symposium on the Interface between
Microbiology and Analytical Chemistry, held June 3 – 7, 1987, at the University of South Carolina,
Columbia, South Carolina

© 1990 Plenum Press, New York
A Division of Plenum Publishing Corporation
233 Spring Street, New York, N.Y. 10013

Printed in the United States of America

PREFACE

The First International Symposium on the Interface between Analytical Chemistry and Microbiology: Applications of Chromatography and Mass Spectrometry was held June 1987 at the University of South Carolina, Columbia, SC, U.S.A. The purpose of the "Interface" meeting was to forge connections between analytical chemists and microbiologists that are using chromatography and mass spectrometry to solve common problems. The goals were admirably fulfilled. Nearly a hundred participants from seven European countries, Japan, and the United States participated in hearing twenty-three plenary talks and thirty-six submitted papers and posters. The papers and discussions displayed the breadth and depth of current research applications and revealed future directions.

This book "Analytical Microbiology Methods: Chromatography and Mass Spectrometry" is loosely based on some of the presentations and discussions at the meeting. Each chapter describes specific methodology and applications in the context of the relevant scientific background. The present book continues the theme of an earlier book, "Gas Chromatography/Mass Spectrometry Applications in Microbiology", edited by G. Odham, L. Larsson, and P-A. Mardh, published by Plenum Press in 1984.

Microbial chemistry has been extremely important in identifying many major structural components of microorganisms (e.g., the teichoic acids, lipopolysaccharides, and peptidoglycans). Structural information has helped provide an understanding of the functional role of many of these molecules in the physiology of the bacterial cell. The uniqueness of many of the monomers present in these macromolecules as well as their natural variability between bacterial species has also provided a framework for chemotaxonomy. The advent of modern analytical chemical methods based on chromatography and mass spectrometry has clarified and expanded on previously known chemical data. Often, this information was not obtainable by more traditional approaches.

The advances in analytical microbiology described in this book have the potential to stimulate a revolution of improved methods for automated and rapid identification of microorganisms, characterization of microbial products and constituents, and trace detection of microbial chemicals. We see the beginnings of a new discipline, "Analytical Microbiology", arising at the interface between analytical chemistry and microbiology.

Alvin Fox and Stephen L. Morgan
Columbia, SC

Lennart Larsson and Goran Odham
Lund, Sweden

December, 1989

CONTENTS

CHAPTER 1

ANALYTICAL MICROBIOLOGY: A PERSPECTIVE

Alvin Fox and James Gilbart

Department of Microbiology & Immunology
School of Medicine
University of South Carolina
Columbia, SC 29208

Stephen L. Morgan

Department of Chemistry
University of South Carolina
Columbia, SC 29208

INTRODUCTION

The term "Analytical Microbiology" describes the application of analytical chemistry to identification, structure elucidation, systematics, diagnosis, and detection in microbiology. The most widely applied instrumental chemical techniques have been the various forms of gas chromatography (GC) and mass spectrometry (MS). These and other analytical techniques (e.g., high performance liquid chromatography, HPLC) and the "hyphenated" methods (including GC-MS, HPLC-MS and MS-MS) can be used to detect trace levels and to identify monomers, oligomers, or polymers derived from microorganisms. Chemical information from these analyses can be of great benefit in medicine, ecology, biotechnology, as well as in the food and pharmaceutical industries. Classical microbiological or biochemical procedures may not be sufficiently sensitive, selective, quantitative, or rapid in some instances. Instrumental approaches, hitherto considered the realm of the chemist, have found application in microbiological analysis. We perceive four major areas in analytical microbiology where chromatography and mass spectrometry are having impact.

First, the monomeric composition of bacteria or secretion of unique metabolites may be used to identify and/or differentiate bacterial species.[1,2] Chromatographic and mass spectrometric methods produce a profile of a selected group of chemical components from the organism. In some cases, more traditional biochemical tests are either inadequate and/or the identification requires secondary growth, which for slow-growing fastidious organisms can be time consuming. As simple and reproducible chemotaxonomic profiling methods are developed they can be applied to other more readily identified organisms. These methods may replace some of the hierarchical identification schemes involving batteries of conventional tests.

Analytical Microbiology Methods
Edited by A. Fox *et al.*
Plenum Press, New York, 1990

1

Second, chromatographic methods (GC or HPLC) for determination of monomeric composition (e.g., fatty acids, carbohydrates, or amino acids), once successfully applied to whole bacteria, can be readily adapted for the analysis of isolated structural polymers such as glycoproteins, glycopeptides, lipopolysaccharides, or cell walls.[3-5] There are many unusual monomers present in bacteria. Traditional chromatographic approaches such as thin layer chromatography (TLC) and paper chromatography are mostly qualitative and often exhibit poor sensitivity and resolution. Spectroscopic methods, usually employed to identify classes of compounds, rarely distinguish between closely related isomers present in a mixture of bacterial constituents and interferences are common. For example, the Elson-Morgan reaction can not differentiate hexosamines such as glucosamine and galactosamine[6]; the Diesche-Shettles method can not distinguish methylpentoses[7] (e.g., fucose and rhamnose); color reactions for ketodeoxyoctonic acid suffer from interferences from other deoxysugars and sialic acid.[8] The use of such classical biochemical methods has provided a basis for understanding cell wall macromolecular composition. The monomeric composition of certain cell wall structures remains unresolved. For example, it is still controversial whether muramic acid (and peptidoglycan) is present within chlamydia.[9,10] Newer analytical approaches have begun to answer some of these questions.

A third application for chromatography and mass spectrometry in analytical microbiology is the trace detection of chemical markers for microbes. GC-MS is the analytical method of choice and can detect bacterial contaminants in complex samples, either without prior culture or with minimal sample handling. Environmental microbiology concerns the nature and distribution of microbial species in the biosphere (air, soil, ground water, and the oceans). Bacteria can be used to process hazardous waste and to clean-up ground water and are also implicated in many degradative processes such as corrosion of metal surfaces and structures.[11,12] GC-MS has considerable potential in biotechnology and molecular biology, e.g., in the rapid measurement of bacterial contamination of pharmacological and medical products.[13] Clinical diagnosis of infectious diseases can be performed by monitoring of characteristic chemical changes or by detection of specific marker compounds in body fluids, both approaches obviating the need to isolate the pathogen.

A fourth GC-MS application is determination of the linkage and sequence of monomers present in microbial oligomers (including carbohydrates, lipids, glycolipids, and peptides). These analyses are often necessary to elucidate relationships between chemical structure and biological function. As discussed in Chapter 10, determining linkage and sequence of intact oligomers involves several steps: isolation by HPLC or TLC, followed by high resolution MS analysis, typically using soft ionization methods (laser desorption and fast atom bombardment). Until recently it has been unclear whether hydroxy fatty acids present in lipid A (the biologically active component of Gram negative bacterial endotoxin) were randomly attached to amino and hydroxyl groups on the glucosamine disaccharide backbone. The structure of lipid A from several different bacterial species has been shown to have specific linkage sites for individual fatty acids.[14] Synthetic analogs of lipid A have been prepared that are useful in immunomodulation (e.g., in non-specifically stimulating the immune system to deal with cancers). MS structural studies of peptidoglycan oligomers may also prove clinically relevant in developing new generations of immunomodulators of known chemical structure.[15]

Some forms of automated and computer-controlled analytical instrumentation have been successfully and commonly applied in microbiology, particularly HPLC. Other analytical techniques including capillary GC and MS have been less readily accepted among microbiologists.[16] Furthermore,

2

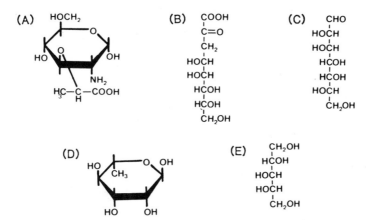

Figure 1. Structures of some carbohydrate chemical markers
for identification and detection of bacteria
and fungi: (A) muramic acid, (B) ketodeoxyoctonic
acid, (C) L-glycero-D-mannoheptose, (D) rhamnose
and (E) arabinitol.

analytical chemists are rarely trained in microbial taxonomy or biochemistry.
This chapter provides an overview of microbial biochemistry, while
anticipating the diverse backgrounds of potential readers in this
interdisciplinary area. Chapter 2 reviews fundamentals of gas chromatography
and mass spectrometry.

THE STRUCTURAL COMPOSITION OF MICROORGANISMS

A chemical marker may be defined as a compound or group of compounds
unique to or prominent in certain groups of microorganisms which can be used
to identify or detect them. Chemical markers may be structural components,
metabolites, or chemically generated products. Some chemical markers (such
as muramic acid or D-alanine) are present in most bacterial pathogens; other
markers (such as rhamnose, ribitol, or arabinitol) are less common. Fatty
acid, amino acid, and carbohydrate monomers are commonly used as chemical
markers when GC or GC-MS is employed. Analytical microbiologists use
chemical markers in the same way that immunologists use antigenic
determinants or molecular biologists use nucleotide sequences. An antigenic
determinant can be recognized using a specific (usually affinity purified or
monoclonal) antibody; a nucleotide sequence may be recognized by
hybridization with a complementary DNA sequence (a gene probe). Some common
chemical markers for bacteria are listed in Table 1. Figures 1-3 provide
chemical structures of some common sugar, lipid, and amino acid markers.

Although bacteria exist in only a limited number of morphological forms
(e.g., rods, cocci, chains, and spirals), their chemical composition varies
considerably, particularly in the cell envelope and other extracellular
structures. Some bacterial cells may be capable of sporulation; capsules may
be present outside some bacterial cell envelopes; some bacteria have an outer
membrane; the thickness of the peptidoglycan layer may vary and, in a few
cases, certain bacteria may not have peptidoglycan at all. Variations in
extraneous growth, treatment, mutation, and sampling conditions influence the
structure and composition of microorganisms. A knowledge of bacterial cell
wall biochemistry promotes an appreciation of the significance of certain
chemical markers. Chapter 7 describes characterization of bacterial
metabolites by GC and MS for industrial and other purposes; Chapter 8
discusses the use of metabolic profiles in clinical identification of
bacteria.

Table 1. Chemical markers for microorganisms

Compound	Source	Microbial Group
Fatty acids (straight and branched, saturated and unsaturated)	Cell membranes	All bacteria (profiles differ among species)
Muramic acid D-amino acids (D-alanine, D-glutamic acid)	Peptidoglycan	Most eubacteria and not in eukaryotic cells
Diaminopimelic acid	Peptidoglycan	Most Gram negative and certain Gram positive bacteria
Mycolic acids	Cell envelope	Mycobacteria, corynebacteria, and nocardiae
Heptoses ketodeoxyoctonic acid, and hydroxy fatty acids (e.g., (β-hydroxymyristic acid)	Lipopolysaccharide	Gram negative bacteria
Dipicolinic acid	Endospores	Certain Gram positive bacteria
Tuberculostearic acid	Cell envelope	Actinomycetales
Rhamnose	Group polysaccharides	Streptococci
Rhamnose	Lipopolysaccharide	Gram negative bacteria
Aminodideoxyhexoses	Lipopolysaccharide	Certain legionellae
Glucitol	Group polysaccharide	Group B streptococci
Arabinitol	Metabolite	*Candida albicans*

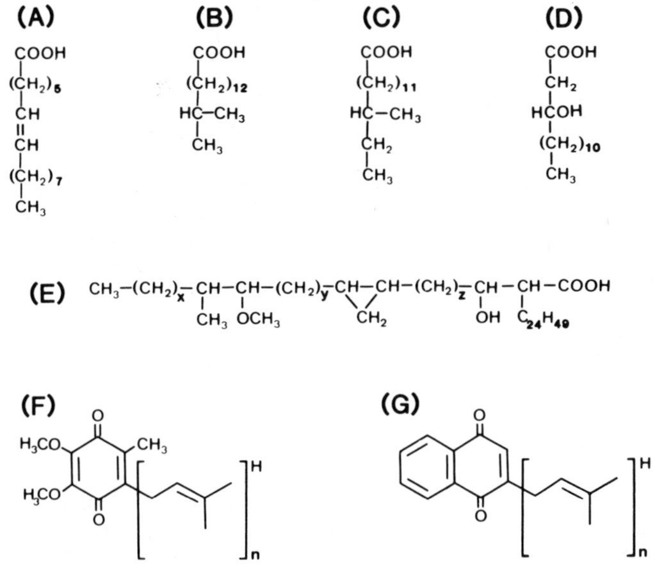

(A)

COOH
|
$(CH_2)_5$
|
CH
‖
CH
|
$(CH_2)_7$
|
CH_3

(B)

COOH
|
$(CH_2)_{12}$
|
HC–CH_3
|
CH_3

(C)

COOH
|
$(CH_2)_{11}$
|
HC–CH_3
|
CH_2
|
CH_3

(D)

COOH
|
CH_2
|
HCOH
|
$(CH_2)_{10}$
|
CH_3

(E) $CH_3-(CH_2)\overline{x}CH-CH-(CH_2)\overline{y}CH-CH-(CH_2)\overline{z}CH-CH-COOH$
                    | | | | |
                   CH_3 OCH_3 CH_2 OH $C_{24}H_{49}$

(F)

(G)

Figure 2.   Structures of some common lipid chemical
            markers: (A) a mono-unsaturated fatty acid
            cis-7-hexadecanoic acid (7 refers to the
            position from the carboxyl end), $C_{16:1\omega9c}$ (ω
            identifies a double bound between carbon n and
            n+1 counting from the hydrocarbon end of the *cis*
            isomer with terminal carbon as carbon 1),
            (B) branched fatty acid with methyl group in *iso*
            position, 14-methyl pentadecanoic acid, $C_{i-16:0}$,
            (C) branched fatty acid with methyl group in
            anteiso position, 13-methyl pentadecanoic acid,
            $C_{a-16:0}$, (D) 3-hydroxy tetradecanoic acid (also
            known as β-hydroxymyristic acid, a chemical marker
            for LPS), $C_{3-OH-14:0}$, (E) generalized mycolic
            acid structure, where x, y, and z may vary,
            (F) ubiquinone (a benzoquinone; in bacteria the
            values for n are 7-15; in mammals, n = 10),
            (G) menaquinone (a naphthoquinone; in bacteria,
            n = 6-14).

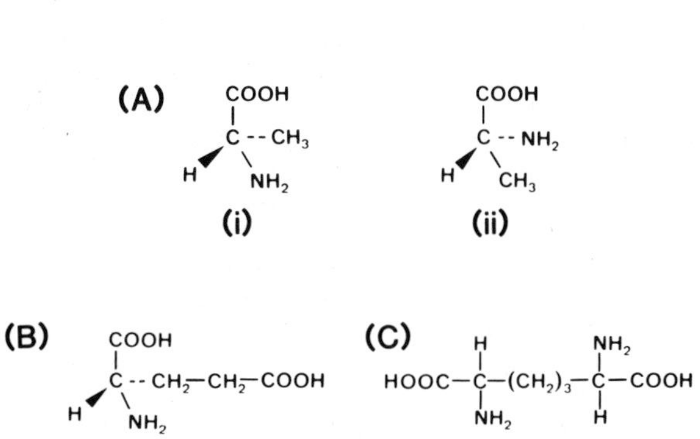

(A)

COOH
|
C -- CH_3
| \
H   NH_2
(i)

COOH
|
C -- NH_2
| \
H   CH_3
(ii)

(B)

COOH
|
C -- CH_2–CH_2–COOH
| \
H   NH_2

(C)

H   NH_2
| |
HOOC–C–(CH_2)_3–C–COOH
| |
NH_2   H

Figure 3.   Structures of some amino acid chemical markers:
            (A) (i) D-alanine and (ii) L-alanine,
            (B) D-glutamic acid, (C) diaminopimelic acid.

5

Bacteria (prokaryotes) lack membranous intracellular structures such as nuclear membranes or mitochondria. Their cytoplasmic make-up is less compartmentalized and simpler than those of animals, plants, and fungi (eukaryotes). Prokaryotic cells have complex and diverse structures outside the cell membrane. These structures include the cell wall and, in some cases, an outer membrane and/or capsules. The cell wall is rigid and protects the cell from osmotic lysis. The outer membrane is the major permeability barrier in Gram negative bacteria.[17] The cell envelope can be defined as the cell membrane and cell wall plus outer membrane if present. The cell wall consists of the peptidoglycan layer and associated structures. Many chemical differences providing taxonomic differentiation between bacteria reside in the cell envelope.

Most cell envelopes fall into one of two major categories: Gram positive or Gram negative. The major component structures of the cell envelope of Gram positive and Gram negative bacteria are shown in Figures 4A and 4B. The so-called "Gram type"[18] of an organism, defined by these structural differences, often but not always correlates with classic Gram staining. After staining with crystal violet, Gram positive cells retain the dark blue dye when washed with ethanol. On washing Gram negative cells, however, the dye is removed and counter-staining with safranin is necessary for microscopic examination. Other cell wall types that are neither Gram positive nor Gram negative are found in certain species (see acid-fastness below).

Peptidoglycan (PG, synonyms murein, mucopeptide) is a single bag shaped, highly crosslinked macromolecule that surrounds the bacterial cell and provides rigidity and protection from osmotic lysis. PG is the cell wall skeleton and is of high molecular weight-- billions of daltons in comparison to proteins which usually are less than a few hundred thousand daltons. PG is the only substance common to almost all eubacteria and not present in non-bacterial matter.[19,20] Strains of *Mycoplasma* are known not to contain peptidoglycan and those of *Chlamydia* have a peptidoglycan that lacks muramic acid. Peptidoglycan consists of a glycan (polysaccharide) backbone that is a repeating polymer of N-acetylglucosamine and N-acetylmuramic acid. Muramic acid (Figure 1A) is unique to bacteria and, because it does not occur elsewhere in nature, is a definitive marker for bacteria. Attached covalently to the lactyl group of muramic acid are tetra- and pentapeptides (composed of repeating L- and D-amino acids) cross-linked by peptide bridges. Diaminopimelic acid (DAP, Figure 3C)) is another chemical marker for peptidoglycan found in many Gram negative bacteria and in certain Gram positive bacteria. DAP is also not found elsewhere in nature. Sequence variations in the peptide side-chains and crossbridges also occur among different bacteria. Other chemical markers in peptidoglycan include the D-amino acids, D-alanine (Figure 3A) and D-glutamic acid (Figure 3B). D-amino acids are sometimes found in other bacterial components, but are not synthesized by mammals. The structure of the peptidoglycan of the group A streptococcus is shown in Figure 5.[21]

The peptidoglycan layer is much thicker in the Gram positive bacterial cell envelope (Figure 4A) than in Gram negative bacteria (Figure 4B). In some Gram positive species, teichoic acids (composed of ribitol or glycerol phosphate backbones with side-chains of variable composition such as D-alanine) are covalently bound to the thick peptidoglycan layer. Teichuronic acids (polymers of glucuronic acid with variable side-chains) or other neutral polysaccharides are also found in this location. Negatively charged teichoic or teichuronic acids concentrate metal ions in the cell.[22] Teichoic acids are implicated in cell growth and division by their control of the action of autolytic enzymes.[23] A typical teichoic acid structure is shown in Figure 6. Although most of the lipoteichoic acid is in the cell membrane,

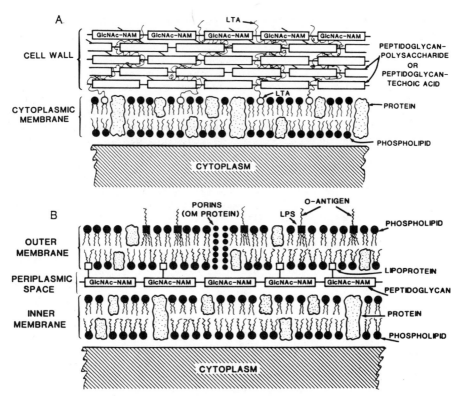

Figure 4.   Generalized structures of bacterial cell
            envelopes: (A) Gram positive, showing thick
            peptidoglycan layer and (B) Gram negative, showing
            double membrane structure and thin peptidoglycan
            layer.

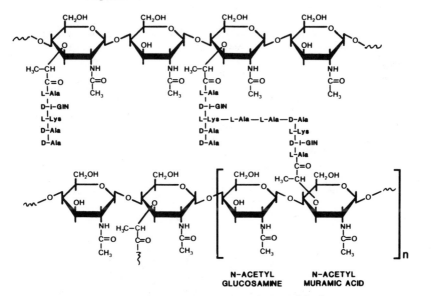

Figure 5.   The group A streptococcal peptidoglycan.
            Degree of peptide cross-linking between the
            glycan backbones is variable, some N-acetyl-
            muramic acid monomers are free, and others
            may be attached to an uncross-linked chain.

some is on the cell surface and involved in bacterial adhesion to epithelial cells, allowing colonization (e.g., *S. pyogenes*).[24,25]

In the Gram negative cell envelope the peptidoglycan is usually a rather thin layer and a lipoprotein is covalently attached (the Braun lipoprotein).[26] This lipoprotein anchors the PG layer to the outer membrane through interaction with phospholipids and other hydrophobic substances. The outer membrane is external to the PG-lipoprotein layer. The inner leaflet of the outer membrane is believed to be similar to other membranes in consisting of phospholipids and proteins. Some of these proteins span across the outer membrane and are referred to as porins since these channels allow passage of hydrophilic nutrients (such as oligosaccharides) into the cell.[27]

Unlike other membranes, the external surface of the outer membrane of Gram negative bacteria consists primarily of a unique lipopolysaccharide (LPS). LPS provides a permeability barrier to hydrophobic compounds and is the primary endotoxin of Gram negative bacteria associated with pyrogenicity. The basic structure of LPS is similar among different Gram negative bacteria and consists of three regions: an outer O antigen, a middle core, and an inner lipid A region. Figure 7 shows an example of a typical LPS structure. The lipid A region consists of a glucosamine disaccharide with covalently bound 2- and 3-hydroxy fatty acids.[13,14,28] Glucosamine is common in nature; 3-hydroxy and 2-hydroxy fatty acids are unusual. The fatty acid composition of LPS varies; however, β-hydroxymyristic acid (Figure 2D) is ubiquitous and thus a useful chemical marker. The core region covalently bound to lipid A in Gram negative bacteria contains two unusual sugars, ketodeoxyoctonic acid (KDO, Figure 1B)) and L-D-glycero-mannoheptose (Figure 1C); in a few organisms, D-D-glycero-mannoheptose is also present. KDO and heptoses are not commonly found in structures other than LPS and are also potential markers for Gram negative organisms. Relative amounts and types of simple sugars, including rhamnose (Figure 1D) and fucose, within the core and O-antigen regions of LPS differ among bacterial species. Carbohydrate profiling is an excellent tool for bacterial differentiation. For example, certain species of *Legionella* may be distinguished by the presence of unusual aminodideoxyhexoses and other sugars.[29,30]

The mycobacteria, nocardiae, corynebacteria, and some related groups have unusual cell envelopes. These cells stain Gram positive but some, particularly mycobacteria and nocardiae, on staining resist decolorization by acid-alcohol (acid-fastness). Acid-fastness is believed to be related to the presence of long chain mycolic acids. These acids are present in all three genera, but those in corynebacteria are relatively short. Mycolic acids are present in various structural forms within the cell envelope and some mycolic acids are covalently bound via a polysaccharide to peptidoglycan. Other mycolic acid containing compounds form a thick waxy layer around the outside of the cell wall. Mycolic acids, are α-alkyl, β-hydroxy fatty acids ranging in size from C_{30} to C_{86} and constitute up to 60% of the dry weight of these bacterial envelopes (Figure 2E). A more detailed discussion of these compounds is given in Chapter 9.

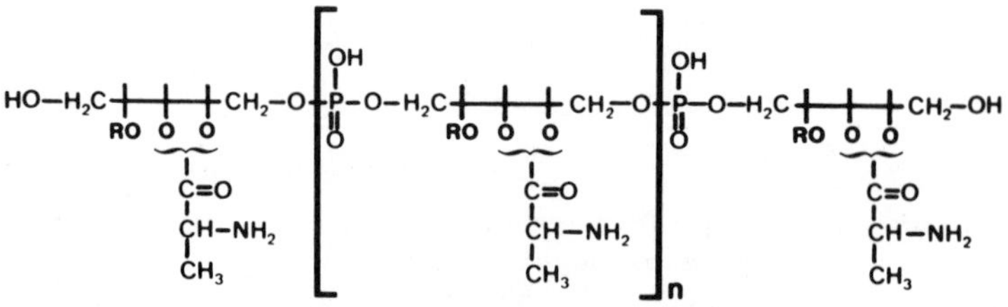

Figure 6. Structure of ribitol teichoic acid found in certain Gram positive bacteria.

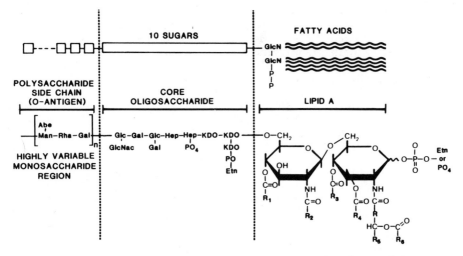

Figure 7. Structure of a typical Gram negative bacterial lipopolysaccharide. The lipid A moiety generally contains 6 or fewer fatty acids. The antigenic type is determined by the sequence of the highly variable polysaccharide side chain.

Under certain circumstances many bacteria produce capsules outside their cell envelopes. These are in some cases referred to as slime layers or the glycocalyx. These layers are not essential to cell viability and species can vary in their ability to produce a capsule. Capsules usually consist of polysaccharide, however certain strains of *Bacillus* have capsules composed of D-glutamic acid polypeptide.[31] Capsules inhibit ingestion of extracellular pathogenic bacteria by phagocytes.

Certain Gram positive bacteria, most notably, strains of *Bacillus* and *Clostridium* produce modified cells (endospores) that can survive adverse environmental conditions. They have an unusual cell envelope that contains inner and outer membranes. Spore PG (sometimes called the cortex) is found between the two membranes. The spore PG differs from the normal vegetative cell in that muramic acid is mostly in the lactam form and the peptidoglycan is less cross-linked by peptides.[32] A keratin coat is located on the cell exterior. The spore also contains large quantities of another potential chemical marker, calcium dipicolinate, which is involved in heat resistance.

CHEMOTAXONOMIC IDENTIFICATION AND DIFFERENTIATION OF MICROORGANISMS

A change in orientation from growth and microscope based traditional methods towards rapid chemical instrumental techniques has not been in the microbiological mainstream. Because bacteria are usually present in low concentration in samples, traditional methods first increase the concentration of the bacteria by culturing on specific agar. This initial isolation also permits selection of a single pure colony of bacteria for further identification. New analytical microbiological methods also rely on isolation of pure colonies before identification.

Traditional methods for identification/differentiation of microorganisms have not been based on the direct analysis of specific chemical components of microorganisms. For example, the classic Gram stain is usually performed on an isolated colony. As noted above, although

9

indirectly related to the chemical composition of the cell envelope, the Gram stain does not provide specific information on chemical structure. The basic appearance of the microorganism under the microscope and its shape and relative size all aid in classifying the organism. These morphological aspects provide taxonomic information but are not directly related to chemical composition.

Following preliminary examination, colonies may be sub-cultured and examined in other ways. Indirect indicators, such as the generation of specific metabolic products or enzymes, utilization of specific growth substrates, or susceptibility to various antibiotics often generate a simple observable change on a culture plate or liquid media. For example, group A streptococci are inhibited in their growth by the antibiotic bacitracin; *Staphylococcus aureus* produces an enzyme (coagulase) that clots plasma; the ability of species of *Enterobacteriaceae* to metabolize certain sugars can be monitored using a pH indicator dye present in the media which is sensitive to acidic metabolic products. Many of these identifying characteristics of microorganisms are more a reflection of metabolic status and the presence or absence of particular enzymes rather than overall chemical composition.

Perhaps the only direct chemical method used routinely in microbiological laboratories is the GC method developed at Virginia Polytechnic Institute (Blacksburg, VA) for identifying metabolic products (short chain alcohols and fatty acids) produced by anaerobic bacteria.[33] After isolation, bacteria are regrown in liquid culture for one to several days before analysis; the broth is extracted with an organic solvent, and an aliquot analyzed by GC. A genus level identification is possible for all the common anaerobes by this method which is frequently used for screening of new isolates before additional biochemical and antibiotic resistance tests are performed. This procedure has been automated by analysis of the headspace vapors above fermentation media. A review of the literature has been published[34] and Chapters 7 and 8 describes recent developments in this area.

Routine clinical laboratories do not commonly perform analyses for structural components of microorganisms. A simple GC method for structural components of microorganisms involves profiling of fatty acids.[1,35] Fatty acid profiles are similar within many bacterial species, but are often quite different between species. Fatty acids are variable in composition and can be unsaturated (Figure 2A) or saturated (Figures 2B and 2C), and straight-chain (Figure 2A) or branched in the *iso* (methyl attached to carbon 1 as in Figure 2B) or *anteiso* (methyl attached to carbon 2 as in 2C) position. An automated GC system for analysis of methyl esters of bacterial fatty acids is commercially available and discussed in Chapter 4. Sample preparation requires a few hours, but could be fully automated. Cellular branched fatty acid profiles allow differentiation of the *Legionellaceae* (including *L. pneumophila*, the causative agent of Legionnaires disease).[35] Although present in smaller amounts, hydroxy-fatty acids present in legionellae provide additional discrimination; unfortunately hydroxy-fatty acids are difficult to detect.[36] Chapter 5 describes the use of fatty acid profiling in bacterial identification.

Fatty acids are useful for differentiation of mycobacteria, where traditional approaches are time consuming. Mycobacteria (including *M. leprae* and *M. tuberculosis*, causative agents of leprosy and tuberculosis) can be identified by GC or GC-MS profiles of derivatized fatty acids.[1] Alcohols and alkanes can also be released from methylated derivatives of mycolic acid by pyrolysis. International interlaboratory studies have compared such fatty acid profiles of mycobacteria and found them to be reproducible.[37] Mycolic acid content has also been investigated using pyrolysis GC without prior derivatization.[38] A series of isomeric alkenes produced by pyrolysis has been shown to differentiate *M. leprae* from other armadillo-derived mycobacteria. Other researchers have analyzed intact long chain mycolic acids by derivatization GC-MS. Species of mycobacteria may be differentiated from one another (and from nocardiae and corynebacteria) by the type of mycolic acids, their chain length, and their degree of unsaturation.[39]

The profiling of other bacterial components has not been exploited extensively for rapid identification of bacteria. Carbohydrates (Figure 1) can be used to differentiate microorganisms and methods for routine characterization of profiles from whole bacterial cells are available.[40] For example, many of the streptococci[41,42] and legionellae[29,30,43] can be differentiated by their sugar profiles. Chapter 5 provides detailed discussion on the analysis of carbohydrates by GC-MS.

Analysis of bacterial isoprenoid quinone composition is valuable for chemotaxonomy and for identification at least to genus and often to species level.[44-46] These compounds are involved in mitochondrial oxidative phosphorylation. As shown in the examples of Figure 2F and 2G, all isoprenoid quinones contain a six carbon ring with oxygens at the 1 and 4 positions and an isoprenoid side chain of variable length. Isoprenoid quinones having n isoprenoid units in the side chain are designated Q_n. There are two major structural groups of these compounds, the benzoquinones and the naphthoquinones, of which ubiquinone (Figure 2F) and menaquinone (Figure 2G) are the most common in bacteria. Gram negative bacteria fall into three groups by quinone content. Aerobes possess only ubiquinone (with minor exceptions) of which Q_8, Q_9 and Q_{10} are most common. The facultative anaerobic genera (such as *Klebsiella* and *Proteus*) have a mixture of ubiquinones and menaquinones. Gram negative obligate anaerobes (such as *Bacteroides* and *Desulfovibrio*) have only menaquinones. Gram positive bacteria also contain only menaquinones, many of which have varying degrees of hydrogenation of the isoprenyl side-chain. Variations in the basic isoprenoid quinone structures commonly occur as minor components in many species and as major components in unusual species, especially those of the Archebacteria. Some species contain other types of quinone, such as dimethylmenaquinone, α-tocopherolquinone, and rhodoquinone.

Isoprenoid quinones can be extracted from bacterial cells by organic solvents, purified, and the resultant homologues separated by reversed phase HPLC or TLC without derivatization. Profiling these compounds is rapid and has proven valuable in many chemotaxonomic studies. Developments in the simultaneous extraction of quinones and fatty acids (as polar lipids[29]) may increase the utility of the technique. In some common bacterial genera, however, there are little or no differences in quinone profiles between species. For example, certain *Pseudomonas* species possess only Q_9 as do many other Gram negative bacteria. In some families, such as the *Legionellaceae*, there are considerable variations in ubiquinone content between many of the species.[45] In this and other unusual groups of bacteria where traditional identification methods are inadequate, isoprenoid quinone profiling has been effective for both identification and recognition of novel strains. Since HPLC applications are not emphasized in this book, the analysis of isoprenoid quinones is not discussed further.

Amino acids (Figure 3) are another major class of structural components with potential for use in bacterial profiling.[47,48] As in the analysis of sugar components, it is important to select unique amino acids that are relevant to the desired taxonomic differentiation. Many L-amino acids are not useful for identification of whole bacterial cells because they are found in all proteins. Analysis of L- and D-amino acids in isolated peptidoglycan or cell walls is of taxonomic value, although the necessity for purified peptidoglycan limits the method.[19] There are some unique amino acids present in bacterial peptidoglycan and not present in proteins including ornithine, L-L-, D-D-, and *meso*-diaminopimelic acid (Figure 3C), D-alanine (Figure 3A), and D-glutamic acid (Figure 3B). Profiling of amino acids present in whole bacterial cells for identification has not been widely investigated.[49] Recent work in this area is reviewed in Chapter 6.

Analytical pyrolysis (thermal degradation in the absence of oxygen) can generate products that provide a chemical profile for identification of microorganisms. Pyrolysis involves dehydration, rearrangements, and bond formation in addition to bond scission; although smaller fragment molecules are commonly produced, higher molecular weight products which retain much of the structure of the original compound are also generated.[50-52] Pyrolysis of carbohydrates, for example, produce anhydrosugars.[53] Groups of bacteria may be differentiated by pyrolysis followed by GC or MS, but the recent more widespread use of GC-MS has created the opportunity to identify specific pyrolytic chemical markers. For example, group B streptococci can be differentiated from other β-hemolytic streptococci by the presence of glucitol identified by GC-MS.[54] Further information on the use of analytical pyrolysis may be found in Chapters 12 and 13.

At present, some profiling techniques are relatively simple and complete automation should soon be possible. In other instances, basic research is needed before such a goal can be accomplished. The prospects are bright, routine applications appear regularly, and growth in this area can clearly be anticipated.

TRACE DETECTION OF MICROBES

Bacteria and bacterial debris can be detected in body fluids, tissues, and other complex matrices without prior culture by searching for chemical markers produced by the organism that are not normally produced by mammals. The analysis of such trace components in a clinical sample is considerably more complex than analysis for a monomer in isolated bacterial cells. Greater selectivity in derivatization and clean-up procedures is required and appropriate MS-based instrumentation and instrumental conditions must be selected.

Tuberculostearic acid is found in the cell envelope of mycobacteria; its presence in patient sputa is diagnostic for tuberculosis.[55,56] Classical methods can take several weeks for a definitive diagnosis, as opposed to a few hours by GC-MS; the time for diagnosis of tuberculosis can be reduced from 3-6 weeks to several hours. The sensitivity and selectivity enhancement achieved by negative ion detection following chemical ionization is also significant. This topic is described in Chapter 14.

D-arabinitol (Figure 1E), a sugar alcohol, is secreted in large amounts by certain fungi and has been employed as a chemical marker for GC-MS detection of *Candida* in serum.[57] Sugar alcohols in normal human serum include ribitol, xylitol, and L- and D-arabinitol. In invasive candidiasis associated with cancer, enhanced levels of arabinitol occur as a result of D-arabinitol secretion by the fungus *Candida albicans*. More information on this topic is found in Chapter 15.

Hydroxy fatty acids such as β-hydroxymyristic acid (Figure 2D) are useful in rapid diagnosis of septic meningitis and potentially useful for detection of many other infections caused by Gram negative bacteria.[13] This technique may be more specific than the more commonly used *Limulus* lysate assay.

Rhamnose (Figure 1D) and muramic acid (Figure 1A) are useful chemical markers for the detection of degradation products of bacterial cell walls in mammalian tissues in animal models. Samples taken from tissues of cell wall injected rats contain the bacterial sugars (rhamnose and muramic acid) in addition to common tissue derived sugars (fucose, mannose, galactose, glucose, glucosamine, and galactosamine). The GC-MS detection of chemical markers for bacteria in inflamed human tissues may answer the question of whether localization of bacterial debris in these sites causes inflammation in man as well as having potential in rapid clinical diagnosis.[58,59] This subject is also discussed in Chapter 5.

As mentioned above, D-alanine (Figure 3A) is another marker for the presence of bacteria and bacterial peptidoglycan.[60] It has been detected in plant roots colonized with bacteria and in animal tissues containing persistent bacterial debris. With large amounts of L-alanine naturally present in mammalian and plant proteins, the racemization of L-alanine to D-alanine during hydrolysis creates a high background of D-alanine which limits the levels at which D-alanine can be detected.[61,62]

CONCLUSION

A number of unique compounds found in microbes are useful for bacterial identification or for detection in complex matrices such as body fluids and tissues. These chemical markers include rhamnose (found in certain Gram positive and negative bacteria), muramic acid, D-alanine and D-glutamic acid (found in most pathogenic bacteria), β-hydroxymyristic acid, L-glycero-D-mannoheptose, D-glycero-D-mannoheptose and ketodeoxyoctonic acid (components of Gram negative lipopolysaccharide), and D-arabinitol (a metabolite of certain fungi). The detection of chemical markers for microbes may, in some instances, eliminate culturing, and could dramatically improve speed of diagnosis and treatment of certain infectious diseases. These markers may also be useful in industrial or biotechnology areas for rapid determination of the presence of microbial contamination with specific bacterial populations or for rapid characterization of biological materials. Rapid identification and detection of certain microbial groups is achievable at present, but further developments in automation and increases in sensitivity will expand the application and utility of these techniques in analytical microbiology.

ACKNOWLEDGEMENTS

This work was supported by grants from the U. S. Army Office of Research and by the National Institutes of Health.

REFERENCES

1. L. Larsson, Gas chromatography and mass spectrometry, in: "Automation in Clinical Microbiology," J. H. Jorgenson, ed., CRC Press Inc., Boca Raton (1987).
2. A. Fox and S. L. Morgan, The chemotaxonomic characterization of microorganisms by capillary gas chromatography and gas chromatography-mass spectrometry, in: "Instrumental Methods for Rapid Microbiological Analysis," W. H. Nelson, ed., VCH, Deerfield Beach (1985).
3. J. F. Collawn, P. Y. Lau, S. L. Morgan, A. Fox and W. W. Fish, A chemical and physical comparison of ferritin subunit species fractionated by high performance liquid chromatography, Arch. Biochem. Biophys. 233:260 (1984).
4. K. Bryn, and E. Jantzen, Analysis of lipopolysaccharides by methanolysis, trifluoroacetylation and gas chromatography on a fused silica capillary column, J. Chromatogr. 240:405 (1982).

5. J. Gilbart, J. Harrison, C. Parks, and A. Fox, Analysis of the amino acid and sugar composition of streptococcal cell walls by gas chromatography-mass spectrometry, J. Chromatogr. 441:323 (1988).

6. N. Boas, Method for determination of hexosamines in tissues, J. Biol. Chem. 204:553 (1953).

7. Z. Dische and L. B. Shettles, A specific color reaction of methylpentoses and a spectrophotometric micromethod for their determination, J. Biol. Chem. 175:595 (1948).

8. Y. D. Karkhanis, J. Y. Zeltner, J. J. Jackson, and D. J. Carlo, A new and improved microassay to determine 2-keto-3-deoxyoctonate in lipopolysaccharide of Gram-negative bacteria, Anal. Biochem. 85:601 (1978).

9. A. J. Garrett, M. J. Harrison, and G. P. Manire, A search for the bacterial mucopeptide component muramic acid, in Chlamydia, J. Gen. Microbiol. 80:315 (1974).

10. A. Barbor, K.-I. Amano, T. Hackstadt, L. Perry, and H. D. Caldwell, Chlamydia trachomatis has penicillin binding proteins but not detectable muramic acid, J. Bacteriol. 151:420 (1982).

11. R. H. Findlay, D. J. W. Moriarty, and D. C. White, Improved method of determining muramic acid from environmental samples, J. Geomicrobiol. 3:135 (1983).

12. D. C. White, Analysis of microorganisms in terms of quantity and activity in natural environments, Soc. Gen. Microbiology. Symposium. 34:37 (1983).

13. S. K. Maitra, R. Nachum, and F. C. Pearson, Establishment of beta-hydroxy fatty acids as chemical marker molecules for bacterial endotoxin by gas chromatography-mass spectrometry, Appl. Environm. Microbiol. 52:510 (1986).

14. E. Rietschel, H.-W. Wollenweber, H. Brade, U. Zahringer, B. Lindner, U. Seydel, H. Bradaczek, G. Barnickel, H. Labischinski, and P. Giesbrecht, Structure and conformation of the lipid A component of lipopolysaccharides, in: "Handbook of Endotoxin, Vol. 1," E. Th. Rietschel, ed., Elsevier Scientific Publishing, Amsterdam (1984).

15. S. A. Martin, M. L. Karnovsky, J. M. Kreuger, J. Pappenheimer, and K. Biemann, Peptidoglycan as promoters of slow wave sleep. I. structure of the sleep promoting factor isolated from human urine, J. Biol. Chem. 259:12652 (1984).

16. J. Goldstein, Future development of automated instruments for microbiology, in: "Automation in Clinical Microbiology," J. Jorgenson, ed., CRC Press Inc., Boca Raton (1987).

17. T. Nakae and H. Nikaido, Outer membrane as a diffusion barrier in Salmonella typhimurium. Penetration of oligo- and polysaccharides into isolated outer membrane vesicles and cells with degraded peptidoglycan layer, J. Biol. Chem. 250:7359 (1975).

18. J. Wiegel, Distinction between the Gram reaction and Gram type of bacteria, Int. J. System. Bacteriol. 31:88 (1981).

19. K. H. Schleifer, Analysis of the chemical composition and primary structure of murein, Meth. in Microbiol. 18:123 (1985).

20. K. H. Schleifer and O. Kandler, Peptidoglycan types of bacterial cell walls and their taxonomic implications, Bacteriol. Rev. 36:407 (1972).

21. E. Munoz, J.-M. Ghuysen and H. Heymann, Cell walls of Streptococcus pyogenes, Type 14, C polysaccharide-peptidoglycan and G polysaccharide-peptidoglycan complexes, Biochemistry 6:3659 (1967).

22. D. C. Ellwood and D. W. Tempest, Control of teichoic acid and teichuronic acid biosynthesis in chemostat cultures of Bacillus subtilus var niger, Biochem. J. 111:1 (1969).

23. A. Tomasz, A. Albino and E. Zanati, Multiple antibiotic resistance in a bacterium with suppressed autolytic system, Nature 227:138 (1970).

24. A. J. Wicken and K. W. Knox, Lipoteichoic acids: a new class of bacterial antigen, Science 187:1161 (1975).

25. E. H. Beachey and I. Ofek, Epithelial cell binding of group A streptococci by lipoteichoic acid on fimbriae denuded of M protein, J. Exp. Med. 143:759 (1976).

26. V. Braun and K. Rehn, Chemical characterization, spatial distribution and function of a lipoprotein (murein-lipoprotein) of the *E. coli* cell wall. The specific effect of trypsin on the cell membrane, Eur. J. Biochem. 10:426 (1969).

27. R. E. W. Hancock, Role of porins in outer membrane permeability, J. Bacteriol. 169:929 (1987).

28. K. Bryn and E. Jantzen, Quantification of 2-keto-3-deoxy octonate in (lipo)polysaccharides by methanolytic release, trifluoroacetylation and capillary gas chromatography, J. Chromatogr. 370:103 (1986).

29. A. Fox, P. Y. Lau, A. Brown, S. L. Morgan, Z.-T. Zhu, M. Lema, and M. D. Walla, Capillary gas chromatographic analysis of carbohydrates of *Legionella pneumophila* and other members of the *Legionellaceae*, J. Clin. Microbiol. 19:326 (1984).

30. M. D. Walla, P. Y. Lau, S. L. Morgan, A. Fox, and A. Brown, Capillary gas chromatography-mass spectrometry of carbohydrate components of legionellae and other bacteria, J. Chromatogr. 288:399 (1984).

31. I. Uchida, T. Sekizaki, K. Hashimoto, and N. Terakado, Association of the encapsulation of *Bacillus anthracis* with a 60 megadalton plasmid, J. Gen. Microbiol. 131:363 (1985).

32. A. D. Warth, Molecular structure of the bacterial spore, Adv. Microb. Physiol. 17:1 (1978).

33. L. V. Holdeman, E. P. Cato, and W. C. Moore, "Anaerobe Laboratory Manual, 4th ed.," Virginia Polytechnic Institute and State University, Anaerobe Laboratory, Blacksburg (1977).

34. L. Larsson and E. Holst, Feasibility of automated head-space gas chromatography in identification of anaerobic bacteria, Acta. Pathol. Microbiol. Scand. Sect. B. 90:125 (1982).

35. C. W. Moss, R. E. Weaver, S. B. Dees, and W. B. Cherry, Cellular fatty acid composition of isolates from Legionaires' disease, J. Clin. Microbiol. 6:140 (1977).

36. W. R. Mayberry, Dihydroxy and monohydroxy fatty acids in *Legionella pneumophila*, J. Bacteriol. 147:373 (1981).

37. L. Larsson, E. Jantzen, and J. Johnsson, Gas chromatographic fatty acid profiles for characterisation of mycobacteria: an interlaboratory methodological evaluation, Europ. J. Clin. Microbiol. 4:483 (1985).

38. G. Wieten, J. J. Boon, D. G. Groothuis, F. Portaels, and D. E. Minnikin, Rapid detection of mycobacterial contamination in batches of whole cells of purified *Mycobacterium leprae* by pyrolysis gas chromatography, FEMS Microbiol. Let. 25:289 (1984).

39. K. Kaneda, S. Naito, S. Imaizumi, I. Yano, S. Mizuno, I. Tomiyasu, T. Baba, E. Kusunose, and M. Kukusone, Determination of molecular species composition of C_{80} or longer chain α-mycolic acids in mycobacterium spp. by gas chromatography-mass spectrometry and mass chromatography, J. Clin. Microbiol. 24:1060 (1986).

40. A. Fox, S. L. Morgan, and J. Gilbart, Preparation of alditol acetates and their analysis by gas chromatography and mass spectrometry, in: "Analysis of Carbohydrates by GLC and MS," C. J. Bierman and G. McGinnis, eds., CRC Press, Boca Raton (1989).

41. D. G. Pritchard, J. E. Colligan, S. E. Speed, and B. M. Gray, Carbohydrate fingerprints of streptococcal cells, J. Clin. Microbiol. 13:89 (1981).

42. D. G. Pritchard, B. M. Gray, and H. C. Dillon, Characterization of the group-specific polysaccharide of group B streptococcus, Arch. Biochem. Biophys. 235:385 (1984).

43. J. Gilbart, A. Fox and S. L. Morgan, Carbohydrate profiling of bacteria by gas chromatography-mass spectrometry: chemical derivatization and analytical pyrolysis, Eur. J. Clin. Micro. 6:715 (1987).

44. M. D. Collins and D. Jones, Distribution of isoprenoid quinone structural types in bacteria and their taxonomic implications, Microbiol. Review. 45:316 (1981).

45. J. Gilbart and M. D. Collins, High performance liquid chromatographic analysis of ubiquinones from new *Legionella* species, FEMS Microbiol. Lett. 26:77 (1985).

46. D. E. Minnikin, A. G. O'Donnell, M. Goodfellow, G. Alderson, M. Athalye, A. Schaal, and J. Parlett, An integrated procedure for the extraction of bacterial isoprenoid quinones and polar lipids, J. Microbiol. Methods 2:233 (1984).

47. S. L. Mackenzie, Amino acids and peptides, in: "Gas Chromatography-Mass Spectrometry Applications in Microbiology," G. Odham, L. Larsson, and P.-A. Mardh, eds., Plenum Press, New York (1984).

48. A. Tunlid and G. Odham, Capillary gas chromatography using electron capture detection or selected ion monitoring detection for the determination of muramic acid, diaminopimelic acid and the ratio of D/L alanine in bacteria, J. Microbiol. Meth. 1:63 (1983).

49. A. G. O'Donnell, D. E. Minnikin, M. Goodfellow, and J. H. Partlett, The analysis of actinomycete wall amino acids by gas chromatography, FEMS Micro. Lett. 15:75 (1982).

50. R. E. Aries, C. S. Gutteridge, and T. W. Ottley, Evaluation of a low-cost, automated pyrolysis-mass spectrometer, J. Anal. Appl. Pyrolysis. 9:81 (1986).

51. H. L. C. Meuzelaar, J. Haverkamp, and F. D. Hileman, "Pyrolysis Mass Spectrometry of Recent and Fossil Biomaterials," Elsevier, Amsterdam (1982).

52. F. L. Bayer and S. L. Morgan, The analysis of biopolymers by analytical pyrolysis gas chromatography, in: "Pyrolysis and GC in polymer analysis," E. Levy and S. A. Liebman, eds., Marcel Dekker, New York (1985).

53. R. J. Helleur, Characterization of the saccharide composition of heteropolysaccharides by pyrolysis gas chromatography-mass spectrometry, J. Anal. Appl. Pyrolysis. 11:297 (1987).

54. C. S. Smith, S. L. Morgan, C. D. Parks, A. Fox, and D. G. Pritchard, Chemical marker for the differentiation of group A and group B streptococci by pyrolysis gas chromatography-mass spectrometry, Anal. Chem. 59:1410 (1987).

55. L. Larsson, G. Odham, G. Westerdahl, and B. Olsson, Diagnosis of pulmonary tuberculosis by selected ion monitoring: Improved analysis of tuberculostearate in sputum using negative-ion mass spectrometry, J. Clin. Microbiol. 25:893 (1987).

56. G. L. French, C. Y. Chan, S. W. Cheung, R. Teoh, M. J. Humphries, and G. O. Mahoney, Diagnosis of tuberculous meningitis by detection of tuberculostearic acid in cerebrospinal fluid, Lancet 2:117 (1987).

57. J. Roboz, D. C. Kappatos, and J. F. Holland, Role of individual serum pentitol concentrations in the diagnosis of disseminated visceral candidiasis, Eur. J. Clin. Microbiol, 6:708 (1987).

58. J. Gilbart, A. Fox, R. S. Whiton, and S. L. Morgan, Rhamnose and muramic acid: chemical markers for bacterial cell walls in mammalian tissues, J. Microbiol. Meth. 5:271 (1986).

59. J. Gilbart and A. Fox, Elimination of group A streptococcal cell walls from mammalian tissues, Infect. Immun. 55:1526 (1987).

60. A. Tunlid and G. Odham, Diastereomeric determination of R-alanine in bacteria in bacteria using capillary gas chromatography and positive/negative ion mass spectrometry, Biomed. Mass. Spectrom. 11:428 (1985).

61. S. Sonnesson, L. Larsson, A. Fox, and G. Odham, Determination of environmental levels of peptidoglycan and lipopolysaccharide using gas chromatography with negative-ion chemical-ionization mass

spectrometry utilizing unique bacterial amino acids and hydroxy fatty acids as marker compounds, J. Chromatogr. Biomed. Appli. 431:1 (1988).

62. K. Ueda, S. L. Morgan, A. Fox, J. Gilbart, A. Sonesson, L. Larsson, and G. Odham, D-alanine as a chemical marker for the determination of streptococcal cell wall levels in mammalian tissues by gas chromatography negative ion/chemical ionization mass spectrometry, Anal. Chem. 61:265 (1989).

CHAPTER 2

GAS CHROMATOGRAPHY AND MASS SPECTROMETRY FOR ANALYTICAL MICROBIOLOGY

Kimio Ueda and Stephen L. Morgan

Department of Chemistry
University of South Carolina
Columbia, SC 29208

Alvin Fox

Department of Microbiology & Immunology
School of Medicine
University of South Carolina
Columbia, SC 29208

INTRODUCTION

Gas chromatography (GC) and mass spectrometry (MS) are powerful analytical techniques for the characterization, identification, and classification of microorganisms. Using different variations or combinations of GC and MS, one can detect and identify natural metabolic products of cells, separate and identify monomeric structural components, profile composition of cells or cell fractions, or detect at a trace level the presence of bacteria using characteristic chemical markers. This chapter reviews basic instrumental aspects of GC, particularly capillary GC, and MS from the perspective of analytical microbiology.

A GC system consists of a source of carrier gas, a sample inlet, a column and column oven, and a detector. The carrier gas carries volatile sample components through the system. The role of the sample inlet is to vaporize a liquid sample or transfer a gas sample rapidly, accurately, and precisely into the chromatographic column. Microbiological liquid samples might include a derivatization mixture of carbohydrates from the cell walls of a microorganism or an extracted aliquot of blood from a patient with a systemic fungal infection; gas samples might result from sampling the volatile metabolites over a fermentation reactor in an industrial process. Ideally, the sample should be introduced to the column as a narrow band; several injection modes have been developed to handle different sampling situations. Many compounds of interest (e.g., sugars, lipids, and amino acids) are not volatile. Because GC requires the sample to be volatile and thermally stable, chemical derivatization or pyrolysis of these compounds is needed prior to GC analysis.

The column is the heart of the GC system: interactions between sample components and the stationary phase in the column cause components (solutes) to migrate at different rates and thus provide separation. The choice of

Analytical Microbiology Methods
Edited by A. Fox *et al.*
Plenum Press, New York, 1990

stationary phase can be based on the nature of the sample components to be separated; although many phases of different polarity and functionality are available, a preferred set of phases suitable for most separations can be listed. The detector generates an electrical signal proportional to the concentration of sample components and, in some cases, provides chemical identity of components.

The typical MS system consists of an inlet, an ion source, a mass analyzer, and a detector. Interfacing GC to MS is discussed in more detail below. Regardless of how sample is introduced, ions are produced from the original molecules in an ion source, separated in the mass analyzer, and their abundance measured in the detector. The ionization mode (e.g., electron impact (EI) or chemical ionization (CI)) determines the type and abundance of ions produced. These ions can be the same molecular weight as the parent compound (molecular ions), fragments of the original structure, or of greater mass produced by reactions between the original compound and an ionizing reagent gas (as in CI) or sample matrix. The plot of ion counts versus ion mass-to-charge ratio (m/z) is called a mass spectrum. Usually only singly charged ions are produced and the m/z ratio is equivalent to ion mass. The abundance and m/z ratios of ions in the mass spectrum provides molecular weight and/or structural information. MS can be used in a direct inlet mode without prior GC for structure elucidation of high molecular weight materials that are not sufficiently volatile to pass through a GC column. Specific ionization and mass analysis modes are reviewed below and an application to the analysis of lipid A from Gram negative bacteria is presented in Chapter 10.

CAPILLARY GAS CHROMATOGRAPHY

A packed GC column is a tube packed with an inert support (such as diatomaceous earth) coated with an organic liquid (the stationary phase). A mixture of volatile solutes introduced at the head of the column is flushed through the system by the carrier gas (the mobile phase); each solute migrates at a different rate and is eluted into the detector as an approximately gaussian peak. Chromatographic efficiency is measured by the ability of the chromatographic system to transmit the sample components from inlet to detector with minimal peak broadening. Sharp peaks are desirable because peak resolution is improved and detector sensitivity is enhanced.

GC performance improved dramatically with the introduction of capillary columns by Golay.[1-3] Golay demonstrated that an open tube of small internal diameter coated with a thin liquid film yields greater efficiency than a packed column. The separation efficiency of packed columns is theoretically limited by multiple flow paths through the packing which lead to peak broadening; this contribution to peak spreading is not present in the open tubular column. More practically, the elimination of packing material lowers the resistance to carrier gas flow and makes possible the use of longer columns to solve more difficult separation problems. The higher resolving power of capillary columns can often be traded for increased speed of separation by using high carrier gas flow rate or shorter column length. The narrower peak widths in capillary GC compared to packed columns provide higher sensitivity due to higher concentration of eluting solute per unit time.

The relative efficiency of a capillary column compared to a packed column is illustrated in Figure 1.[4] Figure 1A shows a chromatogram of a standard mixture of sugar alditol acetates separated on a 6 ft. x 2 mm id packed column containing diatomaceous earth (Chromosorb W, particle size 149 μ) coated with a cyanopropyldimethylsiloxane stationary phase (SP-2330, 3% of the weight of the solid support). Figure 1B shows the same sample separated

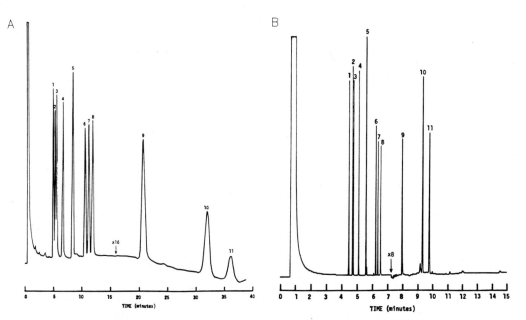

Figure 1. Comparison of packed and capillary column chromatograms of
alditol acetates: (A) 6 ft. packed column containing
Chromosorb W coated with 3% SP-2330; temperature program
starting at 200°C, ramped at 4°C/min to 245°C; (B) 20 m
capillary column coated with SP-2330; temperature program
starting at 100°C for 0.5 min, then ramped at 30°C/min to
245°C. Peak identities: (1) deoxyribose, (2) rhamnose,
(3) fucose, (4) ribose, (5) xylose, (6) mannose, (7) galactose,
(8) glucose, (9) muramic acid, (10) glucosamine, and
(11) galactosamine. Reprinted with permission from reference 4.

on a 20 m x 0.25 mm id capillary column whose interior walls are coated with
a thin film (0.25 μ) of the same stationary phase. Both chromatograms are
produced by programmed temperature GC, with the packed column separation
starting at 200°C and the capillary column separation starting at 100°C, both
ramping to a final temperature of 245°C. The packed column separation is
good, but the first three peaks (deoxyribitol, rhamnitol, and fucitol
pentaacetates) are not completely resolved, and the last peak requires 36 min
to elute. Although decreasing the initial temperature or the program rate
might improve the separation of the early eluting peaks, the overall analysis
time would suffer. The packed column has another disadvantage: the decrease
in the relative area of the amino sugar components in comparison to the
neutral sugars, due to adsorption or decomposition in the chromatographic
system. The capillary column separation of Figure 1B shows an analysis time
under 10 min, near baseline resolution of all components, greater inertness
to the amino sugars, and higher sensitivity. The significance of these
improvements over packed column GC methods for profiling and detecting
bacterial carbohydrates is discussed in Chapter 5.

The introduction of fused silica capillary columns was a major advance
in chromatographic technology.[5] Compared to columns made of borosilicate
glass, the low metal content of fused silica provides a chemically inert
support surface for the stationary phase. The polyimide outside coating on
fused silica columns makes handling and installation of fused silica
capillary columns easier because the columns are flexible and less easily
damaged than glass capillaries. "Bonded phases" are available in which the
stationary phase is cross-linked and immobilized in the column. Cross-

linking imparts greater temperature stability and decreases "bleed" of stationary phase at high temperatures. Because immobilized stationary phases are not readily extractable, when contaminated with sample residue the columns may be washed with solvent.[6] The improved inertness and ease of use of fused silica coupled with high resolution, sensitivity, and speed of separation have contributed to an expansion in applications of capillary GC in analytical microbiology.[7-10]

Capillary columns have some disadvantages. Capillary columns can be easily overloaded, producing poor peak shapes and impaired resolution. Capillary columns are also more demanding of instrumental performance from detectors: rapid response time (due to the sharp peaks) and high detector sensitivity (due to low sample capacity) are required. These difficulties are addressed by the proper design of capillary GC systems and by proper methodology. For laboratories still using packed column instruments, capillary upgrade kits are commercially available and wide bore (greater than 0.5 mm internal diameter) capillary columns may be used in packed column instruments.[11]

CARRIER GAS SYSTEM

In GC the mobile phase plays little role in the separation mechanism and serves only to transport the sample components from the injection system, through the column, to the detector. Nitrogen, helium, and hydrogen are the most commonly used carrier gases. Helium and hydrogen are preferred because their lower molecular weight promotes lower diffusion and produces higher efficiency. Hydrogen and helium also permit faster analysis because efficiency decreases less rapidly than with nitrogen at higher flow rates.[12] The choice of carrier gas can also be based on detector characteristics: the carrier is the background from which the detector must distinguish sample components and the choice of gas affects sensitivity of some detectors. Mass spectrometers typically require hydrogen or helium as a carrier gas; nitrogen may be used with flame ionization detectors. Although hydrogen may be preferred for fast analysis, its explosion potential and possible reactions with metallic oxides are of concern.[13] Carrier gas should be purchased at highest purity and further purified with in-line traps to scrub oxygen, water, and trace organic impurities.

SAMPLE INTRODUCTION SYSTEMS FOR GC

In capillary GC, most samples are introduced as liquids by syringe. The amount of sample that can be injected on capillary columns is less than that of conventional packed columns (by a few orders of magnitude); injection of a relatively large sample volume can easily overload the column. The low carrier gas flow rate in capillary GC also creates a broad initial band. Various injection methods, such as split, splitless, and cold on-column injection methods have been developed to deal with these problems. Two other sampling techniques, pyrolysis and headspace sampling, have also been employed in analytical microbiology and are described here.

Split injection technique, first introduced by Desty et al.,[14] is perhaps the most popular method of GC sample introduction. Practical aspects and theoretical reasoning for split, splitless, and on-column injections have been detailed.[15] During split injection, the evaporated sample is mixed with carrier gas and divided into two unequal streams. The smaller proportion is directed to the column and a larger proportion is vented through the split exit to the atmosphere. The sample amount entering the column is controlled by the ratio of the vent to column flow (the split ratio). Splitting allows the inlet vaporization chamber to be swept clear of sample by a high flow rate and a small portion of the injected sample is introduced to the column as a narrow band. Splitting ensures the sample volume transferred to the column is sufficiently small to avoid overloading the low amount of stationary phase in capillary columns.

The challenge of split injection is to divide the sample into two unequal streams of identical composition. Incomplete transfer and/or decomposition of higher molecular weight components in a sample may invalidate the analysis. Mass discrimination occurs as a result of fractional distillation in the injector, passing lower molecular weight components to the column in different proportions than higher molecular weight components. Raising the injection port temperature or loosely packing the injection volume (or, more commonly, an glass insert) with glass beads or glass wool to promote mixing of sample components and carrier gas may reduce discrimination, although catalytic decomposition of sensitive sample components may occur.[16] For example, GC analysis of mycolic acids may be irreproducible because of problems relating to differential degradation of these thermally sensitive compounds in hot injection ports.[17,18] Poor quantitation and poor reproducibility due to high molecular weight discrimination with split injection methods may often be improved by using appropriate internal standards.[4] Split injection methods are universally applicable for all but analyses in which sample components are present in low abundance.

In splitless injection, introduced in 1969,[19] the sample is injected into the heated injection port and vapor is slowly transferred to the analytical column by carrier gas flow. After a short time, the split valve is opened and a high flow is used to vent any sample and solvent remaining in the injection port. An injection port designed for split/splitless operation is shown in Figure 2. The high sensitivity of splitless sampling is vital for trace analysis of chemical markers in body fluids or tissues.

In splitless mode most of the sample is transferred to the analytical column while most of the solvent is vented. Splitless injection is often used to take advantage of the "solvent effect".[20-22] During splitless sampling, a large plug of solvent condenses on the head of the capillary column and traps sample components. Migration of components along the column does not occur until this solvent layer has evaporated. As the column temperature increases, trapped sample components are released in a sharp band. The initial wide injection band is refocussed on the head of the column before separation takes place. The strength of the solvent effect is affected by the column temperature, the solvent volatility, and the solvent amount. The initial column temperature should be 20-30°C less than the boiling point of the solvent which in turn should be less than the boiling point of the sample components. Too long a sampling period or too low a column temperature may cause excessive solvent condensation and result in peak broadening or stripping of stationary phase from wall coated columns that are not cross-linked. Temperature sensitive compounds may also be degraded by the longer residence time in the hot injection port.

Keeping the first 50-80 cm of the capillary column free of stationary phase produces a "retention gap" that accelerates migration and reconcentrates sample components at the beginning of the stationary phase film.[23,24] Rapid vaporization of the sample is not required and sample degradation is minimized.

Splitless methods have been compared to split injection methods for profiling of bacterial fatty acid methyl esters.[25] The chromatograms shown in Figure 3 from that work do not show the solvent effect because the column temperature was 50°C above the boiling point of the solvent. Cold-trapping of methyl esters does, however, occur under the conditions used because the

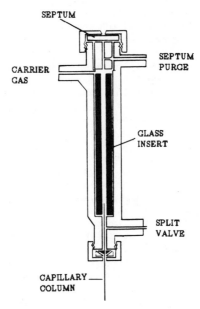

Figure 2.   Schematic of a capillary GC injection
            port designed for split/splitless operation.

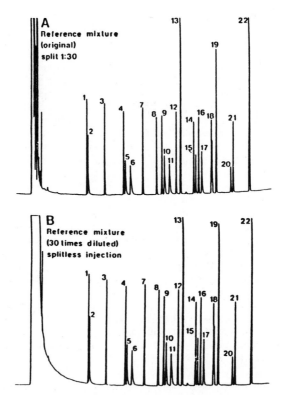

Figure 3.   Split injection methods (A) compared to splitless injection
            methods (B) for profiling a mixture of 22 fatty acid methyl
            esters.   Reprinted with permission from reference 25.

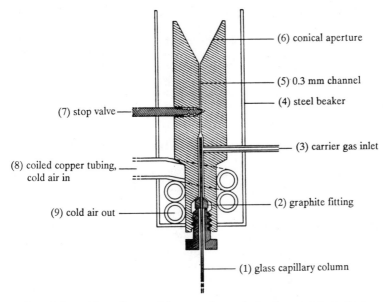

(6) conical aperture

(5) 0.3 mm channel

(4) steel beaker

(7) stop valve

(3) carrier gas inlet

(8) coiled copper tubing, cold air in

(9) cold air out

(2) graphite fitting

(1) glass capillary column

Figure 4.   Schematic of a cold on-column injector system for
            capillary GC.   Reprinted with permission from reference 27.

initial column temperature was 100°C below that of the first eluting peak.
The resulting chromatogram shows excellent sensitivity.

Injection problems such as sample dilution and band broadening
following injection, discrimination in favor of low molecular weight
compounds, degradation of thermally labile compounds at higher injection port
temperature are partially solved by the use of cold on-column injection
techniques which place the liquid sample directly on top of the capillary
column without heating or mixing with carrier gas.[26] Figure 4 shows a
schematic of a cold on-column injector.[27] The conical top of the injector
guides the syringe needle into the capillary column. The bottom of the
injector is cooled by circulating cold air. A syringe with a narrow diameter
needle which can fit into the capillary column is needed. Both stainless
steel or fused silica needles are available.

Cold on-column injection eliminates sample loss during vaporization,
maximizes sample transfer to the column, and reduces or eliminates sample
discrimination. Excellent quantitative precision of individual sample
components and elimination of thermal or catalytic decomposition have been
reported. However, the difficulty of interfacing to standard automated
injection systems make this septum-free injection technique somewhat unsuited
for high throughput routine analyses.

Programmed temperature injection utilizing a cold splitless injection
technique followed by programmed heating of the injection port has been
recently introduced,[15] but these techniques have not been thoroughly
evaluated in microbiological applications.

Headspace GC (HSGC) analyzes vapors from a closed vessel containing a
liquid sample, such as an anaerobic bacterial culture. If the vapor phase is
allowed to reach thermodynamic equilibrium with the liquid (or solid) sample
in a closed system, the concentration of components in the head-space vapor
is directly proportional to concentrations in the original solution.
Information on the chemical composition of the volatile components in the
liquid sample (e.g., alcohols, fatty acids, and amines) can be obtained by
this method without derivatization or extraction.[28,29]

25

Figure 5 shows a HSGC apparatus with a gas sampling valve. The sample is placed in a thermostatted container until equilibrium is established. The valve sample loop is filled with the volatile headspace gas by syringe and then injected into the column. HSGC automates the introduction of a volatile sample free of non-volatile contaminants which would normally have to be removed by some form of sample clean-up or solvent extraction. The sensitivity of HSGC can be improved by raising the temperature of the sample or by otherwise enhancing the distribution of solute(s) in the vapor over the liquid phases.

For trace analysis of headspace volatiles, techniques such as cryogenic trapping or columns packed with sorbents may be used to condense and concentrate gaseous samples.[30] Cold traps are generally maintained at a low temperature using liquid N_2 or dry ice.

The chemical characterization of volatile compounds present in headspaces above cultures of clostridia for clinical, biochemical, and industrial purposes is discussed in Chapter 7. Anaerobic bacteria from clinical isolates can be identified by the profiling of metabolites in spent culture media.[31-34] HSGC for the identification of bacteria is also discussed in Chapter 8. Direct diagnosis of anaerobic infections by analysis of body fluids for volatiles (e.g., low molecular weight fatty acids) without prior culture has been performed.[35]

GC requires the analyte to be volatile and thermally stable. Unless substituent groups confer exceptional volatility and are stable at high temperature (such as fluorinated derivatives), GC is generally limited to molecular weights less than 300-500, usually monomers or dimers. Thus, the first step in the GC analysis of monomeric chemical markers in bacteria involves their release by chemical means such as hydrolysis, saponification, or methanolysis.

Monomers released from the macromolecular structure of bacteria are often non-volatile, thermally labile, and possess complex functional groups; the second step in their GC analysis often involves <u>chemical derivatization</u>.

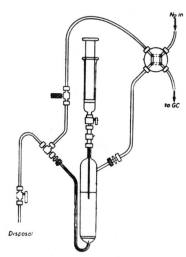

Figure 5.   Schematic of a head-space GC sampling system.
The sample loop (3) of a gas sampling valve (2)
is filled by syringe (4) from a sampling vessel (1).
Reprinted with permission from reference 124.

For example, because of hydrogen bonding and ionic interactions, most carbohydrates and amino acids are not volatile. Other compounds such as fatty acids are not normally analyzed by GC without derivatization because their polarity may cause adsorption effects resulting in poor chromatographic peak shapes. Derivatization inhibits these interactions and ideally produces a single, volatile, stable compound. Fatty acids are almost always analyzed as methyl esters (Chapter 4) in which the carboxyl group is reacted with methanol. For the analysis of carbohydrates, the alditol acetate derivatization method is popular (Chapter 5). In this procedure, the carbonyl group is reduced using $NaBH_4$ and polyols are reacted with acetic anhydride, producing a single peak for each derivatized sugar. Amino acids have both amino and carboxyl groups, often requiring derivatization steps for each: the carboxyl group may be esterified with butanol or propanol, the amino group may be acylated using halogenated anhydrides (Chapter 6).

Analytical pyrolysis[36,37] extends the utility of GC to allow direct introduction of samples that are not volatile due to high molecular weight. Whole bacterial cells may be profiled or polar materials such as bacterial polysaccharides can be analyzed for compositional information. Thermal fragmentation of a sample in the absence of oxygen produces a complex chromatogram (pyrogram), consisting of contributions from all structural components of the original sample. In some cases, pyrolysis may circumvent the need for derivatization and increase speed of analysis. The disadvantage of pyrolysis is the difficulty of interpreting the complex mixture of pyrolysis products and relating this pattern to the structure or composition of the sample. Validation and use of pyrolytically generated chemical markers and some microbiological applications of pyrolysis are described in greater detail in Chapters 12 and 13.

Two types of pyrolyzers are in common use: resistively heated pyrolyzers, and Curie-point pyrolyzers.[38] Resistively heated pyrolyzers usually consist of platinum ribbons or coils across which a capacitor is discharged (Figure 6A). Although manufacturers cite linear heating rates as fast as 20,000°C/s, actual temperature rates achieved by the sample are lower. With coil probes, samples are positioned within a quartz tube placed inside the coil; with ribbon probes, the sample is deposited or coated on the ribbon. In Curie-point pyrolyzers, a ferromagnetic wire coated with the sample is inductively heated by a radio-frequency field to the Curie-point temperature, where the alloy becomes paramagnetic and ceases to adsorb energy (Figure 6B). The Curie temperature is characteristic of the wire composition and may range from 300-1000°C. Either type of pyrolyzer can be connected directly to or mounted inside the injection port of the GC or MS for introduction of pyrolysates.[39]

STATIONARY PHASES FOR GC

If only nonpolar interactions occur between the components of a sample and a stationary phase, the elution order of the components will be the same as that of their boiling point. GC stationary phases are classified by the interactions promoted with solute molecules to be separated. These interactions are determined by the many intramolecular forces such as dispersion forces between nonpolar moieties, hydrogen bonding between hydrogen and electronegative atoms (e.g., O, F, N), polarity and polarizability causing either permanent or induced dipoles, and charge transfer (electron transfer).[40] Some common GC stationary phases are listed in Table 1 along with their relative polarity indexes and upper temperature limits; representative chemical structures are given in Figure 7.

In packed column GC, the selection of a stationary phase greatly influences the success of a separation. With high efficiency fused silica capillary columns, the selectivity of the stationary phase may not be as critical. Tuning of chromatographic selectivity is usually still required for difficult problems such as the separation of mono- and di-saturated fatty acids from unsaturated fatty acids (Chapter 4), the separation of enantiomeric D- and L-isomers of amino acids and sugars (Chapter 6 and Chapter 15 respectively), and the separation of sugar epimers such as galactosamine and mannosamine (Chapter 5).

Methyl silicones (polysiloxanes) are the most commonly used phases in capillary GC.[41] Polysiloxanes possess high thermal stability, a wide operating temperature range, resistance toward oxidation, and excellent film forming ability. The chemical structure of methyl silicone stationary phases is shown in Figure 7A. Methyl silicone phases are nonselective: compounds are separated by nonpolar or dispersive interactions and elution occurs in boiling point sequence.

Separations of most fatty acid mixtures can be achieved using methyl silicone stationary phases; however, purified methyl silicones such as OV-1 or SP-2100 with greater temperature stability are preferred and long (up to 50 m) capillary columns may be needed. Separations of heptafluorobutyryl-n-butyl ester derivatives of amino acids mixtures have also been successfully performed on methyl silicone columns.[42] Problems such as separations of mixtures of neutral and amino sugars typically require polar phases. Many polar phases are built on the nonpolar polysiloxane backbone by adding polar substituents.

The polar stationary phase, trifluoropropyl silicone, was introduced for the analysis of steroids and other natural compounds.[43] The most common stationary phase of this type (such as OV-210) is shown in Figure 7B and consists of 50:50 methyl and trifluoropropyl substituents. Because of the electron donating character of fluoro groups, this phase selectively interacts with dipolar functional groups such as carbonyl groups.

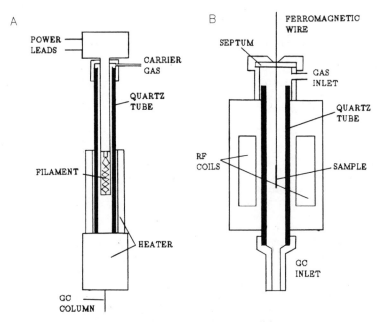

Figure 6. The two major types of pyrolysis interfaces: (A) a resistively heated pyrolyzer, and (B) a Curie-point pyrolyzer.

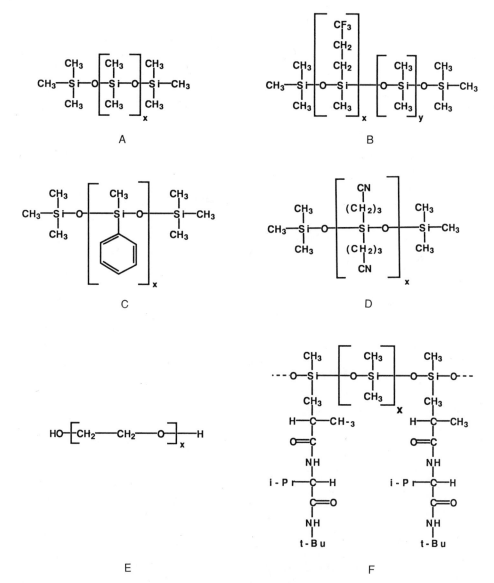

Figure 7. Chemical structures for some common GC stationary phases: (A) methyl silicone, (B) trifluoropropyl silicone; (C) a phenyl silicone; (D) cyanopropyl silicone; (E) Carbowax-20M[tm]; (F) Chirasil-val.

Table 1. Common GC stationary phases

Phase	Chemical composition	Upper temperature, °C	Relative Polarity index[a] (Squalane = 0%, OV-275 = 100%)
OV-101	methyl silicone fluid	260	5.0
OV-1 (SE-30)	methyl silicone gum 100% methyl	350	5.0
QF-1 (OV-210)	methyl silicone fluid 50% trifluoropropyl- 50% methyl	250	36.0
SE-52	methyl silicone gum, 5% phenyl-95% methyl	300	8.2
OV-17	methyl silicone polymer, 50% methyl-50% phenyl	350	21.0
SP-2330	methyl silicone fluid, 68% bis cyanopropyl- 32% dimethylsiloxane	250	83.0
Silar 10C (CP Sil 88)	100% cyanopropyl silicone	275	87.0
Carbowax 20M[tm]	polyethylene glycol	225	55.0
Chirasil-Val	chiral polysiloxane	240	
SE-54	methyl silicone polymer, 1% vinyl-5% phenyl	300	8.0
OV-1701	methyl silicone polymer, 7% cyanopropyl-7% phenyl- 86% dimethylsiloxane	250	18.0
OV-225	methyl silicone polymer, 25% cyanopropyl- 25% phenyl-50% methyl	265	42.0
Squalane	C_{30} alkane	100	0.0
OV-275	100% Cyanoethyl silicone oil	275	100.0

[a]information supplied by Supelco (Bellefonte, PA) and Chrompack (Bridgewater, NJ).

Phenyl silicones with 25% phenyl substitution were first used by James and Martin[44] in 1952. Phenyl silicones with differing ratios of phenyl to methyl groups are available (Figure 7C). Because phenyl groups are polarizable and have some electron donating capacity, this liquid phase exhibits enhanced polarity in comparison to methyl silicones. Fused silica coated with SE-52 is standard for separations of polyaromatic hydrocarbons.[45] OV-17 columns have been employed for rapid diagnosis of tuberculosis using separations of esterified tuberculostearic acid from patients' sputum.[46] Hewlett-Packard (Palo Alto, CA) introduced an automated fatty acid profiling system that employed a methyl-phenyl fused silica column (Chapter 4).

Methyl silicones containing cyanopropyl groups for enhanced polarity were first introduced in 1962.[47] One explanation for the selectivity exhibited by this liquid phase is interactions of the sort between nitrile groups and the π-electrons of olefins and aromatic compounds.[48] Complex mixtures of neutral sugars are not well resolved on less polar columns such as SE-54 or OV-17. If separations of mixtures of both neutral and amino sugars are desired (e.g., in analysis of bacterial hydrolysates) capillary columns coated with a polar selective phase such as SP-2330 (68 mole % cyanopropyl and 32 mole % methyl) are needed (Figure 7D).[4,49]

Polyethylene glycol (PEG) phases are the next most widely used of the polar phases (Figure 7E). Carbowax-20M[tm], with a molecular weight around 20,000, is the most commonly used polyglycol partly because it is one of the few polar phases which can be coated in a thin uniform film onto smooth glass surfaces to give an efficient column. PEG phases preferentially retain compounds which can hydrogen bond. Lack of temperature stability can be a disadvantage, however, and PEG phases cannot tolerate temperatures greater than 230°C for long periods of time. The Superox[tm] series, highly purified PEG's of higher molecular weight, give slightly better temperature stability (upper temperature around 250°C).[50] Free fatty acids, as well as free amines, are analyzed on polar phases such as Carbowax-20M[tm] coated on fused silica.[5,51]

For the separation of enantiomers, several chiral stationary phases have been prepared by modification of silicones.[52-54] Chirasil-L-Val, a chiral stationary phase for the separation of amino acid isomers, was synthesized by connecting L-valine-tert-butylamide to the carboxyl group of the copolymer of dimethyl- and carboxyalkylmethyl polysiloxane (Figure 7F). The principle, "like dissolves like", suggests that L-amino acids will interact more strongly with Chirasil-L-Val than will D-amino acids, and thus D-amino acids elute first. Formation of transient diastereoisomeric association complexes between solute and chiral phase due to hydrogen bonding may play an important role in the separation.[55] Chirasil-L-Val has been widely used for the separation and quantitative determination of amino acids isomers derived from bacterial peptidoglycan.[56-58]

Phases with intermediate polarities can be obtained by mixing non-polar and polar phases in different proportions and methods to calculate the optimal composition of the phase mixture for a particular separation are available.[59] For example, OV-1701 phases (7 mole % cyanopropyl, 7 mole % phenyl, and 86 mole % dimethyl silicones) and OV-225 phases (25% cyanopropyl, 25% phenyl, and 50% dimethylsiloxane) have been widely used for analysis of polar materials.

GC DETECTORS

GC detectors can be described in terms of sensitivity, minimum detectable quantity, selectivity, and linear dynamic range.

Detector _sensitivity_ is defined as the change in the measured signal
resulting from the change in detected quantity; i.e., sensitivity is the
slope of the calibration curve. Sensitivities depend upon detector design as
well as the compounds to be detected. The _minimum detectable quantity_ (MDQ)
is the minimum amount of sample detectable with a signal-to-noise ratio of 2.
Detectors with high sensitivities provide better MDQ. Detector sensitivity
should be matched with chromatographic operating conditions. If a detector
is not very sensitive, the capillary column has to be overloaded to obtain
detectable peaks.

Selectivity is the ability of the detector to respond to only certain
compound types, such as those containing nitrogen or phosphorus. Selectivity
improves analyses of complex samples by eliminating extraneous background
peaks from the chromatogram.

The _linear dynamic range_ of a detector is the range of concentration
(or mass flow rate) of the sample over which the proportionality between the
signal and concentration (or mass flow rate) is constant. Detectors should
not be operated in their nonlinear range to prevent peak distortion and
errors in quantitation.

The _flame ionization detector_ (FID) measures variations in the
ionization current as components pass through a hydrogen-oxygen flame. A
schematic of a FID detector is shown in Figure 8. The carrier gas and
eluting sample components are mixed with hydrogen and the mixture burns at
the tip of the jet with air ($2100°C$). Positive ions, negative ions, and
electrons produced in the flame give an ion current which is measured by
establishing an electrical field (400 V) between the negatively charged jet
tip and the positively charged collector electrode. The FID is a destructive
detector.

The linear dynamic range of the FID is 10^6 to 10^7. FID sensitivity is
dependent on ionization efficiency and the number of oxidizable carbon atoms
in an organic compound. The FID MDQ for methane is in the range of 10^{-11} to
10^{-12} g/s. Carbonaceous compounds are universally detected by FID; compounds
such as H_2O, N_2, and O_2 have little FID response. The FID is popular because
of its low cost, ease of use, relatively high sensitivity, and universal
detection capabilities for organic compounds.

The presence of alkali metal (such as rubidium) in an FID flame between
the collector electrode and burner jet increases the ionization efficiency of
organic compounds containing phosphorus or nitrogen.[60] Although the
detection mechanism is still incompletely understood, the _nitrogen-phosphorus_
detector (NPD) provides selective detection of these compounds. This
detector is also called the alkali FID or thermionic ionization detector. A
schematic diagram of the NPD, originally designed by Kolb, is shown in Figure
9.[61] The alkali source is separately heated to minimize noises due to flame
shape, size, and temperature.

Although the linear dynamic range varies depending on the compound of
interest, usually the NPD has a quite acceptable linearity of 10^3 to 10^4.
The NPD MDQ for nitrogen is about 10^{-13} g/s and for phosphorous about 10^{-14}
g/s. The NPD is 10^4 to 10^5 times more sensitive to nitrogen or phosphorous
than organic carbon. The NPD has been used in the analysis of nitrogen
containing compounds of microbiological significance such as amino acids and
amino sugars. Despite its sensitivity, the NPD has not been widely employed,
perhaps because the practical difference between the absolute sensitivity of
the FID and the NPD is not dramatic, especially if the number of carbons in
the compound of interest is much larger than the number of nitrogens. Using

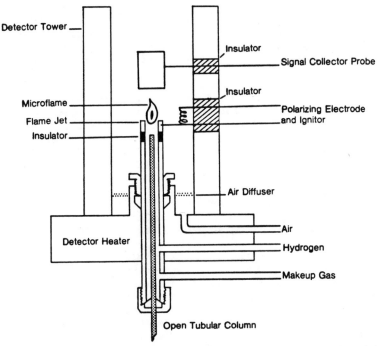

Figure 8.   Schematic of a flame ionization detector (FID).   Reprinted with permission from reference 13.

an NPD does provide selective information on a sample and thus may allow trace detection in situations of high background signal.

The <u>electron capture detector</u> (ECD) is based on the ability of eluting compounds to form negative ions by capturing electrons.[62,63] Derivatization is not required for compounds with high electron affinities such as those containing halogens. Unfortunately, few compounds of microbiological significance contain halogens and derivatization is usually required.

A schematic of a typical ECD detector is shown in Figure 10. Carrier gas is ionized by β electrons from ^{63}N or ^{3}H; slow thermal electrons and positive ions are formed. The potential (90 V) applied between two electrodes, either continuously or pulsed, allows a "standing" electric current to flow. Eluting molecules with high electron affinities will capture the thermal electrons in the detector. The change in standing current due to the loss of thermal electrons is proportional to the concentration of the sample.

The linear dynamic range of the ECD is 10^4 and the detection limit 10^{-13} g. The ECD is selective in the sense that only those compounds with high electron affinities in a mixture are detected. The detector response is dependent on both sample concentration and electron affinities. Proper quantitative analysis require accurate calibration. Although highly sensitive to halogenated derivatives and less expensive than a mass spectrometer, the ECD is not stable over longer periods and is subject to contamination. Unreacted halogenated derivatizing reagent or side-reaction products, if not removed prior to GC, can cause excessive background.

Increased sensitivity for profiling bacterial fatty acids has been demonstrated by using halogenated derivatives with the ECD.[64] Chapter 14 discusses such applications in greater detail. An automated approach for the preparation of halogenated derivatives has been described.[65] The sensitivity of the ECD has potential in the direct detection of microorganisms in body fluids.[66]

Fourier-transform infrared (FTIR) spectrometry measures absorption of infrared energy by molecules that change their dipole moment during vibration. By providing information on the presence of functional groups (such as carbonyl groups), FTIR spectra can contribute to elucidating molecular structure. The sensitivity of FTIR is low with an MDQ around 50 ng. Geometrical isomers can be differentiated by FTIR, but homologous compounds can not be distinguished without additional information from MS.[67] GC-MS coupled with FTIR has been developed[68,69] and is commercially available. FTIR can be a powerful tool for structure elucidation when MS fails to yield a complete structure; for example, the presence of a lactam in the aldononitrile acetate derivative of muramic acid was suggested by GC-FTIR.[70]

The microwave induced plasma detector (MIPD) can provide information on the presence of specific elements and can potentially determine the empirical formula of unknown molecules. A plasma is an ionized gas at a temperature of 4,000-10,000°K and is able to excite elements in a sample to induce characteristic emission. The sensitivity of the MIPD is adequate for trace analysis of biological compounds containing metal atoms, halogens, and elements such as C, H, N, P, O, and S. The MIPD is also now commercially available and applications should be forthcoming.[71-73]

MASS SPECTROMETRIC INSTRUMENTATION

Mass spectrometry is particularly advantageous for the analysis of complex samples encountered in analytical microbiology. In terms of

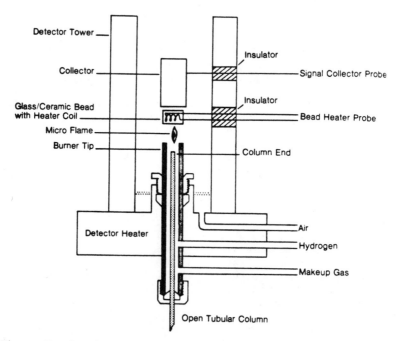

Figure 9. A schematic of a nitrogen-phosphorus detector (NPD). Reprinted with permission from reference 13.

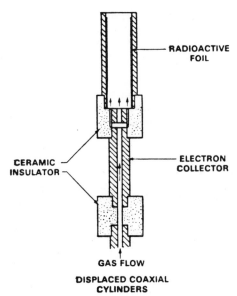

RADIOACTIVE
FOIL

CERAMIC
INSULATOR

ELECTRON
COLLECTOR

GAS FLOW

DISPLACED COAXIAL
CYLINDERS

Figure 10. Schematic of an electron capture detector (ECD).
Reprinted with permission from reference 125.

versatility, selectivity, sensitivity, and information content MS is rapidly
becoming the detector of choice for GC. Lower cost benchtop mass
spectrometers with quadrupole mass analyzers (Figure 11) are available. MS
is in transition from a specialized research tool to an indispensable adjunct
to routine GC. Microbiological applications of GC-MS to the analysis of
carbohydrates (Chapter 5 and 15), amino acids (Chapter 6), and fatty acids
(Chapters 4, 14 and 16) are discussed in this text.

Pure standards for microbial components are often not available and
qualitative identification of compounds by MS is an important driving force
for its use. The mass spectrum of a compound is often sufficient for
chemical identification. Even if the mass spectrum cannot be fully
interpreted or matched to a library spectrum or standard, characteristic
fragmentation patterns and masses often suggest structural features. Even
with the high resolving power of capillary GC, peaks may not represent pure
components. A single chromatographic peak consisting of a mixture of
coeluting components may often be deconvoluted using MS.[8] Although sample
preparation is often required, the instrumental measurement of a mass
spectrum is rapid. Ionization of sample molecules, ion separation, and ion
detection require only about 10^{-2}-10^{-5} s. When used as a selective detector
for ion masses characteristic of particular chemical markers, MS possesses
exquisite sensitivity for trace analysis.

Besides its use as a GC detector, MS is an important tool as a stand-
alone instrument for structural elucidation, particularly of higher molecular
weight materials. Large oligomers may not pass through a GC system or may
not volatilize or ionize efficiently with EI. Structural characterization of
oligomers and polymers by MS requires off-line purification (usually by
liquid chromatography), followed by direct probe introduction, soft
ionization, and mass analysis. Besides EI and CI, other ionization modes
include field desorption, plasma desorption, fast atom bombardment, and laser
desorption. If higher mass analysis is desired or more mass resolution for

accurate mass determination is needed, different mass analysis modes (such as double-focusing) are appropriate. Another increasingly popular MS mode described below is tandem-MS in which ions separated by an initial MS stage are further fragmented and directed into a second MS. Chapters 10 and 11 provide examples of some these applications.

Practical considerations related to MS vacuum systems, sample introduction, ionization modes, mass analysis, and ion detection are reviewed in the following sections.

MS VACUUM SYSTEMS

A vacuum system is crucial in MS to avoid ion-molecular collisions which will alter fragmentation and make interpretation of spectra difficult. Furthermore, any material introduced into the MS must be removed to minimize ion source contamination and background noise. Several vacuum pumping systems, including rotary pumps, diffusion pumps, and turbomolecular pumps are available.[74-76]

Rotary pumps are used to reduce the forepressure of high vacuum systems that employ diffusion or turbomolecular pumps; rotary pumps can be run directly against atmospheric pressure.[74] Gas molecules are drawn into the pump housing and pushed forward by rotor vanes; as the gas is compressed, the pressure opens an outlet valve, and gas escapes through an oil trap which lubricates and seals. Single-stage rotary pumps can achieve pressures of 10^{-2} torr; two-stage rotary pumps can produce 10^{-4} torr. Rotary pump speeds are in the range of 20-200 L/min of air, depending on size.

In diffusion pumps, the pump fluid is vaporized by electrical heating and emerges with supersonic speed through nozzles.[74] Air or any gas molecules diffused into this vapor stream are brought toward the bottom of the pump and removed by a backing pump. Cooling is accomplished by circulating water through coils. Commonly used pumping fluids include polyphenyl ether or silicone oils, and mercury; the vapor pressure of the pumping oil determines the obtainable pressure. Diffusion pumps can not operate against atmospheric pressure and need to be backed up by other pumps such as rotary vacuum pumps. The working range of diffusion pumps is from 10^{-3} to 10^{-11} torr. Diffusion pumps can typically remove 100-1000 L/s, depending on pump size at 10^{-5} torr; large pumps with capacities of more than 50,000 L/s are available.

In a turbomolecular pump, gas molecules collide with the surface of rapidly moving (50,000 rpm) rotor blades and are removed from the pump.[75-76]

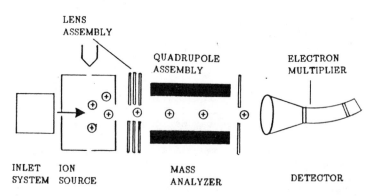

Figure 11. Schematic of a quadrupole mass spectrometer.

At this speed the blades easily disintegrate if hit by small particles, and the gas must be filtered before entering the housing. The working pressure range and pumping speed of typical turbomolecular pumps is about the same as that of diffusion pumps. Turbomolecular pumps reach their operating pressure faster than diffusion pumps and allow more rapid maintenance and return to operating conditions.

MS SAMPLE INLET SYSTEMS

Samples are generally introduced to the ion source as gases. Ion source temperatures are typically 150-250°C. Ion source pressures should be less than 10^{-6} torr to avoid collisions of ions with other molecules. To keep ion source pressure low and constant, the amount of sample introduced must be small and produce a sample pressure less than about 10^{-2} torr. Inlet systems are designed to satisfy these requirements.

Gases, liquids, or solids can be volatilized in a heated storage reservoir (1-5 L capacity) and passed into the MS ion source through porous material or pinholes. Solid or liquid samples can also be introduced directly into the ion source chamber by means of a heated direct insertion probe.

GC carrier gas flows are usually between 1 and 70 mL/min at 1 atmosphere pressure; commonly used pumps can handle 500 L/s at 10^{-5} torr. A modern MS system with high capacity vacuum pumps can easily handle the 1-2 mL/min flow rate from a typical capillary GC column. Direct capillary column interfacing allows all the GC effluent to enter the MS ion source and maximizes sensitivity. Packed GC column flow rates are 20-30 times too high for direct interfacing; a variety of separation devices (such as effusion, membrane, and jet separators) are available for removing excess carrier gas prior to introduction to the MS.[76-78] As most modern applications of GC-MS in analytical microbiology involve capillary columns we will not discuss interfacing packed column GC to MS further.

MS IONIZATION METHODS

In _electron impact_ (EI) sample molecules are bombarded with electrons to produce ionized fragments.[79] EI is a fairly energetic process which tends to fragment sensitive compounds severely, often to the extent that the molecular ion is not observed in the mass spectrum. Electrons are produced from a heated tungsten or rhenium filament and are accelerated at 70 eV in a high vacuum toward a trap (anode). Sample molecules are introduced perpendicular to the electron beam, ionized, and positively ionized molecules are pushed toward the mass analyzer by a repeller and/or draw-out potential. Ionization efficiencies are typically low, on the order of one per thousand molecules.[80] Although 70 eV accelerating voltage is standard, lower voltages may produce more molecular ions and fewer fragment ions, thus providing molecular weight information.

Neutral sample molecules (M) that collide with sufficiently energetic electrons (e^-) can produce positive radical ions ($M^{+\cdot}$) with loss of an electron:

$$M + e^- \rightarrow M^{+\cdot} + 2 e^-$$

Other processes can also occur, such as electron capture,

$$M + e^- \rightarrow M^-$$

or the formation of doubly charged ions,

$$M + e^- \rightarrow M^{2+} + 3 e^-$$

depending on the nature of the sample molecules.[80] The energy required to ionize a neutral organic molecule is typically under 20 eV and the excessive electron energy tends to produce further fragmentation of the molecular ion to smaller charged or neutral species. The plot of abundances of all ions produced as a function of mass-to-charge ratio (the mass spectrum) can be interpreted on the basis of simple cleavage reactions or intramolecular rearrangements that occur in predictable patterns.[81] The EI fragmentation pattern is a signature that can identify an unknown compound by matching with a library spectra or, even better, matching with the mass spectrum of a pure standard.

Chemical ionization (CI) uses collisions with a reagent gas (e.g., methane, isobutane, or ammonia) to induce ionization and tends to produce less fragmentation and a more abundant molecular ion peak. In positive chemical ionization[82], the reagent gas (CH_4) at about 1 torr pressure is ionized by a stream of energetic electrons (240 eV) to give the reactive species $CH_4^{+\cdot}$:

$$CH_4 + e^- \rightarrow CH_4^{+\cdot} + 2 e^-$$

which can undergo ion molecule reactions to produce $C_2H_5^+$:

$$CH_4^{+\cdot} + CH_4 \rightarrow CH_5^+ + CH_3\cdot$$

$$CH_3^+ + CH_4 \rightarrow C_2H_5^+ + H_2$$

Although CH_5^+ and $C_2H_5^+$ are the major ions, small amounts of ions such as $C_3H_5^+$ and other ions are also formed. These three ions (at m/z 17, 29, and 41) are often used to tune MS instruments for chemical ionization. Reagent ions further react with sample molecules through a variety of ion-molecule reactions such as proton transfer,

$$CH_5^+ + MH \rightarrow MH_2^+ + CH_4$$

hydride abstraction,

$$C_2H_5^+ + MH \rightarrow M^+ + C_2H_6$$

and alkyl transfer with charge exchange,

$$CH_5^+ + MH \rightarrow MHCH_5^+$$

Differences between EI and CI spectra are illustrated in Figure 12, showing mass spectra of the alditol acetate of a aminodideoxyhexose (X1) found in a sample of *Fluoribacter (Legionella) bozemanae*.[8] The EI mass spectra (Figure 12A) is similar to that of glucosaminitol hexaacetate, except that the highest mass is 302 instead of 360; this 58 mass unit shift indicates replacement of acetate with hydrogen on one of the carbon atoms. The CI spectrum (Figure 12B) of the same peak is simpler. The peak at m/z 376 represents the protonated molecular ion (M + H). The loss of acetic acid (M + H - 60) and ketene (M + H - 42) and the addition of C_2H_5 (M + 29) are also observed at m/z 316, 334, and 404. Other fragmentation is minimal in the CI spectrum. For structural identification and molecular weight determination both EI and CI methods are often necessary: EI because the fragmentation pattern provides structural detail, CI because the prominent molecular ion peak provides molecular weight information.

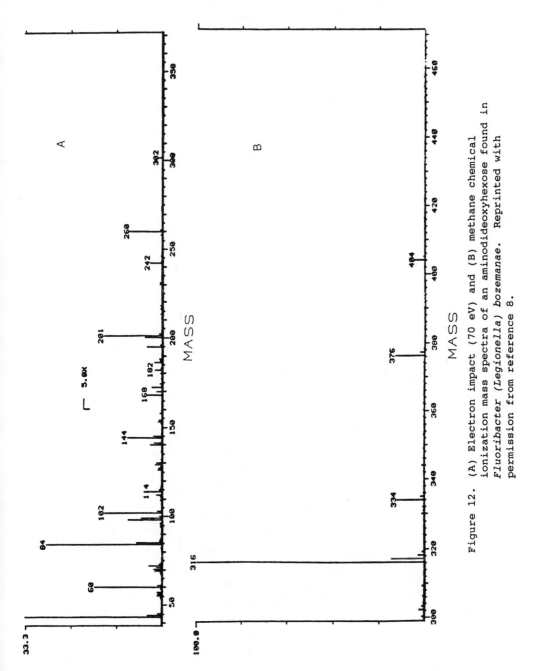

Figure 12. (A) Electron impact (70 eV) and (B) methane chemical ionization mass spectra of an aminodideoxyhexose found in *Fluoribacter (Legionella) bozemanae*. Reprinted with permission from reference 8.

Even in EI, negative ions (M⁻) are formed by electron capture. In positive ion EI, these ions are usually discharged by collisions or move in the direction opposite to the positive ions. These negative ions may be detected by reversing the electrical and/or magnetic fields in the ion source region.[83-86] At source pressures employed in CI (1 torr), formation of negative ions dominates over that of positive ions because low energy electrons ("thermal electrons") produced by collisions with reagent gas (e.g., methane) are easily captured by highly electronegative molecules.[83] Negative ions can be formed by the interaction between electrons and sample molecules in several mechanisms[80]:

$$MX + e^- \rightarrow MX^- \quad \text{(Resonance capture)}$$

$$SF_6 + e^- \rightarrow SF_6^- \quad (0.08 \text{ eV})$$

$$MX + e^- \rightarrow M^- + X^\cdot \quad \text{(Dissociative resonance capture)}$$

$$HBr + e^- \rightarrow Br^- + H^\cdot \quad (0.8 \text{ eV})$$

$$MX + e^- \rightarrow M^+ + X^- + e^- \quad \text{(Ion-pair production)}$$

$$CH_3Cl + e^- \rightarrow CH_3^+ + Cl^- + e^- \quad (10 \text{ eV})$$

The energy of the electron determines which of these processes predominates. In general, resonance capture takes place when electron energy is near 0 eV, dissociative resonance capture is dominant in the 0-15 eV range, and ion-pair production only occurs with electrons of more than 10 eV.[85] Under EI conditions, negative ions are formed primarily by ion-pair production; under CI conditions, they are predominantly formed by resonance capture and dissociative resonance capture. Formation of negative ions by resonance capture and dissociative resonance capture in CI strongly depends on the electron affinities of sample molecules: aromatics and halogenated compounds easily form negative ions. Negative ion detection enhances selectivity and sensitivity over positive ion detection for some molecules.

In analytical microbiology, negative ion chemical ionization (NICI) has been used for trace detection of halogenated derivatives of marker compounds. Selectivity is also increased since underivatized materials are not detected. Minimal fragmentation of the sample insures production of a molecular ion and gentler ionization decreases background signal.

Figure 13A shows a NICI mass spectrum of a derivatized standard of D-alanine.[56] The molecular weight of the butyl heptafluorobutyryl derivative of D-alanine is 341; a prominent peak is present at m/z 321, due to loss of HF. Figure 13B shows the mass spectrum of a D-alanine peak at the same nominal GC retention time in a hydrolysate of rat liver tissue taken 24 hr after the animal was injected with a bacterial cell wall preparation. The ions at mass m/z 321, originating from bacterial D-alanine, predominate. Hydrolysis with deuterated HCl to release D-alanine from peptidoglycan forms labelled D-alanine by racemization of protein L-alanine, producing another smaller peak at m/z 322. Figure 13C shows the spectrum taken at the same retention time in a control liver hydrolysate. In this instance the base ion is at m/z 322 since most of the D-alanine is derived from racemization and labelled with deuterium. Chapter 6 discusses this application in greater detail.

In field ionization, positive ions are produced from molecules subjected to intense electric fields of about 108 V/cm between a sharp point or edge (anode or emitter) and an electrode (cathode).[87] Molecules in high electric gradients produce positive ions by quantum tunnelling of electrons from molecules to the positively charged emitter. Gas phase compounds may be

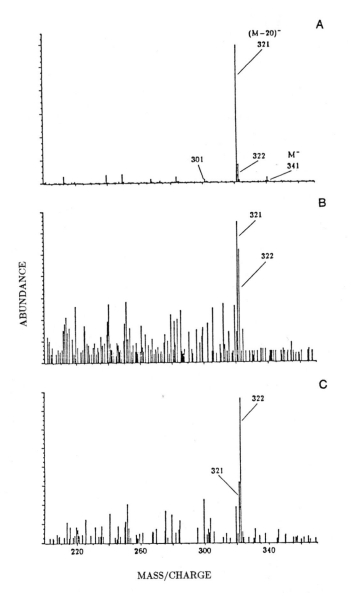

Figure 13. Mass spectra of butyl heptafluorobutyryl
derivatives of D-alanine from (A) a standard sample
of D-alanine, (B) a liver hydrolysate of an animal
injected 24 h previously with streptococcal cell
walls, and (C) a liver hydrolysate from a control
animal. Reprinted with permission from reference 56.

ionized by passing close to the electric field without being deposited onto
surfaces. Since excess internal energy is not transferred to molecules,
field ionization spectra exhibit fewer fragment ions.

In _desorption ionization_, gas phase components of samples are generated
and ion source pressures are kept low. Heat is commonly employed to vaporize
samples, although thermally labile or nonvolatile compounds can be chemically
derivatized before analysis. Derivatization usually requires additional
clean-up steps, introduces background noise, and increases the molecular

weight of the analyte molecule. To overcome these problems, several desorption ionization techniques have been developed in which sample is desorbed from the surface as an ion without decomposition using methods such as particle bombardment, high electric fields, and lasers.[88]

Fast atom bombardment (FAB) MS has been used to analyze large biomolecules and to sequence peptides.[89,90] High energy particles (usually 8 keV Xe or Ar atoms) from a particle beam gun strike a sample dissolved or suspended in a liquid phase and the resulting secondary ions are detected.[91] Secondary ion yields are dependent on the coverage of sample by the primary particles, the mass of the primary particles, the angle of the impact, and the distance between the gun and sample.[92] Dissolving or suspending the sample in liquid matrices permits continual refreshment of the sample surface which reduces damage to the sample and allows the use of higher energy particle beams for increased secondary ion current. Chemical properties of the liquid matrices (such as pH and the presence of additives) determine the ion abundance.[92] Low volatility of the matrices is important so that they remain intact at high temperature and low source pressures. Among several liquid phases, thioglycerol and glycerol (positive ion mass spectra) and diethanolamine (negative ion mass spectra) are commonly used. FAB MS can be used to obtain molecular weight and structural information on nonvolatile and thermally unstable compounds in complex biological matrices without prior chemical ionization or sample clean-up.[93] Caprioli and Smith[94] used FAB MS to study enzymatic reactions and calculated kinetic parameters, which were later verified by GC-MS. Heller et al.[95] differentiated Gram positive bacteria from Gram negative bacteria and characterized different species by their composition of polar lipids using FAB MS. Combining FAB with tandem MS may also prove valuable for obtaining structural information on biomolecules.[96]

Secondary ion (SI) MS is closely related to FAB MS: secondary ions from a sample are created by bombardment with energetic ions (such as Ar^+, Xe^+, and Cs^+) instead of neutral particles.[97] Samples are usually supported on liquid phases such as glycerol although coating on thin metal films are also employed.

Plasma desorption (PD) MS[98,99] has provided molecular weight information on molecules up to 34,633 daltons. A ^{252}Cf source produces about 180 MeV/atom; more than 2000 fission fragments per second pass through a metal film on which the samples are deposited. Spectra resulting from desorbed positive and negative ions, electrons, and neutral molecules are accumulated for long times (many hours) to avoid statistical fluctuations from low ion yields. Samples are generally deposited as a thin film on mylar or Nafion (a perfluorinated cation-exchange polymer). Nitrocellulose membranes[100] and glutathione sample additives[101] have been found to improve molecular ion yields and enhance formation of multiply charged ions. For large molecules, PD MS contains mainly molecular ions with few interpretable fragmentations; for smaller molecules, interpretable fragment ions are often generated. Garcia-Bustos et al.[102] used PD MS to study peptide moieties from the peptidoglycan of the Gram positive bacterium, *Streptococcus pneumoniae*. A structural analysis of the lipid A portion of lipopolysaccharide from the cell wall of the Gram negative *bacterium E. coli* was reported by Cotter et al.[103] Nisin (a polycyclic peptide antibiotic), intact cecropin B (an antibacterial protein), and interleukin-1 and -2 have been also studied by PD MS.[104,105]

Rapid heating of a sample coated on a thin wire induces desorption or evaporation before decomposition takes place and is the basis for laser desorption methods.[106] Pulsed lasers are often used to obtain high heating rates for thermal desorption and ionization of large biomolecules.[107] In laser desorption, sample is deposited on a metal support and a pulse of infrared radiation from a CO or UV radiation from a neodymium YAG laser is focused to a small spot (less than 1 μ^2). The absorption of the laser energy by the metal support results in a rapid temperature rise and causes desorption and ionization of the sample. Time of flight is often employed for mass analysis. Laser beam power density can be controlled to obtain either molecular ions (less than 108 W/cm^2) or small fragment ions (higher than 109 W/cm^2).

In <u>field desorption</u> (FD), sample is coated directly onto a metal surface of removable electrode and high field is applied; in high vacuum, positively charged particles are ejected and analyzed by MS.[108] FD is a gentle ionization method and molecular ions dominate spectra. Desorption may also be combined with CI as was done by Dutton et al.[109] in sequencing bacterial oligosaccharide antigens.

MASS SEPARATION TECHNIQUES

In the <u>magnetic sector</u>, or single-focusing analyzer, ions generated in the ion source are accelerated by a uniform electrostatic field through focussing slits toward a magnetic sector. In the uniform magnetic field ions experience a centrifugal force which curves their path. The radius of curvature of an ion is dependent on the mass-to-charge ratio and can be controlled by either varying the magnetic field or the accelerating potential. Because sensitivity decreases with lower accelerating voltage, the acceleration voltage is usually kept constant and the magnetic field is varied to scan the mass spectrum. Ions with curved paths matching the radius of curvature of the analyzer tube in the magnetic field pass through and are focussed on an exit slit for detection. Transmission efficiency of the single focusing analyzer is only 1-4%.

The spread of translational energies in the ion beam from the heated ion source filament give rise to a divergent beam of ions from the accelerator and impedes accurate mass measurement. The resolving power of MS analyzers may be quantitated by referring to the ratio of the mass to be measured (M) divided by the difference in mass (ΔM) of two mass peaks that are resolvable to 10% of their heights:

$$R = M / \Delta M$$

Since ions leaving the ion source are not focused in terms of kinetic energy, only low resolution spectra (R = 1,000-5,000) can be obtained on single-focusing instruments.

<u>Double focusing</u> mass analyzers offer higher mass resolution by focussing ions by an electrostatic analyzer according to their kinetic energy and then refocussing the ions using a magnetic analyzer according to their mass-to-charge ratio. The resolving power of double-focusing MS is at least 20,000-30,000, and mass analysis as high as 2500 is possible.

<u>Time of flight</u> (TOF) mass analyzers, accelerate ions electrostatically through slits; ions of different masses acquire different velocities, depending on their mass-to-charge ratio. After allowing these ions to separate in their travel down a drift tube, they arrive at a detector at different times. Although TOF resolving power is low, this analysis mode is widely used in studies of gas phase reactions because of the exceptionally fast response time (10^{-4} s). Transmission efficiencies are high, around 50%. Because TOF ion sources are open, CI can not be performed.

<u>Quadrupole</u> mass analyzers (Figure 11) pass the ion beam from the ion source into four precisely arranged parallel rods. Only ions within a narrow mass region can go through the rods, depending on the ratios of a DC voltage

to a superimposed radio-frequency (rf) potential applied between each pair of rods.[110-111] With the rf/dc voltage ratio held constant, the voltage magnitudes are varied to achieve ion separation according to mass.

Many features of the quadrupole mass analyzer make it ideal as a detector for capillary GC. Quadrupole mass analyzers are compact, the entire mass range (up to m/z 1200) can be scanned rapidly (milliseconds), and they have a high angle of acceptance. The quadrupole's compactness is important because the short path length between the ion source and the detector allows ion transmission at relatively high pressure (10^{-5} torr) without many collisions. Single- or double-focussing magnetic analyzers cannot be scanned at rates faster than about 0.1 s across the mass range from 40 to 400 mass units. The fast scan capability of the quadrupole is advantageous if compounds elute in narrow bands from chromatographic columns. The high angle of acceptance means that a higher percentage of ions impact the detector (transmission efficiency of about 10%), another feature that is important for trace analysis. The maximum mass range and resolution are determined by a variety of factors (geometry, ion source conditions, and rf parameters); mass ranges up to 1000 m/z and mass resolutions of 2000 are common in commercial quadrupole instruments. Quadrupoles are, however, insensitive to metastable ions.

The quadrupole ion trap (QIT) has been widely used as a detector for GC because of its compactness: an ionizer assembly is not required.[112] A typical QIT consists of a ring electrode and two end-cap electrodes. An rf voltage is applied to the ring electrode, while the two end-cap electrodes are grounded. Sample is introduced through one of the end-cap electrodes and ionized by EI. Under appropriate conditions, all ions whose m/z ratios are greater than a minimum cut-off value (set by the magnitude of the rf voltage) are trapped; smaller ions are ejected through the other end-cap electrode and detected with an electron multiplier. A mass spectrum is obtained by increasing the rf voltage and detecting the ions coming out from the end-cap electrode.

An ion moving in the direction perpendicular to a magnetic field experiences a constant acceleration perpendicular to its direction of motion, which will cause the ion to travel in small circular orbits.[111-114] The frequency of the revolution (cyclotron frequency) is inversely proportional to the mass-to-charge ratio of the ion; the radius of the circular trajectory is proportional to ion velocity and inversely proportional to frequency of revolution. When an rf electric field is applied perpendicular to both the magnetic field and the initial direction of motion of the ion, and when its frequency matches the cyclotron frequency, the ion will absorb energy and begin to spiral (a resonance condition). The current induced by the cyclotron motion of ions in resonance is detected and an ion cyclotron resonance (ICR) mass spectrum is obtained by sweeping either the magnetic field or the rf electric field.[115] The amplitude of the current is related to the number resonance ions and the frequency of the signal is related to the mass-to-charge ratio of the ions.

In Fourier transform (FT) ICR MS, a rapidly scanned pulse of frequencies is applied to excite all ions simultaneously in an analyzer and the signal, due to all resonant ions, is measured.[113-115] FT ICR MS provides high mass resolution and rapid scanning capability. High mass resolution (1,000,000 at m/z 100) is required for the analysis of high molecular weight biological compounds. The rapid scanning capability of FT ICR MS reduces the time required to record a mass spectrum, which is useful in capillary GC-MS or tandem MS. FT ICR MS is still in development but will soon be a more attractive method for microbiological applications.

In <u>tandem mass spectrometry</u> (MS-MS), parent ions separated in the first MS are further fragmented; daughter ion fragments from these parent molecules are then analyzed in a second stage mass analyzer.[75,116] Tandem MS may be employed following a GC separation or following direct introduction of a solid, liquid, or gaseous sample. Separations and identification of components of a mixture can be performed by tandem MS. For example, a magnetic analyzer (MS-I) could be used to separate molecular ions or parent ions of a given compound or a mixture of compounds, produced by a soft ionization technique such as CI or FD, according to their mass-to-charge ratios. Parent ions are then fragmented by collisions with neutral molecules (e.g., helium). Finally, a mass spectrum is produced by measuring the kinetic energy of each fragment, which is proportional to mass, by an electrostatic analyzer (MS-II).

In general, addition of a second mass analyzer enhances the resolution and sensitivity of MS and various versions of MS-MS exist. For instance, an magnetic analyzer can be added to MS-II to bring ions of the same mass but different kinetic energy to a common focus, increasing sensitivity and resolution; an additional electrostatic analyzer could be added to MS-I. Yost and Enke constructed quadrupole MS-MS instrumentation in which a selected parent ion (from MS-I) can be fragmented by collisions with neutral molecules in a center quadrupole (an rf-only quadrupole trap) used to focus scattered ions and efficiently transmit them into MS-II.[117-118]

Recently, the quadrupole ion trap was evaluated for use in MS-MS.[119] In this MS/MS system, selected parent ions generated by electron impact are trapped and undergo collision activated dissociation (CAD), while other ions are ejected from the trap. This system only requires a single mass analyzer which makes the instrument small compared with other systems. Ion trapping and collision steps can be repeated indefinitely, effectively providing an MS instrument raised to the nth power with associated enhanced signal-to-noise ratios.

Tandem MS offers advantages over GC- or liquid chromatography-MS: time-consuming sample preparation prior to analysis may not be necessary. MS-MS is a rapid procedure for analysis of individual components in complex biological mixtures. In conjunction with capillary GC, MS-MS may be useful for rapid identification, classification and characterization of microorganisms.

MS DETECTION METHODS

The <u>electron multiplier</u> provides high sensitivity and rapid response for ion detection.[120] A typical electron multiplier consists of 15-20 Cu-Be dynodes. An ion beam hits the first dynode and secondary electrons are emitted to the next dynode where the process is repeated, amplifying the signal by up to 10^6.

The <u>Faraday-cup collector</u> was developed to detect low ion current (less than 10^{-9} A).[121] The detection mechanism is based on a long metallic cylinder (Faraday-cup) oriented so that ions can travel directly into the cup. Electron suppressor slits and screens located at the entrance of the cylinder also focus ions into the cup. Ion currents as low as 10^{-15} A are detectable. For detection of ion currents below 10^{-15} A, electron multipliers are better.

<u>Photographic detection</u>[122] is often used with the classical Mattauch-Herzog geometry[123] because the complete spectrum can be simultaneously recorded at high resolution. The instrument requires calibration before use because the response is not linearly related to the number of ions, and photographic plates are less convenient.

MS sensitivity can be increased by monitoring those prominent mass ions in the spectrum of the desired analyte. A mass spectrometer operating in selected ion monitoring (SIM) mode is a versatile, stable, and selective detector. With an appropriate selection of the monitored ions, all components of interest may be detected at trace levels and the background ignored. SIM is not limited to a particular class of compounds (as is the NPD) and does not require specific chemical derivatives (as does the ECD). To enhance selectivity, characteristic high mass analyte ions are often chosen for their uniqueness and absence from the background spectra. Detection limits in the range of 10^{-12} g have been reported for SIM GC-MS.

CONCLUSIONS

This chapter has reviewed instrumental options for analytical microbiologists interested in applying modern GC and MS to their problems. Capillary GC and MS has added new dimensions to the selectivity and sensitivity with which microbiological compounds can be separated and identified.

GC instruments equipped with a glass capillary inlet system, a single fused silica capillary column, and a flame ionization detector are adequate for many tasks. Polar phases appropriate for separation of chemical markers of microbiological significance are now available as cross-linked or "bonded" phases; fused silica columns coated with these phases have great durability, low column bleed, and permit high resolution separations at high speed. The routine analysis of multiple samples is facilitated by automating the sample injection. For sample pretreatment prior to GC, not much automation is currently available; several manufacturers have recently introduced workstations and robotic systems for automating routine steps such as addition of solvents and reagents, evaporation, liquid-liquid and solid-phase extraction, and mixing.

Determining chemical composition of microbiological samples or elucidating structure of bacterial components requires low resolution GC-MS at a minimum. Lower cost benchtop units using fixed 70 eV EI and quadrupole or ion trap mass analysis are available providing unit mass resolution up to 1,000 daltons. The cost of such instruments is typically 5-10 times the cost of a GC instrument, however, the additional capabilities and benefits far outweigh this cost. User-friendly operating software with integrated graphics, data processing, and report generation are usually integrated together. Benchtop GC-MS instruments are also applicable for trace analysis using SIM. A MS in the SIM mode is a versatile, stable, and selective detector. Through appropriate selection of the monitored ions, all components of interest may be detected and background from the sample matrix may be ignored. SIM GC-MS is not limited to a particular class of compounds (as is the nitrogen phosphorus detector) and does not require specific chemical derivatives (as does the electron capture detector). CI may be added as an option to some instruments at this level.

Structural elucidation of large microbial oligomers or polymers cannot be performed by GC or by benchtop GC-MS and require a higher level of instrumental sophistication and, commensurately, higher cost. High resolution systems with multiple soft ionization and desorption modes (e.g., FD or FAB) and multiple mass analysis modes (e.g., MS-MS) typically involve 5-10 times the cost of benchtop instruments.

A barrier still remaining that inhibits the application of chromatographic and mass spectrometric methods in analytical microbiology is the steep learning curve. Simple and rapid sample preparation procedures must be chosen to present the compound(s) of interest to the instrument in an appropriate form and chromatographic or mass spectrometric conditions for analysis must be chosen. The development of methods that specify, standardize, and integrate together the complete sequence of steps to be carried out in the laboratory, from sample preparation to data reporting, will aid in the more widespread use of these methods in analytical microbiology.

ACKNOWLEDGEMENTS

This work was supported by grants from the U. S. Army Office of Research and by the National Institutes of Health.

REFERENCES

1. M. J. E. Golay, Vapour Phase Chromatography and the Telegrapher's Equation, Anal. Chem. 29:928 (1957).
2. M. J. E. Golay, "Theory of chromatography in open and coated tubular columns with round and rectangular cross-sections," in: "Gas Chromatography" D. H. Desty, ed., Academic Press, New York (1958).
3. M. J. E. Golay, "Theory and practice of gas liquid partition chromatography with coated capillaries," in: "Gas Chromatography", V. J. Coates, H. J. Noebles, and I. S. Fagerson, eds., Academic Press, New York (1958).
4. A. Fox, S. L. Morgan, J. R. Hudson, Z. T. Zhu, and P. Y. Lau, Capillary gas chromatographic analysis of alditol acetates of neutral and amino sugars in bacterial cell walls, J. Chromatogr. 256:429 (1983).
5. R. D. Dandeneau and E. H. Zerenner, An investigation of glasses for capillary chromatography, HRC&CC 2:351 (1979).
6. K. Grob and G. Grob, Capillary columns with immobilized stationary phases II. Practical advantages and details procedure, J. Chromatogr. 213:211 (1981).
7. K. Bryn and E. Jantzen, Analysis of lipopolysaccharides by methanolysis, trifluoroacetylation, and gas chromatography on a fused-silica capillary column, J. Chromatogr. 240:405 (1982).
8. M. D. Walla, S. L. Morgan, P. Y. Lau, and A. Fox, Capillary gas chromatography-mass spectrometry of carbohydrate components of legionella and other bacteria, J. Chromatogr. 288:399 (1984).
9. J. Gilbart, A. Fox, R. S. Whiton, and S. L. Morgan, Rhamnose and muramic acid: chemical markers for bacterial cell walls in mammalian tissues, J. Microbiol. Meth. 5:271 (1986).
10. J. Roboz, D. C. Kappatos, and J. F. Holland, Role of individual serum pentitol concentrations in the diagnosis of disseminated visceral candidiasis, Eur. J. Clin. Microbiol. 6:708 (1987).
11. R. T. Wiedemer, S. L. McKinley, and T. W. Rendl, Advantages of wide-bore capillary columns, Amer. Lab. 18:110 (1986).
12. W. Jennings, "Gas Chromatography with Glass Capillary Columns", 2nd ed., Academic Press, New York (1980).
13. M. L. Lee, F. J. Yang, K. D. Bartle, "Open Tubular Gas Chromatography", John Wiley, New York (1984).
14. D. H. Desty, A. Goldup, and B. A. F. Whymann, Potentialities of coated capillary columns for gas chromatography in the petroleum industry, J. Inst. Petroleum. 45:287 (1959).
15. K. Grob, "Classical Split and Splitless Injection in Capillary GC", Huethig, Heidelberg (1986).
16. W. Jennings, "Gas Chromatography with Glass Capillary Columns", 2nd ed., Academic Press, New York (1980), p. 63-77.
17. L. Larsson, E. Jantzen, and J. Johnsson, Gas chromatographic fatty acid profiles for characterization of mycobacteria: an interlaboratory methodological evaluation, Eur. J. Clin. Microbiol. 4:483 (1985).

18. M. A. Lambert, C. W. Moss, V. A. Silcox, and R. C. Good, Analysis of mycolic acid cleavage products and cellular fatty acids of Mycobacterium species by capillary gas chromatography, J. Clin. Microbiol. 23:731 (1986).
19. K. Grob and G. Grob, Splitless injection on capillary columns, Part I. The basic technique; steroid analysis as an example, J. Chromatogr. Sci. 7:584 (1969).
20. W. E. Harris, Some aspects of injection of large samples in gas chromatography, J. Chromatgr. Sci. 11:184 (1973).
21. K. Grob and K. Grob, Jr., Splitless injection and the solvent effect, HRC & CC 1:57 (1978).
22. K. Grob and K. Grob Jr., Characteristics of donor-acceptor properties of amines and cyclic nitrogen-containing substances determined by their retention on stationary phases of different types, J. Chromatgr. 94:184 (1974).
23. K. Grob Jr., 'Band broadening in space' and the 'retention gap' in capillary gas chromatography, J. Chromatgr. 237:15 (1982).
24. K. Grob Jr. and R. Muller, Some technical aspects of the preparation of a 'retention gap' in capillary gas chromatography, J. Chromatogr. 244:185 (1982).
25. L. Larsson and G. Odham, Injection principles in capillary gas chromatographic analysis of bacterial fatty acids, J. Microbiol. Methods, 3:77 (1984).
26. A. Zlatkis and Q. Walker, Direct sample introduction for large bore capillary columns in gas chromatography, J. Gas Chromatgr. 1:9 (1963).
27. K. Grob and K. Grob, Jr., On-column injection onto glass capillary columns, J. Chromatgr. 151:311 (1978).
28. H. Hachenberg and A. P. Schmidt, "Gas Chromatographic Head-space Analysis," Heyden and Son Ltd., London, (1977).
29. B. V. Ioffe and A. G. Vitenberg, "Head-space Analysis and Related Methods in Gas Chromatography", John Wiley, New York (1982).
30. J. M. Purcell and P. Magidman, Rapid desorption-injection system for gas chromatography of preconcentrated headspace samples, Anal. Lett. 16:465-483 (1983).
31. L. Larsson and E. Holst, Feasibility of automated head-space-gas-chromatography in identification of anaerobic bacteria, Acta Pathol. Microbiol. Scand. Sect. B. 90:125 (1982).
32. L. Larsson, E. Holst, C. G. Gemmell, and P. A. Mardh, Characterization of Clostridium difficile and its differentiation from Clostridium sporogenes by automatic head-space gas chromatography, Scand. J. Infect. Dis., Suppl. 22:37 (1980)
33. A. J. Taylor, "The Application of Gas Chromatographic Head-Space Analysis to Medical Microbiology," Kolb, B., ed., Heyden, London (1980).
34. N. J. J. Hayward, Head-space gas-liquid chromatography for the rapid laboratory diagnosis of urinary tract infections caused by Enterobacteria, J. Chromatogr. 274:27 (1983).
35. L. Larsson, P.-A. Mardh, and G. Odham, Volatile metabolites in identification of microbes and diagnosis of infectious diseases, in: "Gas Chromatography/mass Spectrometry Applications in Microbiology", Plenum, New York (1984), p. 207.
36. W. J. Irwin, "Analytical Pyrolysis," Marcel Dekker, New York (1982).
37. H. L. C. Meuzelaar, J. Haverkamp, and F. D. Hileman, "Pyrolysis Mass Spectrometry of Recent and Fossil Biomaterials," Elsevier, Amsterdam (1982).
38. I. Tyden-Ericson, A new pyrolyzer with improved control of pyrolysis conditions, Chromatographia. 6:353 (1973).
39. R. S. Whiton and S. L. Morgan, Modified interface for pyrolysis gas chromatography with capillary columns, Anal. Chem. 57:778(1985).

40. T. J. Stark, P. A. Larsson, and R. D. Dandeneau, Selective phases for wall-coated open tubular columns, J. Chromatogr. 279:31 (1983).
41. J. K. Haken, Developments in polysiloxane stationary phases in gas chromatography, J. Chromatogr. 300:1 (1984).
42. C. F. Poole and M. J. Verzele, Separation of protein amino acids as their N(O)-acyl alkyl ester derivatives on glass capillary columns, J. Chromatogr. 150:439 (1978).
43. W. J. A. Vanden Henvel, E. O. A. Haachte and E. C. Horning, A new liquid phase for gas chromatographic separations of steroids, J. Amer. Chem. Soc. 83:1513 (1961).
44. A. T. James and A. J. P. Martin, Gas-liquid partition chromatography: the separation and micro-estimation of volatile fatty acids from formic acid to dodecanoic acid, Biochem. J. 50:679 (1952).
45. K. D. Bartle, M. L. Lee, S. and Wise, Modern analytical methods for environmental polycyclic aromatic compounds, Chem. Soc. Rev. 10:113 (1981).
46. L. Larsson, P. A. Mardh, G. Odham, and G. J. Westerdahl, Detection of tuberculostearic acid in biological specimens by means of glass capillary gas chromatography-electron and chemical ionization mass spectrometry, using selected ion monitoring, J. Chromatogr. Biomed. Appl. 182:402 (1980).
47. H. Rotzsche, in: "Gas Chromatography 1962," M. van Swaay, ed., Butterworth, London (1962).
48. F. Bayer, K. P. Hope, and H. Mack, in: Proc. Int. Symp. Advances in Gas Chromatography, Houston, A. Zlatkis, ed., Preston Technical Abstracts, Niles (1963).
49. A. Fox, P. Y. Lau, A. Brown, S. L. Morgan, Z. -T. Zhu, and M. Lema, "Capillary Gas Chromatographic Analysis of Carbohydrates of Legionella Pneumophila and Other Members of the Family Legionellaceae," J. Clin. Micro. 19:326 (1984).
50. R. F. Arrendale, R. F. Severson, and O. T. Chortyk, Preparation of fused silica polar stationary phase wall-coated open tubular columns, J. Chromatogr. 254:63 (1983).
51. G. Becher, Glass capillary columns in the gas chromatographic separation of aromatic amines. II. Application to samples from workplace atmospheres using nitrogen-selective detection, J. Chromatogr. 211:103 (1981).
52. E. Gil-Av, B. Feibush, and R. Charles-Singler, Separation of enantiomers by gas-liquid chromatography with an optically active stationary phase, in: "Gas Chromatography 1966", A. B. Littlewood, ed., Institute of Petroleum, London, p. 227 (1967).
53. H. Frank, G. J. Nicholson, and E. Bayer, Gas chromatographic-mass spectrometric analysis of optically active metabolites and drugs on a novel chiral stationary phase, J. Chromatogr. 146:197 (1978).
54. W. A. Konig, I. Benecke, and S. Siever, New results in the gas chromatographic separation of enantiomers of hydroxy acids and carbohydrates, J. Chromatogr. 217:71 (1981).
55. R. H. Liu, and W. W. Ku, Chiral stationary phases for the gas- liquid chromatographic separation of enantiomers, J. Chromatogr. 271:309 (1983).
56. K. Ueda, S. L. Morgan, A. Fox, J. Gilbart, A. Sonesson, L. Larsson, and G. Odham, D-alanine as a chemical marker for the determination of streptococcal cell wall levels in mammalian tissues by gas chromatography/negative ion chemical ionization mass spectrometry, Anal. Chem., 61:265 (1989).
57. A. Tunlid, G. Odham, D. C. Findlay, and D. C. White, Precision and sensitivity of the measurement of 15N enrichment in D-alanine from bacterial cell wall using positive/negative ion mass spectrometry, J. Microbiol. Meth. 3:237 (1985).

58. A. Tunlid and G. Odham, Capillary gas chromatography or selected ion monitoring detection for the determination of muramic acids, diaminopimelic acids and the ratio of D/L alanine in bacteria, J. Microbiol. Meth. 1:63 (1983).

59. R. J. Laub, Global optimization strategy for gas-chromatographic separations," in: "Physical Methods of Modern Chemical Analysis," T. Kuwana, ed., Academic Press, New York, Vol. 3, p. 249 (1983).

60. R. C. Hall, The nitrogen detector in gas chromatography, CRC Crit. Rev. Anal. Chem. 8:323 (1978).

61. B. Kolb and J. Bishoff, A new design of a thermionic nitrogen and phosphorus detector, J. Chromatogr. Sci. 12:625 (1974).

62. J. E. Lovelock, Electron absorption detectors and technique for use in quantitative and qualitative analysis by gas chromatography, Anal. Chem. 35:474 (1963).

63. A. Zlatkis and C. F. Pool, eds., "Electron Capture. Theory and Practice in Chromatography," Elsevier, Amsterdam (1981).

64. L. Larsson, A. Sonesson, and J. Jimenez, Ultrasensitive analysis of microbial fatty acids using gas chromatography with electron capture detection, Eur. J. Clin. Microbiol. 6:729 (1987).

65. J. M. Rosenfeld, O. Hammerberg, and M. C. Orvidas, Simplified methods for preparation of microbial fatty acids for analysis by gas chromatography with electron-capture detection, J. Chromatogr. Biomed. Appl. 378:9 (1986).

66. J. B. Brooks, Gas-liquid chromatography as an aid in rapid diagnosis by selective detection of chemical changes in body fluids, in: "The Direct Detection of Microorganisms in Clinical Samples," J. D. Coorod, J. J. Kunz, and M. J. Ferraro, eds., Academic Press, New York (1983).

67. P. R. Griffiths, J. A. de Haseth, and L. V. Azarraga, Capillary GC/FTIR, Anal. Chem. 55:1361A (1983).

68. C. L. Wilkins, Hyphenated techniques for analysis of complex organic mixtures, Science. 222:291 (1983).

69. C. L. Wilkins, Linked gas chromatography infrared mass spectrometry, Anal. Chem. 59:571A (1987).

70. R. H. Findlay, D. J. W. Moriarty, and D. C. White, Improved method of determining muramic acid from environmental samples, Geomicrobiology J. 3:135 (1983).

71. T. H. Risby and Y. Talmi, Microwave induced electrical discharge detectors for gas chromatography, CRC Crit. Rev. Anal. Chem. 14:231 (1983).

72. J. W. Carnahan, K. J. Mulligan, and J. A. Caruso, Elemental-selective detection for chromatography by plasma emission spectrometry, Anal. Chim. Acta. 130:227 (1981).

73. R. Baum, Atomic-emission detector for gas chromatography introduced, Chem. Eng. News. 67(3):37 (1989).

74. R. R. LaPelle, "Practical Vacuum Systems," McGraw-Hill, New York (1972).

75. F. W. McLafferty, Introduction to tandem mass spectrometry, in: "Tandem Mass Spectrometry", F. W. McLafferty, ed., John Wiley, New York (1983).

76. G. M. Message, "Practical Aspects of Gas Chromatography/Mass Spectrometry," John Wiley, New York (1984), pp 123-140.

77. J. T. Watson, Gas chromatography and mass spectrometry, in: "Ancillary Techniques of Gas Chromatography, L. S. Ettre and H. McFadden, eds., Wiley-Interscience, New York (1969), pp 145-225.

78. H. H. Willard, L. L. Merritt. Jr., J. A. Dean, and F. A. Settle, Jr., "Instrumental Methods of Analysis", 7th ed, Wadsworth, Belmont, CA, p. 466 (1988).

79. E. M. Chait, Ionization sources in mass spectrometry, Anal. Chem. 44:77A (1972).

80. G. Odham and L. Larsson, Mass spectrometry, in: "Gas Chromatography/mass Spectrometry Applications in Microbiology", G. Odham, L. Larsson, and P.-A. Mardh, Eds., Plenum, New York, p. 207 (1984).

81. F. W. McLafferty, "Interpretation of Mass Spectra", 3rd ed., University Science Books, Mill Valley (1980).

82. B. Munson, Chemical ionization mass spectrometry: ten years later, Anal. Chem. 49:772A (1977).

83. K. F. Faull and J. D. Barchas, Negative-ion mass spectrometry fused-silica capillary gas chromatography of neurotransmitters and related compounds, in: "Methods of Biochemical Analysis", D. Glick, ed., John Wiley, New York (1983), Volume 29, pp. 325-383.

84. D. F. Hunt and F. W. Crow, Electron capture negative ion chemical ionization mass spectrometry, Anal. Chem. 50:1781 (1978).

85. D. F. Hunt, G. C. Stafford, Jr., F. W. Crow, and J. W. Russell, Pulsed positive negative ion chemical ionization mass spectrometry, Anal. Chem. 48:1976 (1976).

86. C. H. Melton, Negative ion mass spectra, in: "Mass Spectrometry of Organic Ions", F. W. McLafferty, ed., Academic Press, New York, (1963), pp. 163- 205.

87. M. Anbar and W. H. Aberth, Field ionization mass spectrometry: a new tool for the analytical chemist, Anal. Chem. 46:59A (1974).

88. C. J. McNeal, Symposium on fast atom and ion induced mass spectrometry of nonvolatile organic solids, Anal. Chem. 54:43A (1982).

89. M. Barber, R. S. Bordori, R. D. Sedgwick, and A. N. Tyler, fast atom bombardment of solids (FAB)-- a new ion-source for mass spectrometry, J. Chem. Soc., Chem. Commun. 325 (1981).

90. D. J. Surman and J. C. Vickerman, Fast atom bombardment quadrupole mass-spectrometry, J. Chem. Soc., Chem. Commun., 324 (1981)

91. M. Berber, R. S. Bordoli, G. J. Elliot, R. D. Sedgwick, and A. N. Tyler, Fast atom bombardment mass spectrometry, Anal. Chem. 54:645A (1982).

92. C. Fenselau and R. Cotter, Chemical aspects of fast atom bombardment, Chem. Rev. 87:501 (1987).

93. Y. Tondeur, R. C. Moschel, A. Dipple, and S. R. Koepke, Fast atom bombardment and collisional activation mass spectrometry as probes for the identification of positional isomers in a series of benzylated guanosines, Anal. Chem. 58:1316 (1986).

94. R. M. Caprioli, and L. Smith, Determination of K_m and V_{max} for tryptic peptide hydrolysis using fast atom bombardment mass spectrometry, Anal. Chem. 58:1080 (1986).

95. D. N. Heller, R. J. Cotter, C. Fenselau, and O. M. Uy, Profiling of bacteria by fast atom bombardment mass spectrometry, Anal. Chem. 59:2806 (1987).

96. L. M. Mallis and D. H. Russell, Fast atom bombardment-tandem mass spectrometry studies of organo-alkali metal ions of small peptides, Anal. Chem. 58:1076 (1986).

97. R. J. Day, S. E. Unger, and R. G. Cooks, Molecular secondary ion mass spectrometry, Anal. Chem. 52:557 (1980).

98. R. D. MacFarlane, Californium-252 plasma desorption mass spectrometry, Anal. Chem. 55:1247A (1983).

99. R. J. Cotter, Plasma desorption mass spectrometry: coming of age, Anal. Chem. 60:781A (1988).

100. P. J. Gunnar, A. B. Hedin, P. L. Hakansson, B. U. R. Sundqvist, B. G. S. Save, P. F. Nielsen, P. Roepstorff, K. Johansson, I. Kamensky, and S. L. Lindberg, Plasma desorption mass spectrometry of peptides and proteins adsorbed on nitrocellulose, Anal. Chem. 58:1084 (1986).

101. M. Alai, P. Demirev, C. Fenselau, and R. J. Cotter, 1986, Glutathione as a Matrix for Plasma Desorption Mass Spectrometry of Large Peptides, Anal. Chem. 58:1303 (1986).

102. J. F. Garcia-Bustos, B. T. Chait, and A. Tomasz, Structure of the peptide network of pneumococcal peptidoglycan, J. Biol. Chem. 262:1540 (1987).

103. R. J. Cotter, J. Honovich, N. Qureshi, and K. Takayama, Structural determination of lipid A from Gram negative bacteria using laser desorption mass spectrometry, Biomed. Environ. Mass Spectrom. 14:591 (1987).

104. A. G. Craig, A. Engstrom, H. Bennich, and I. Kamensky, Plasma desorption mass-spectrometry coupled with conventional peptide sequencing techniques, Biomed. Environ. Mass Spectrom. 14:669 (1987).

105. I. Jardine, G. F. Scanlan, A. Tsarbopoulous, and D. J. Liberato, Plasma desorption mass spectrometry of peptides adsorbed on nitrocellulose from a glutathione matrix, Anal. Chem., 60:1086 (1988).

106. R. J. Cotter, Mass spectrometry of nonvolatile compounds: desorption from extended probes, Anal. Chem. 52:1589A (1980).

107. R. J. Cotter, Lasers and mass spectrometry, Anal. Chem. 56:485A (1984).

108. W. D. Reynolds, Field desorption mass spectrometry, Anal. Chem. 51:283A (1979).

109. G. G. S. Dutton, G. Eigendorf, Z. Lam, and A. V. S. Lim, Bacteriophage endoglycanases and desorption chemical ionization mass spectrometry (DCI-MS) to sequence bacterial antigens, Biomed. Environ. Mass Spectrom. 15:459 (1988).

110. P. E. Miller and M. B. Denton, The quadrupole mass filter: basic operating concepts, J. Chem. Educ. 63:617 (1986).

111. P. H. Dawson, "Quadrupole Mass Spectrometry and its Applications," Elsevier, New York (1976).

112. J. M. Henis, Analytical implications of ion cyclotron resonance spectroscopy, Anal. Chem. 41:22A (1969).

113. C. L. Wilkins, Fourier transform mass spectrometry, Anal. Chem. 50:493A (1978).

114. T. A. Lehman and M. M. Bursey, "Ion cyclotron resonance spectrometry," John Wiley, New York (1976).

115. R. T. McIver and W. D. Bowers, Fourier-transform ion cyclotron resonance instrument for tandem mass spectrometry, in: "Tandem Mass Spectrometry", F. W. McLafferty, Ed., John Wiley, New York (1983), p. 287.

116. F. W. McLafferty, Tandem mass spectrometry, Science 214:280 (1981).

117. R. A. Yost and C. G. Enke, Triple quadrupole mass spectrometry for direct mixture analysis and structure elucidation, Anal. Chem. 51:1251 (1979).

118. R. A. and C. G. Enke, Tandem quadrupole mass spectrometry, in: "Tandem Mass Spectrometry", F. W. McLafferty, ed., John Wiley, New York (1983), p. 175.

119. N. J. Louris, R. G. Cooks, J. E. Syka, P. E. Kelley, Stafford, Jr., and J. F. Todd, Instrumentation, applications, and energy deposition in quadrupole ion-trap tandem mass spectrometry, Anal. Chem. 59:1677 (1987).

120. H. V. Malmstadt, C. G. Enke, and S. R. Crouch, "Electronics and Instrumentation for Scientists," Benjamin/Cummings, MA (1981), p. 97.

121. M. C. Hamming and N. G. Foster, "Interpretation of Mass Spectra of Organic Compounds," Academic Press, New York (1962), p. 17.

122. W. J. McMurray, Photographic techniques, in: "Organic-High Resolution Mass Spectrometry, in Mass Spectrometry: Techniques and Applications," G. W. Milne, ed., Robert E. Krieger, New York (1979), p. 11.

123. M. E. Rose and R. A. W. Johnson, "Mass Spectrometry for Chemists and Biochemists," Cambridge University Press, Cambridge, (1982), p. 30.

124. R. Otson and D. T. Williams, Headspace chromatographic determination of water pollutants, Anal. Chem., 54: 942 (1982).

125. P. L. Patterson, Pulse-modulated electron capture detection with nitrogen carrier gas, J. Chromatogr., 134:25 (1977).

CHAPTER 3

THE CHEMOTAXONOMY OF CORYNEFORM ORGANISMS: MORPHOLOGY VS. CHEMICAL ANATOMY

Cecil S. Cummins

Department of Anaerobic Microbiology
Virginia Polytechnic Institute and State University
Blacksburg, VA 24061

This short contribution describes no new test or method which will revolutionize bacterial taxonomy and classification. To judge from the other contributions to the First International Symposium on the Interface between Analytical Chemistry and Microbiology, these matters are already in very capable hands, and have produced an impressive mass of information which has already helped to clarify many of the problems of bacterial taxonomy.

Instead of discussing a new technique I would like to put in a plea for a very old one, namely the examination of morphology, and to discuss briefly its failures and its present uses.

The inadequacies of morphology in bacterial taxonomy and classification are many and are nowhere better illustrated than in the successive attempts to produce a rational classification of the coryneform group of organisms. I was first introduced to the morphology of this group in the 1940's when I worked for a time in a clinical laboratory and had to look at 10-20 stained slides every morning to determine whether or not *Corynebacterium diphtheriae* was present. This certainly did focus one's attention on morphology, and one came to recognize the long, curved, irregularly stained cells of *mitis* strains, with many metachromatic granules, as opposed to the shorter, much less irregular *gravis* strains with few granules, and both of these had to be distinguished from non-pathogens like *C. pseudodiphtheriticum*, which at that time we happily called *C. hoffmanii*. Of course, we had to be cautious in reporting these results. "Organisms present morphologically resembling *C. diphtheriae*" is what we said, if I remember correctly. Nevertheless, as judged by the subsequent cultural results, our presumptive reports were correct more often than not, showing that at this level of investigation, morphology can be a very useful criterion.

Having been introduced to coryneform organisms in this very practical way, I have never found any particular difficulty with the words "coryneform" or "diphtheroid". I think of them both as meaning essentially the same thing: a Gram positive organism that shows some or all of the morphological features of that classic coryneform *C. diphtheriae* var. *mitis*. The major feature is irregularity in all aspects of morphology: the cells may be straight or curved, of irregular width with uneven staining, and are often club or wedgeshaped or arranged at angles to each other. Without laboring the point further, I feel that both terms are very useful and should certainly not be discarded as long as they are used non-taxonomically. We do not, after all, get into trouble with "bacillus" with a small b as a general term for a rod shaped organism, despite the fact when we write it with a capital B it has a specific taxonomic meaning. Forty years ago, J. H. Conn, at the 47th meeting of the American Society for Microbiology, protested against the misuse of the generic name *Corynebacterium*,[1] and indeed *Corynebacterium* was, as Conn rightly complained, used as a dumping ground for a large number of Gram positive organisms of irregular outline and staining. For example, the 7th edition of Bergey's Manual contained 33 species of corynebacteria (largely human, animal and plant parasites, and pathogens). These had a wide range of physiological properties and their only common feature seemed to be that they were straight to slightly curved Gram positive rods with irregularly stained segments and sometimes granules. It might be noted in passing that *Mycobacterium*, although its members show a typical coryneform morphology, has never suffered from the same influx of unrelated organisms as did *Corynebacterium*, probably because in the case of *Mycobacterium* the special and distinctive property of acid-fastness was attached to its description from an early period.

In Chester's Manual of Determinative Bacteriology[2] matters were simple: the coryneform organisms were in the genus *Mycobacterium*, and were put into group I or II depending on whether or not they were easily decolorized in the Ziehl Niehlsen stain (Nocardias were in the genus *Streptothrix*). Would that things remained that simple! In the first five editions of Bergey's Manual,[3-7] the two genera *Corynebacterium* and *Mycobacterium* were in *Mycobacteriaceae*, while *Nocardia* was in *Actinomycetaceae*; although these were different families, at least they are in the same order-- *Actinomycetales*. In the 6th and 7th editions of the manual, even greater separation was proposed. *Corynebacterium* was now in *Eubacteriales*, while *Mycobacterium* and *Nocardia* (then designated *Actinomyces*) were in *Actinomycetales*.[7-9]

These problems, of course have their routes in the interpretation of morphology and staining reactions. Is the organism acid-fast or partially acid-fast? Does it show real branching? Does it grow in the form of a mycelium and does the mycelium break up into rod-like or coryneform elements? And, more difficult, what do these things mean in terms of relationships, phylogenetic or otherwise? One can see all these points argued at great length by Jensen,[10,11] complete with a tentative phylogenetic scheme of some complexity with a hypothetical coccus as a starting point.

It is interesting to re-read Jensen's reviews, for it brings home clearly the immense changes that bacterial taxonomy has undergone since his time. To see this one only needs to read the articles on the Symposium on Coryneform Bacteria published 10 years ago[12] or the contributions to the present symposium. Nevertheless, Jensen's instincts were sound on many points. He was clear in stating, for example, that *Corynebacterium* and *Mycobacterium* were closely related to *Nocardia*, that *Microbacterium* should be regarded as a group of strains within *Corynebacterium*, and that *C. acnes* should be included in *Propionibacterium*. This is not really a bad record in the light of recent developments.

The seventeen year (1957-1974) gap between Bergey's 7th and 8th editions[9,13] corresponded to the period in which a number of methods involving the chemical anatomy of bacteria began to be applied to bacterial taxonomy, starting with cell wall analysis. By the time of the 8th edition, it was possible to begin to make sense out of the taxonomy of the great mess of species that sheltered under the umbrella of *Corynebacterium*. Using crude cell wall criteria such as the types of sugars present in wall hydrolysates, and the diaminoacid revealed by acid hydrolysis, it was evident, as had long been suspected, that the organisms allocated to *Corynebacterium* were a very heterogeneous group.[14] But the results also indicated in a positive way the probable affinities of some of them. For example, cell wall analyses gave positive evidence that *Corynebacterium*, *Mycobacterium*, and *Nocardia* did indeed share important characteristics. Similarly, there was evidence that the "anaerobic coryneforms" including our old friend "*Corynebacterium acnes*" had little in common with *C. diphtheriae* but shared characteristics with the propionibacteria. The results also showed that, whatever *C. pyogenes* and *C. hemolyticum* were, they did not belong in a genus where *C. diphtheriae* was the type species, and that likewise the plant pathogenic corynebacteria represented yet another group.

Since then, of course, the list of useful characters has grown apace, and we can now call on mycolic acids, cord factors, and other glycolipids as distinguishing factors, as well as the presence of acetyl or glycolyl groups in peptidoglycan, quite apart from the menaquinones and fatty acids of the cell membrane.

In the case of coryneform organisms, outline morphology (the morphology and staining as seen in the Gram stained smear) has been obviously misleading, and chemical anatomy has given us what appears to be a much more rational taxonomy. One might conclude from this that morphology is a pretty useless character to consider as a taxonomic criterion, and this point of view has been put quite forcibly on a number of occasions.

Nevertheless, whether we like it or not, the usages of bacterial taxonomy are still very heavily influenced by morphology. Consider some names:

coccus -- a grain or berry

bacillus -- from bacillum, a small rod

sarcina -- a package or bundle

clostridium -- a small spindle

arachnia -- from arachne, a spider

spirochaeta -- a coiled hair

The list is endless, and, as new names proliferate, it is interesting to see how many of them are based on some suitable classical derivation of a morphological feature. It is inevitable that this is so. How far would we get in identifying an organism if we were allowed to use only touch, hearing and smell, but not sight? It is going to be a very long time before we do not have to rely on cultural appearances and microscopic examination of Gram stained preparations to make a preliminary taxonomic decision, or in clinical work, a preliminary identification.

Moreover, morphology, whether examined in simple stained smear with the optical microscope, or in more elaborate preparations in the electron microscope, is the result of the interactions of a substantial part of the genome, and therefore likely to be more significant than examination of a single feature such as fermentation of a single sugar, or the presence of a single marker substance.

My message then is a simple one. We will need to use morphology for a long time to come to enable us to make a rapid allocation to broad groups, and this is something at which the human computer, relying on visual input, is still very efficient. By the use of sophisticated tests, we can learn to avoid the more serious deficiencies of the morphological approach, but it is quite inconceivable that we will be able to do without the simple cultural and microscopic appearances of microbial cultures as a starting point.

Table 1. Distinctive cell wall components of coryneform organisms[a]
(modified from Bergey's Manual, 8th Ed., Table 17.1, p. 600)

Organisms	Sugars	Amino Acids in Peptidoglycan
C. diphtheriae, and related organisms	Arabinose galactose	*meso*-DAP[b]
Plant pathogenic, and corynebacteria	Rhamnose, mannose, galactose, glucuronic acid, xylose, fucose	Homoserine, DABA[c], or ornithine
"*C. pyogenes*", and "*C. hemolyticum*"	Rhamnose, glucose	Lysine
"*C. acnes*" and anaerobic diphtheroids	Galactose, Glucose, Mannose	LL-DAP[b] (occ. *meso*-DAP[b])

[a]Published by permission of William and Wilkins, Baltimore.
[b]DAP, diaminopimelic acid.
[c]DABA, diaminobutyric acid.

Morphology, in that sense, will always be with us and we should learn to make as much use of it as we can.

Finally, we should pay tribute to that first of chemical taxonomists, Christian Gram, for his elegant and simple technique published a little over a hundred years ago[15] that has stood the test of time so well. One only has to compare the cell wall organization of Gram positive and Gram negative bacteria to see what a fundamental observation he made. The Gram stain, like simple morphological descriptions, will be also with us for a very long time, and of course the two will be combined, as they usually are, in the simple technique of the heat fixed, Gram stained smear.

So let us by all means use all the sophistication we can in examining bacteria and obtaining more information about them, but do not let us forget to use the simple tests first. They might even spare us the embarrassment of finding that we have been examining the wrong organism.

REFERENCES

1. H. J. Conn, A protest against the misuse of the generic name *Corynebacterium*, J. Bacteriol., 54:10 (1947).
2. F. D. Chester, "A Manual of Determinative Bacteriology," The Macmillan Co., New York (1901).
3. D. H. Bergey, F. C. Harrison, R. S. Breed, B. W. Hammer, and F. M. Huntoon, "Bergey's Manual of Determinative Bacteriology, 1st Ed.," Williams and Wilkins, Baltimore (1923).
4. D. H. Bergey, F.C. Harrison, R. S. Breed, B. W. Hammer, and F. M. Huntoon, "Bergey's Manual of Determinative Bacteriology, 2nd Ed.," Williams and Wilkins, Baltimore (1925).
5. D. H. Bergey, F. C. Harrison, R. S. Breed, B. W. Hammer, and F. M. Huntoon, "Bergey's Manual of Determinative Bacteriology, 3rd Ed.," Williams and Wilkins, Baltimore (1930).

6. D. H. Bergey, R. S. Breed, B. W. Hammer, F. M. Huntoon, E. G. D. Murray, and F. C. Harrison, "Bergey's Manual of Determinative Bacteriology, 4th Ed.," Williams and Wilkins, Baltimore (1934).

7. D. H. Bergey, R. S. Breed, E. G. D. Murray, and A. P. Hitchens, "Bergey's Manual of Determinative Bacteriology, 5th Ed.," Williams and Wilkins, Baltimore (1939).

8. R. S. Breed, E. G. D. Murray, and A. P. Hitchens, "Bergey's Manual of Bacteriology, 6th Ed.," Williams and Wilkins, Baltimore (1948).

9. R. S. Breed, E. G. D. Murray, and N. R. Smith, "Bergey's Manual of Bacteriology, 7th Ed.," Williams and Wilkins, Baltimore (1957).

10. H. L. Jensen, The coryneform bacteria, Ann. Rev. Microbiol. 6:77 (1952).

11. H. L. Jensen, The genus *Nocardia* (or *Proactinomyces*) and its separation from other *Actinomycetales* with some reflections on the phylogeny of the Actinomycetes, in: Symposium on *Actinomycetales*, 6th Int. Congress of Microbiology, Fond. Emanuele Paterno, Rome (1953).

12. I. J. Bousefield and A. G. Callely, eds., "Coryneform Bacteria", Academic Press, New York (1978).

13. R. E. Buchanan and N. E. Gibbons, "Bergey's Manual of Determinative Bacteriology, 8th Ed.," Williams and Wilkins, Baltimore (1974).

14. M. Rogosa, C.S. Cummins, R. A. Lelliott and R. M. Keddie, Coryneform group of bacteria, in: "Bergey's Manual of Determinative Bacteriology, 8th Edition," R. E. Buchanan and N. E. Gibbons, eds., Williams and Wilkins, Baltimore (1974).

15. C. Gram, Uber die isolierte Farburg der Schizomyceten in Schnitt-und-Trocken Praparaten, Fortschr. Med., 2:185 (1884).

CHAPTER 4

THE USE OF CELLULAR FATTY ACIDS FOR IDENTIFICATION OF MICROORGANISMS

C. Wayne Moss

Division of Bacterial Diseases
Centers for Disease Control
1600 Clifton Road
Atlanta, GA 30333

Use of trade names is for identification only and does not imply endorsement by the Public Health Service or by the U.S. Department of Health and Human Services.

INTRODUCTION

During the past 22 years, data from this laboratory and others throughout the world have firmly established the value of cellular fatty acids as an aid in rapid identification and classification of microorganisms. Applications of fatty acids and other chemical data for taxonomy of a variety of microorganisms are presented in Bergey's Manual of Systematic Bacteriology.[1] Gas liquid chromatography (GLC) has been the analytical method of choice for fatty acid analysis. Continued advancements in GLC instruments, columns, and computerized analysis of raw data have developed to the point where microbial fatty acid analysis can be accomplished with accuracy and precision outside the research laboratory. Thus, this review is directed to technical laboratory personnel with little or no experience in GLC techniques for chemical analysis of microorganisms. The methods and procedures currently used in this laboratory for microbial fatty acids are outlined in detail with examples of the usefulness of the data.

CULTURE AND GROWTH CONDITIONS

Pure cultures are required for cellular fatty acid analysis. After primary isolation, the culture is inoculated onto growth media and incubated under optimum conditions to produce good to heavy growth. The growth medium and the time and temperature of incubation for good growth differ among bacterial species and groups. Most nonfermentative, Gram negative bacteria and other organisms isolated in the clinical laboratory give good growth after 24-48 hr incubation at 32-35°C on trypticase soy agar or heart infusion agar supplemented with 5% rabbit blood.[2-4] However, these media will not support the growth of *Legionella* species that require a special formulation such as charcoal yeast extract agar which contains L-cysteine and iron salts.[5] Thus, selection and use of a standardized medium which produces good growth of the organisms under study is an important consideration.

Length of incubation is also a major factor since the cellular fatty acid composition of an organism is most stable and reproducible during the stationary phase of growth. Physiological age of the cell rather than exact time of incubation is the important parameter; efforts should be made to analyze the cells when they are in the stationary phase. For example, when examining many strains within a species, one or more strains may be slow growing and require 48 hr to obtain equivalent growth to other strains incubated for 24 hr. Cells of the slow growing strain will most likely contain relatively larger amounts of unsaturated fatty acids compared with physiologically older cells. The fatty acid composition of the slow growing strain at 48 hr (or until good growth) will be most similar to the other strains which give good growth at 24 hr. Thus, from the practical viewpoint, familiarity with the growth characteristics of the organisms under study will lead to the most consistent and reproducible fatty acid results for all strains within the group. Obviously, for meaningful comparison, all organisms within the study or comparison group should be grown on the same medium and under similar conditions.

HYDROLYSIS OF CELLS

The lipids of microorganisms are found in the cell wall/cell membrane fractions where the fatty acids are chemically bonded to proteins, carbohydrates, and other chemical entities. Most but not all lipids are removed by extraction with an organic solvent. However, to determine total cellular fatty acids, the cells (or extracted lipids) must be hydrolyzed with acid or base to liberate the unit fatty acid. The hydrolysis procedure used is critical to the final results as acid hydrolysis degrades cyclopropane acids and base hydrolysis fails to liberate all the amide-linked hydroxy acids.[6]

We have developed a simple and rapid base hydrolysis procedure which gives accurate and reproducible results with either lyophilized or fresh whole cells (see Figure 1). Growth from the surface of an agar plate or slant is removed by adding approximately 0.5 mL of sterile distilled water and gently scraping. The turbid cell suspension is placed in a screw-capped tube (13 by 100 mm) fitted with a teflon-lined cap and saponified by heating at 100°C for 30 min after adding 1 mL of 15% NaOH in 50% aqueous methanol. The sample is cooled to ambient temperature, 1.5 mL of ether:hexane (1:1) is added, and the contents are mixed by shaking. The phases are allowed to separate by standing 1 to 2 min, and the lower (aqueous) layer is carefully removed with a Pasteur pipette and discarded (or saved to test for amide-linked acids). One mL of phosphate buffer (pH 11.0) is added, the contents are mixed by shaking, and the phases are allowed to separate by standing 2 to 3 min. About two thirds of the top (organic) layer containing the fatty acid methyl esters (FAME) is removed to a clean test tube or a septum capped sample vial for subsequent analysis by GLC.

Amide-linked hydroxy acids, if present, are not totally released with this saponification procedure. These acids, which remain in the methanolic aqueous layer after the methylation step in the above procedure, can be completely released by adding 1 ml of concentrated HCl to the aqueous layer and heating at 100°C for 1 hr. The resulting methyl esters of these acids can be combined and analyzed with the FAME from the regular saponification procedure. Essentially all of the fatty acid data generated in our laboratory have been obtained with the saponification procedure without subsequent acid hydrolysis.

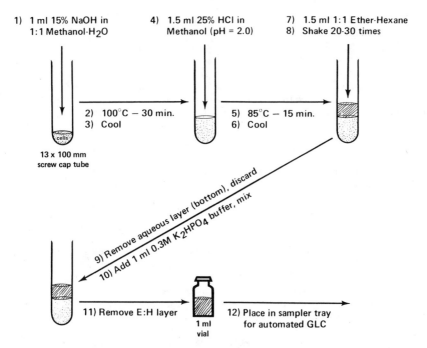

1) 1 ml 15% NaOH in   4) 1.5 ml 25% HCl in   7) 1.5 ml 1:1 Ether-Hexane
   1:1 Methanol-H₂O   Methanol (pH = 2.0)   8) Shake 20-30 times

2) 100°C − 30 min.   5) 85°C − 15 min.
3) Cool   6) Cool

13 x 100 mm
screw cap tube

9) Remove aqueous layer (bottom), discard
10) Add 1 ml 0.3M K₂HPO₄ buffer, mix

11) Remove E:H layer   1 ml   12) Place in sampler tray
                            vial   for automated GLC

Figure 1.   Saponification procedure for fatty acid analysis of
            microorganisms.

GAS LIQUID CHROMATOGRAPHY

Analysis of bacterial fatty acids has been performed almost exclusively
by GLC because of speed, sensitivity, and resolution of the technique. In
early studies, GLC analysis was done on 6 ft or 8 ft by 4 mm (inside
diameter) glass columns packed with a nonpolar (SE-30, OV-1 or OV-101) or a
polar stationary phase (EGA or OV-17). GLC retention time (RT) comparisons
to known standards on both polar and nonpolar phases were essential for
tentative identification of the individual fatty acid components;
confirmation of identity required additional ancillary techniques (i.e.,
hydrogenation, acetylation, mass spectrometry).

Since the early 1980s, most GLC FAME analyses have been done using the
fused silica capillary column.[7-9] This column, which is typically 25 to 50 m
in length with an internal diameter of 0.2 to 0.4 mm, gives superior
resolution and separating efficiency compared with packed columns. Most of
the fatty acids known to occur in bacteria can be resolved with a 50 m OV-1
capillary column, and the accuracy of retention time (RT) measurements with
current gas chromatographs is routinely greater than 0.005 min. Thus,
precise RT matches of FAME peaks from the analytical sample with those of
known standards under standard chromatographic conditions give a high
probability of identification. The accuracy of identification by RT
measurements is increased by hydrogenation of unsaturated acids and
acetylation of hydroxy acids with subsequent GLC analysis under identical
conditions as the original sample. Thus, peaks of unsaturated acids will
disappear from the chromatogram as these will be converted by hydrogenation
to their saturated homolog (i.e., $C_{16:1}$ to $C_{16:0}$), and the size of the
saturated homolog peak will be increased in proportion to the amount of
unsaturated component in the original sample. With acetylation, hydroxy acid
methyl esters are converted to a more volatile diester derivative which
elutes faster from a nonpolar column than the original methyl ester. With

these combinations of GLC analyses, essentially all the fatty acids in a given microorganism can be identified by RT alone. Other analytical techniques (IR, NMR, MS) are useful for confirmation and are required for identification of any component not identified by RT measurements. Some fatty acids known to occur in bacteria are not available commercially for use as standards, but the RT's of these relatively unusual acids (i.e., $C_{2-OH-13:0}$, $C_{i-3-OH-17:0}$, $C_{a-17:1}$, etc.) can be obtained by consulting the literature and then analyzing microorganisms known to contain these acids.[3,4,10,11]

ACETYLATION AND HYDROGENATION OF FAME

The results obtained by acetylation and hydrogenation are illustrated in Figures 2-4 with a strain of *Neisseria subflava*. The chromatogram in Figure 2 shows peaks of FAME of this organism processed with the saponification procedure described above. Tentative identification of peaks in the chromatogram was made on the basis of RT comparison to known standards and recorded on the chromatogram in shorthand designation (i.e., $C_{12:0}$ is a 12-carbon saturated straight-chain acid, $C_{16:1}$ is a monounsaturated 16-carbon straight-chain acid). To confirm the presence of the hydroxyl group in the three peaks tentatively identified as hydroxy acids ($C_{3-OH-12:0}$, $C_{3-OH-14:0}$, $C_{3-OH-16:0}$), the sample was treated with trifluoroacetic anhydride as described previously[12-15] and then reanalyzed on the same column under identical GLC conditions as in Figure 2.

A chromatogram of the acetylated FAME of *N. subflava* is shown in Figure 3. By comparing the chromatograms, it is clear that the peaks of hydroxy FAME at 6.4 min, 9.5 min, and 13.0 min in Figure 2 are not present in Figure 3 and that each of these peaks has been converted to a more volatile diester derivative which elutes at 5.3 min, 8.0 min, and 11.3 min, respectively, in Figure 3. These data which show identical RT matches with two derivatized functional groups (-COOH, -OH) give strong support for positive identification. It is also apparent that no peaks other than the hydroxy acids were affected by the acetylation step as these peaks have the same RT in both chromatograms.

The hydrogenated FAME of *N. subflava* is shown in Figure 4. All labelled peaks of unsaturated acids in Figure 2 ($C_{14:1}$, $C_{16:1\omega7c}$, $C_{18:2}$, $C_{18:1\omega9c}$, $C_{18:1\omega7c}$) are absent in the hydrogenated sample shown in Figure 4 since each unsaturated acid has been converted to its saturated homolog. The increase in the amount of the saturated homolog is proportional to the amount of unsaturated component(s) present as illustrated by comparing the relative peak heights of $C_{14:0}$, $C_{16:0}$, and $C_{18:0}$ in Figure 2 with those in Figure 4. The area of the $C_{18:0}$ peak in Figure 4 represents the total of all unsaturated 18-carbon acids in Figure 2 ($C_{18:2}$, $C_{18:1\omega9c}$, $C_{18:1\omega7c}$) as well as the area of $C_{18:0}$ in the original sample; the $C_{14:1}$ and $C_{16:1}$ peaks are not present in Figure 4 since these have been converted to $C_{14:0}$ and $C_{16:0}$, respectively. These data give additional evidence for identification of unsaturated (as well as saturated) acids. The fact that the hydroxy acids were unaffected by hydrogenation indicates that these acids were saturated and contained only a hydroxyl group. Unsaturated hydroxy acids are occasionally present in bacteria; these require the combined analysis as FAME, acetylated FAME, and hydrogenated FAME for accurate identification. Organisms with relatively complex fatty acid profiles such as *Xanthomonas* (*Pseudomonas*) *maltophila* and *Flavobacterium meningosepticum* also require these techniques for accurate peak identification.[2,16]

IDENTIFICATION

After the individual fatty acids are identified, the fatty acid composition of the unknown organism is compared with culturally or

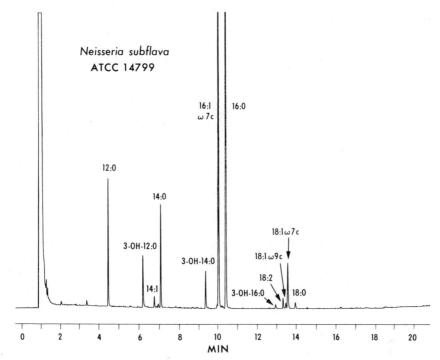

Figure 2. Gas chromatogram of the cellular fatty acids (as methyl esters) of *Neisseria subflava* analyzed on a 25 m by 0.2 mm methyl phenyl silicone fused silica capillary column.

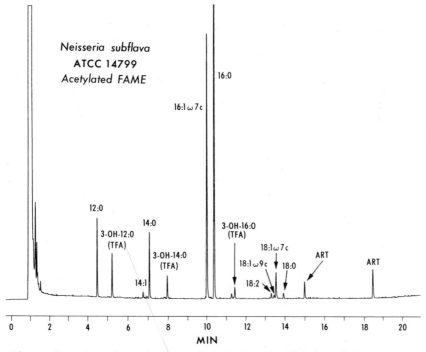

Figure 3. Gas chromatogram of acetylated cellular fatty acids (as methyl esters) of *Neisseria subflava* analyzed on a 25 m by 0.2 mm methyl phenyl silicone fused silica capillary column.

biochemically related organisms whose fatty acid contents are known. Visual comparisons are made on the basis of qualitative and quantitative similarities or differences in fatty acid composition. Visual comparison requires time, experience, and familiarity with the fatty acid composition of many bacterial genera and species. As the number of species for comparison grows, the difficulty of visual identification increases because of the large amount of data that must be correlated. Use of a microcomputer for identification is a logical and necessary step for widespread use of cellular fatty acid data outside research or specialty laboratories.

THE HEWLETT-PACKARD (HP) 5898A MICROBIAL IDENTIFICATION SYSTEM

     In early 1985, the Hewlett-Packard Company (Avondale, PA) introduced an automated GLC computer system (HP 5898A MIS) for analysis of the cellular fatty acid content of microorganisms. This system consists of a gas chromatograph containing a 25 m x 0.2 mm methyl phenyl silicone fused silica capillary column, an automatic sampler, a reporting integrator, a computer, and a printer. To use the system, bacterial FAME are prepared as described above and loaded into the tray of the automatic injector. The numbers assigned to the analytical samples are entered into the computer through a keyboard and analysis is initiated by pressing the "start sequence" key. At this time, the complete system is under full control of the computer which initiates sample injection, automatic sample sequencing, and GLC temperature-programmed analysis (170-280°C at 5°C/min). The integrator plots the signal (chromatogram) from the GLC flame ionization detector and calculates peak data. At the end of the GLC run, the raw data are automatically transferred from the integrator to the computer for identification and the final report from the printer.

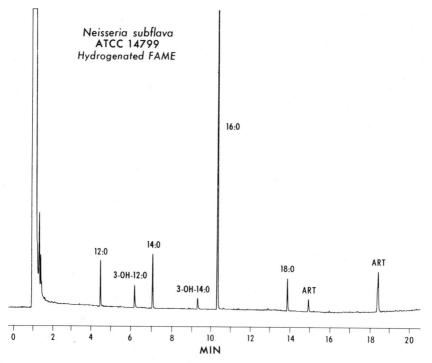

Figure 4. Gas chromatogram of hydrogenated cellular fatty acids
          (as methyl esters) of *Neisseria subflava* analyzed on a 25 m
          by 0.2 mm methyl phenyl silicone fused silica capillary column.

64

```
--------------------------------------------------------------------------------------------------
ID:    1          CAL                                          Date of run: 19-NOV-87 00:08:19
Bottle: 1         CALIBRATION [AEROBE]                                    /

  RT     Area    Ar/Ht Respon   ECL       Name          %     Comment 1         Comment 2
------  --------- ------ ------  ------  -------------- ------  -------------     ---------------
 1.573 37306000  0.074  . . .   7.032  SOLVENT PEAK . . . . .  . . .  < min rt
 1.827     2106  0.034  . . .   7.569  . . . . . . . . .       . . .  < min rt
 1.888     2228  0.029  . . .   7.698  . . . . . . . . .       . . .  < min rt
 2.504    50668  0.031  1.116   9.000  9:0 . . . . . . . .     5.09
 2.977   104230  0.032  1.081  10.000  10:0 . . . . . . . .   10.15  Peak match  0.0000
 3.632    54648  0.033  1.052  11.000  11:0 . . . . . . . .    5.18  Peak match -0.0009
 3.768    21993  0.033  1.048  11.156  10:0 20H . . . . . .    2.08  Peak match  0.0006
 3.999    11007  0.034  1.041  11.421  10:0 30H . . . . . .    1.03  Peak match  0.0017
 4.504   110650  0.035  1.028  12.000  12:0 . . . . . . . .   10.24  Peak match -0.0010
 5.606    56413  0.037  1.008  13.000  13:0 . . . . . . . .    5.12  Peak match  0.0002
 6.926   112330  0.039  0.992  14.000  14:0 . . . . . . . .   10.04  Peak match  0.0002
 8.427    57774  0.042  0.980  15.000  15:0 . . . . . . . .    5.10  Peak match -0.0021
 8.757    23912  0.044  0.978  15.202  14:0 20H . . . . . .    2.10  Peak match  0.0038
 9.223    11236  0.046  0.975  15.488  Sum In Feature 3 . .    0.99  Peak match  0.0006   14:0 30H/16:1 ISO I
10.059   115750  0.044  0.970  16.000  16:0 . . . . . . . .   10.11  Peak match -0.0010
11.764    59381  0.047  0.963  17.000  17:0 . . . . . . . .    5.15  Peak match -0.0017
12.169    23776  0.048  0.961  17.232  16:0 20H . . . . . .    2.06  Peak match  0.0025
13.506   119840  0.047  0.957  18.000  18:0 . . . . . . . .   10.33  Peak match -0.0008
15.245    59458  0.049  0.952  19.000  19:0 . . . . . . . .    5.10  Peak match  0.0000
16.966   118900  0.050  0.948  20.000  20:0 . . . . . . . .   10.15
17.951      933  0.052  . . .  20.572  . . . . . . . . . . .  . . .
*******    11236 . . . . . . .  . . .  SUMMED FEATURE 3 . .    0.99  12:0 ALDE ?      unknown 10.928
******* . . . . . . . . . . . .  . . .  . . . . . . . . . .   . . .  16:1 ISO I/14:0 30H  14:0 30H/16:1 ISO I


Solvent Ar  Total Area  Named Area  % Named  Total Amnt  Nbr Ref  ECL Deviation  Ref ECL Shift
----------  ----------  ----------  -------  ----------  -------  -------------  -------------
 37306000    1111966     1111966    100.00    1110596       0    . . . . . . . . . . . . . . .
GOOD PEAK MATCHING: PEAK POSITION MATCHING ERROR (RMS) IS 0.0015.
--------------------------------------------------------------------------------------------------
```

Figure 5. Computer printout of a fatty acid methyl ester calibration mixture analyzed with the Hewlett-Packard 5898A microbial identification system.

Over the last two years, our experience with the HP-MIS has shown that the system gives a high level of reproducibility, accuracy, and precision in analysis and identification of individual fatty acid components in the analytical sample. This capability is due not only to optimized column and chromatographic parameters but also in large part to careful selection of a standard FAME mix that is run initially and then reanalyzed at selected intervals along with samples. The HP standard FAME consists of a quantitative mixture of odd and even chain saturated FAME ranging from 9-20 carbons in length as well as a homologous series of hydroxy FAME with a free hydroxyl group at the 2- or 3-carbon atom (Figure 5). In the GLC temperature program run, RT data of the components in the standard FAME mixture are used to determine FAME equivalent chain length (ECL) values. The ECL, which is an index of the FAME carbon number, indicates where the FAME elutes from the column with respect to a series of FAME whose ECL's are known. Since ECL plots of FAME in a given homologous series (i.e., $C_{14:0}$, $C_{16:0}$, $C_{18:0}$, $C_{14:1}$, $C_{16:1}$, $C_{18:1}$; $C_{2-OH-14:0}$, $C_{2-OH-16:0}$, $C_{2-OH-18:0}$, $C_{3-OH-14:0}$, $C_{3-OH-16:0}$, $C_{3-OH-18:0}$, etc.) tend to be linear, the individual components in the analytical sample are identified by the computer which calculates the ECL and correlates these values with known ECL's. Peaks are identified when they lie within a window of approximately 0.005 ECL units. The MIS computer printout of the standard FAME mixture or "calibrator" in Fig. 5 shows that each component of the mixture is identified as indicated under the "name" column. For example, the report shows that the peak eluting from the column at RT 2.974 min has an ECL of 10.000 and thus is identified as the methyl ester of decanoic acid ($C_{10:0}$), a saturated straight-chain acid with a 10-carbon chain.

In addition to ECL's, the computer also calculates, monitors, and records several other parameters during the chromatographic run as indicated in the report (Figure 5). In addition to RT and peak identification, the computer calculates peak area (area), peak shape (area/height) and percentage (%) composition of each peak by dividing individual peak area by the total area. The peak shape (area/height) calculation allows for accurate determination of true peaks and thus eliminates any broad (from contamination) or narrow (i.e., dirty detector, electrical spike) peaks from the final report. Calculation of quantitative amounts of each component in the standard mixture allows for continuous monitoring of the column and overall system performance. If the quantitative value of any peak is outside the expected value, a warning is printed on the report and the calibration standard is rerun. If the system fails to calibrate after two consecutive runs, the error message is repeated and the sample sequence is aborted. The most frequent reason for failure to calibrate is the loss of hydroxy acids due to a dirty injector liner which should be replaced after receiving the error message from the computer. Slight losses in hydroxy acids that are easily detected with the MIS are generally not apparent from visual inspection of the chromatogram. Thus, the overall precision and accuracy in identification and quantitation of bacterial fatty acids with the MIS is far superior to hand injection and analysis without a computer.

The MIS identifies the unknown or test organism by comparing its fatty acid composition with that of known library entries. The library entries contain data from analysis of several known or reference strains of a given bacterial species and searchable library entries are generated through multivariate gaussian analysis of this data. Within 1 min after completing the chromatographic run and transfer of data from the integrator, the computer analyzes the unknown fatty acid data, does the library search, and prints the report. In the report, the MIS prints the names or the abbreviated names of the most likely matches and a similarity index for each match. The similarity index is a numerical value that expresses how closely the composition of the library matches. This index value is a calculation of the unknown organism's distance in n-dimensional space from the mean profiles of the closest library entries. If the search results in more than one possible match, the suggested identities are listed in descending probability. The most probable genus is listed first followed by the species.

Shown in Figure 6 is a complete printout of the fatty acid composition and computer identification of a strain of *Legionella oakridgensis* (OR-10) processed with the MIS. The top portion of the printout contains the data and time of analysis and the sample name and number. The middle portion contains the qualitative and quantitative data and peak identification as discussed above for the calibrator or reference standard. The library search report, which is listed just before the report summary, shows that *L. oakridgensis* strain OR-10 was identified to the genus (LGN) and species (*L. oakridgensis*) level with a very high similarity index match of 0.926 (1.0 is the highest possible value). This high match would be expected for *L. oakridgensis* as the fatty acid composition of this organism is not only unique for the genus but is sufficiently distinct from the other 23 named *Legionella* species and is easily identified.[7,17,18] The similarity index of some strains of other *Legionella* species are much lower, but in all cases, strains of each of the 23 species are always identified to the genus level and are also correctly listed as the first choice at the species level. Close similarity index values simply reflect close similarities in fatty acid composition which is often the case for species within a genus (i.e., *Legionella*, *Streptococcus*) but which may also occur among species of different genera such as has been observed in the *Enterobacteriaceae*. At present, the MIS library of aerobic bacteria contains approximately 75 genera

```
-------------------------------------------------------------------------------------------
ID:    284      LGN-OAKRID(OR-10,5/81,72H,ATCC33761,TS)              Date of run: 28-APR-87 16:49:25
Bottle: 35      SAMPLE [AEROBE]                                      Date edited: 09-JUL-87 11:29:15
```

RT	Area	Ar/Ht	Respon	ECL	Name	%	Comment 1	Comment 2
1.596	28301000	0.056	...	7.036	SOLVENT PEAK	. . .	< min rt	
6.538	1150	0.041	0.989	13.618	14:0 ISO	0.46	ECL deviates 0.000	Reference -0.003
7.048	654	0.038	0.984	13.999	14:0	0.26	ECL deviates -0.001	Reference -0.004
8.129	4425	0.042	0.977	14.712	15:0 ANTEISO	1.76	ECL deviates 0.001	Reference -0.001
8.348	3044	0.043	0.976	14.857	15:1 B	1.21	ECL deviates 0.001	
8.563	1808	0.044	0.975	14.999	15:0	0.72	ECL deviates -0.001	Reference -0.004
9.320	3637	0.047	0.972	15.459	16:1 ISO H	1.44	ECL deviates -0.002	
9.596	69666	0.045	0.972	15.627	16:0 ISO	27.58	ECL deviates 0.001	Reference -0.002
9.911	37298	0.047	0.971	15.818	16:1 CIS 9	14.75	ECL deviates 0.001	
10.073	815	0.049	0.970	15.917	16:1 C	0.32	ECL deviates 0.009	
10.208	25600	0.046	0.970	15.999	16:0	10.12	ECL deviates -0.001	Reference -0.004
10.363	1175	0.063	...	16.090	. . . .	. . .		
11.108	2410	0.055	0.969	16.524	17:1 ANTEISO C . . .	0.95	ECL deviates -0.001	
11.447	13203	0.047	0.969	16.722	17:0 ANTEISO	5.21	ECL deviates -0.000	Reference -0.003
11.734	28861	0.049	0.969	16.889	17:0 CYCLO	11.39	ECL deviates 0.001	Reference -0.002
11.925	7475	0.049	0.969	17.001	17:0	2.95	ECL deviates 0.001	Reference -0.002
13.029	5156	0.052	0.970	17.631	18:0 ISO	2.04	ECL deviates -0.001	Reference -0.004
13.238	1662	0.049	...	17.751	. . . .	. . .		
13.674	33536	0.051	0.971	18.000	18:0	13.27	ECL deviates -0.000	Reference -0.003
14.461	1583	0.098	...	18.451	. . . .	. . .		
14.948	1974	0.054	0.975	18.730	19:0 ANTEISO	0.78	ECL deviates 0.001	
15.419	6605	0.055	0.976	19.000	19:0	2.63	ECL deviates -0.000	Reference -0.004
17.144	5337	0.056	0.985	20.000	20:0	2.14	ECL deviates 0.000	Reference -0.004

Solvent Ar	Total Area	Named Area	% Named	Total Amnt	Nbr Ref	ECL Deviation	Ref ECL Shift
28301000	257074	252654	98.28	245420	13	0.002	0.003

```
        CDC3-A [Rev 1.0]  Legionella . . . . . . . . . . . . . . . . . . . . 0.926
                          L. oakridgensis . . . . . . . . . . . . . . . 0.926
```

Figure 6. Computer printout of the cellular fatty acids
(as methyl esters) of *Legionella oakridgensis*
analyzed with the Hewlett-Packard 5898A microbial
identification system.

and 300 species; a library of mycobacteria and an updated one of anaerobic
bacteria are near completion. Also, software for generating one's own
library will soon be available.

SUMMARY

For the past several years, the use of cellular fatty acid data has
become an integrated part of the identification procedures of many
laboratories throughout the world. The fatty acid compositions of a large
number of bacterial species have been reported.[10,13-15,19-29] Moreover,
cellular fatty acid data are included in the description of essentially all
new or recently discovered organisms.[30] Routine fatty acid analysis in the
clinical or diagnostic laboratory is now feasible with current GLC
instrumentation and computer technology, which permits automated analysis
using technical personnel who have little or no experience or knowledge with
the technique. Although fatty acid data are quite specific for many bacteria
such as *Legionella*[11] and *Francisella*[12], it will not distinguish all species or
groups. However, the data will significantly reduce the possibilities to
several best choices. Then, in many instances, the correct choice can be
made from a few simple conventional tests, while in other cases the complete
battery of microbiological tests will be required. Thus, the microbiologist
should use cellular fatty acid data as the first step in the identification

procedure as this information will save time and money by significantly reducing the number of subsequent tests required for identification.

ACKNOWLEDGMENTS

The author gratefully acknowledges Patricia Lynn Wallace for preparation of the figures and Ellen Lamb for typing the manuscript.

REFERENCES

1. N. R. Krieg and J. G. Holt, eds., "Bergey's Manual of Systematic Bacteriology," Vols. 1 and 2, Williams and Wilkins, Baltimore (1984).
2. C. W. Moss and S. B. Dees, The cellular fatty acids of *Flavobacterium meningosepticum* and *Flavobacterium* species group IIb, J. Clin. Microbiol. 8:772 (1978).
3. S. B. Dees, G. M. Carlone, D. Hollis, and C. W. Moss, Chemical and phenotypic characteristics of *Flavobacterium thalpophilum* compared with those of other Flavobacterium and *Sphingobacterium* species, Int. J. Syst. Bacteriol. 35:16 (1985).
4. S. B. Dees, C. W. Moss, D. G. Hollis, and R. E. Weaver, Chemical characterization of *Flavobacterium odoratum*, *Flavobacterium breve*, and *Flavobacterium*-like groups IIe, IIh, and IIf, J. Clin. Microbiol. 23:267 (1986).
5. P. H. Edelstein, Improved semiselective medium for isolation of *Legionella pneumophila* from contaminated clinical and environmental specimens, J. Clin. Microbiol. 14:298 (1981).
6. M. A. Lambert and C. W. Moss, Comparison of the effects of acid and base hydrolysis on hydroxy and cyclopropane fatty acids in bacteria, J. Clin. Microbiol. 18:1370 (1983).
7. C. W. Moss, W. F. Bibb, D. E. Karr, and G. O. Guerrant, Chemical analysis of the genus *Legionella*: fatty acids and isoprenoid quinones, INSERM, 114:375 (1983).
8. C. W. Moss, A. Kai, M. A. Lambert, and C. M. Patton, Isoprenoid quinone content and cellular fatty acid composition of *Campylobacter* species, J. Clin. Microbiol. 19:772 (1984).
9. L. M. Teixeira, C. W. Moss, and R. R. Facklam, Gas-liquid chromatography of the fatty acids of *Streptococcus faecalis* with a fused silica capillary column, FEMS Microbiol. Lett. 17:257 (1983).
10. D. B. Drucker, "Microbiological Applications of Gas Chromatography," Cambridge University Press, New York (1981).
11. E. Yabuuchi and and C. W. Moss, Cellular fatty acid composition of strains of three species *Sphingobacterium* gen *nov.* and *Cytophaga johnsonae*, FEMS Microbiol. Lett. 13:87 (1982).
12. C. W. Moss, Gas-liquid chromatography as an analytical tool in microbiology, J. Chromatogr. 203:337 (1981).
13. M. A. Lambert, D. G. Hollis, C. W. Moss, R. E. Weaver, and M. L. Thomas, Cellular fatty acids of nonpathogenic *Neisseria*, Can. J. Microbiol. 17:491 (1971).
14. M. A. Lambert, C. W. Moss, V. A. Silcox, and R. C. Good, Analysis of mycolic acid cleavage products and cellular fatty acids of *Mycobacterium* species by capillary gas chromatography, J. Clin. Microbiol. 23:731 (1986).
15. M. A. Lambert, C. M. Patton, T. J. Barrett, and C. W. Moss, Differentiation of *Campylobacter* and *Campylobacter*-like organisms by cellular fatty acid composition, J. Clin. Microbiol. 25:706 (1987).
16. C. W. Moss, S. B. Samuels, J. Liddle, and R. M. McKinney, Occurrence of branched-chain hydroxy acids in *Pseudomonas maltophila*, J. Bacteriol. 114:1018 (1973).

17. C. W. Moss, R. E. Weaver, S. B. Dees, and W. B. Cherry, Cellular fatty acid composition of isolates from Legionnaires' disease, J. Clin. Microbiol. 6:140 (1977).

18. L. H. Orrison, W. B. Cherry, R. L. Tyndall, C. B. Fliermans, S. B. Gough, M. A. Lambert, L. K. McDougal, W. F. Bibb, and D. J. Brenner, *Legionella oakridgensis*: unusual new species isolated from cooling tower water, App. Environ. Microbiol. 45:536 (1983).

19. E. Jantzen, K. Bryn, T. Bergan, and K. Bovre, Gas chromatography of bacterial whole cell methanolysates, V, Fatty acid composition of neisseriaceae and moraxellae, Acta. Path. Microbiol. Scand., Sect. B, 82:767 (1974).

20. E. Jantzen, K. Bryn, T. Bergan, and K. Bovre, Gas chromatography of bacterial whole cell methanolysates, VII, Fatty acid composition of *Acinetobacter* in relation to the taxonomy of neisseriaceae, Acta. Path. Microbiol. Scand., Sect. B, 83:569 (1975).

21. E. Jantzen, B. P. Berdal, and T. Omland, Cellular fatty acid composition of *Francisella tularensis*, J. Clin. Microbiol. 10:928 (1979).

22. C. W. Moss, New methodology for identification of non-fermentation, in: "Glucose nonfermenting gram-negative bacteria in clinical microbiology," G. L. Gilardi, ed., CRC Press, West Palm Beach, FL, pp. 188-196 (1978).

23. M. A. Lambert and C. W. Moss, Cellular fatty acid composition of *Streptococcus mutans* and related streptococci, J. of Dental Research 55:A96-A102 (1976).

24. B. C. Mayall, Rapid identification of mycobacteria using gas liquid chromatography, Pathology 17:24 (1985).

25. P. A. Tisdall, D. R. De Young, G. D. Roberts, and J. P. Anhalt, Identification of clinical isolates of mycobacteria with gas-liquid chromatography: a 10-month follow-up study, J. Clin. Microbiol. 16:400 (1982).

26. P. Valero-Guillen, F. Martin-Luengo, L. Larsson, J. Jimenez, I. Juhlin, and F. Portaels, Fatty and mycolic acids of *Mycobacterium malmoense*, J. Clin. Microbiol. 26:153 (1988).

27. W. R. Mayberry, Hydroxy fatty acids in *Bacteroides* species: D-(-)-3-hydroxy-14-methylhexadecanoate and its homologs, J. Bacteriol. 143:582 (1980).

28. K. Suzuki and K. Komagata, Taxonomic significance of cellular fatty acid composition in some coryneform bacteria, Int. J. Syst. Bacteriol. 33:188 (1983).

29. H. N. Shah and M. C. Collins, Genus *Bacteriodes* a chemotaxonomical perspective, J. Appl Bacteriol. 55:403 (1983).

30. W. L. Thacker, H. W. Wilkinson, B. B. Plikaytis, A. G. Steigerwalt, W. R. Mayberry, C. W. Moss, and D. J. Brenner, Second serogroup of *Legionella feeleii* strains isolated from humans, J. Clin. Microbiol. 22:1 (1985).

CHAPTER 5

PROFILING AND DETECTION OF BACTERIAL CARBOHYDRATES

Alvin Fox and James Gilbart

Department of Microbiology & Immunology
School of Medicine
University of South Carolina
Columbia, SC 29208

Stephen L. Morgan

Department of Chemistry
University of South Carolina
Columbia, SC 29208

INTRODUCTION

Carbohydrates are a major class of structural components in bacterial
cell envelopes. Sugar profiles can differentiate and identify isolated
bacteria. Carbohydrates can also serve as chemical markers for direct
detection of bacteria in complex matrices such as mammalian body fluids and
tissues. This chapter reviews the composition of some bacterial
polysaccharides and their characteristic sugars. Preparation and analysis of
derivatives of these sugar monomers by gas chromatography (GC) and mass
spectrometry (MS) is described. A modified alditol acetate procedure
developed in our laboratories is presented along with some examples of its
use.

CARBOHYDRATE STRUCTURES PRESENT WITHIN BACTERIA

Nomenclature

Carbohydrates are polyhydroxyl organic compounds with exceptional
variability in composition. The backbone of simple sugars, varying in chain
length from 3 to 7 carbons, are designated as trioses, tetroses, pentoses,
hexoses, and heptoses respectively. The various classes and structures of
carbohydrates are illustrated in Figure 1.

The D- or L- specification of a sugar denotes which of two possible
optical isomers results from the configuration of the asymmetric carbon at
the penultimate position (carbon 5 in hexoses). In Fischer projections (left
side Figure 1A), this hydroxyl is drawn to the right in D-sugars. The
terminal carbon (carbon 6 in hexoses) is above the ring shown in the Haworth
projection (right side Figure 1A). In L-sugars, the hydroxyl is to the left
and the terminal carbon is below the ring.

Analytical Microbiology Methods
Edited by A. Fox *et al.*
Plenum Press, New York, 1990

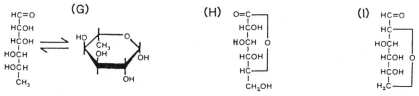

Figure 1.   Classes of sugars, and and an example of each:
(A) the aldose, D-glucose; (B) the ketose, D-fructose;
(C) the alditol, D-glucitol; (D) the amino sugar,
D-galactosamine; (E) the alduronic acid, D-glucuronic
acid; (F) the aldonic acid, D-gluconic acid; (G) the
methyl sugar, L-rhamnose; (H) the lactone, 1,5-
gluconolactone; (I) the anhydrosugar, 2,6-anhydro-D-
glucose.

Aldoses contain an aldehyde group on carbon 1. Six carbon aldoses (e.g., hexoses such as glucose) mostly form pyranose rings (Figure 1A), but can also form furanose rings. In Figure 1, where ring forms are shown, the orientation of the hydroxyl on carbon 1 is not indicated since this group can either be below the ring (α) or above it (β); these possibilities constitute two possible anomers for each sugar. Ketoses, such as fructose, contain a ketone group on carbon 2 and tend to form only furanose rings. Pentoses (such as ribose, Figure 1B) also form furanose rings since the pyranose form of these sugars is unstable.

Alditols are straight chain sugar alcohols (such as glucitol, Figure 1C); because they do not contain carbonyl groups, pyranose or furanose rings can not form. Some sugars exist naturally as alditols in bacteria; alditols can also be formed by reduction of an aldose. Amino sugars (such as galactosamine, Figure 1D) contain one, or less commonly, two amino groups in place of hydroxyls. Alduronic acids (such as glucuronic acid, Figure 1E) contain a carboxyl moiety on carbon 6 in addition to an aldehyde on carbon 1; these compounds can exist in the ring form. Aldonic acids contain a carboxyl group instead of the aldehyde on carbon 1 and cannot form pyranose rings (Figure 1F). Aldaric acids have carboxyl moieties at both ends of the sugar. In methyl sugars (such as rhamnose, Figure 1G), a methyl group replaces a hydroxyl, while in deoxysugars (such as deoxyribose) a hydrogen replaces a hydroxyl. Under dehydrating conditions, acidic sugars can form lactones, or sugars with an internal ester ring (Figure 1H). Neutral sugars can lose water to become anhydrosugars containing an internal ether ring (Figure 1I). The ester or ether ring in lactones or anhydrosugars can be in several different positions, but more commonly the ether ring links carbons which are 4 or 5 places apart (1,5; 2,6; 3,6; etc.).

Sources of Microbial Sugars

The major sources of carbohydrates useful for microbial chemotaxonomy are different in Gram positive and Gram negative bacteria.[1-3] For Gram positive bacteria, the neutral or acidic polysaccharides and teichoic acids associated with peptidoglycan as well as membrane lipoteichoic acids are most commonly employed to distinguish among related genera and species. For Gram negative bacteria, the core and O-antigen region of lipopolysaccharides are commonly used for differentiation. Capsular polysaccharides found in Gram positive and Gram negative bacteria are also characteristic of certain species. Certain species can vary dramatically in capsular polysaccharides synthesized; this intraspecies variability may limit usefulness of those carbohydrates for bacterial identification or detection. Proteins are not glycosylated in prokaryotic cells as they are in eukaryotic cells.

Deoxyribose and ribose are major microbial sugars but, because they are present in DNA and RNA respectively in all microorganisms, these sugars are not generally useful for chemotaxonomy. Deoxyribose is unstable under some hydrolysis conditions and may not appear in chromatograms of bacterial hydrolysates. Muramic acid and glucosamine, components of the glycan backbone of peptidoglycan, are present in almost all bacterial chromatograms.[4,5] Glucosamine is commonly found in other prokaryotic and eukaryotic structures. Muramic acid is not believed to occur elsewhere in nature. The presence of glucosamine or muramic acid is therefore not characteristic of particular bacterial species. However, Gram positive bacteria contain substantially higher levels of peptidoglycan than Gram negative bacteria. The core region of most Gram negative lipolysaccharides contain L-D-glyceromannoheptose and ketodeoxyoctonic acid (KDO). Structures of KDO and heptose, along with muramic acid, are shown in Figure 1 of Chapter 1. These two sugars can be used to distinguish Gram negative from Gram positive bacteria. Less commonly present is D-D-glyceromannoheptose which can be used to distinguish among certain Gram negative bacteria.[6,7] In some capsular polysaccharides, KDO is a major constituent. Many Gram positive bacteria possess teichoic acids that usually contain ribitol or glycerol[8]. Glycerol is also a component of bacterial glycerides which can be found in both Gram positive and Gram negative bacteria. The presence of ribitol indicates the organism is likely to be Gram positive; the presence of glycerol may not be as informative. The composition of cell envelope macromolecules, including polysaccharides, has been described in detail in Chapter 1.

Most early use of carbohydrates in bacterial taxonomy involved qualitative determination of sugar profiles by paper or thin layer chromatography. Qualitative profiling of bacterial carbohydrates has provided a framework for the chemotaxonomic differentiation of many organisms, but quantitative comparisons are now becoming more commonplace. In addition to identification of carbohydrate monomers, GC or GC-MS methods can provide quantitative composition. Carbohydrate composition is usually expressed as a percentage of total carbohydrate peak areas or, perhaps more meaningfully, as a percentage of the dry weight.

CARBOHYDRATE PROFILING IN CHEMOTAXONOMY

Gram Positive Bacteria

The differentiation of propionibacteria by thin layer chromatography is among the earliest applications of carbohydrate profiling in microbial taxonomy. For example, the cell wall polysaccharide of *Propionibacterium acnes* serotype 1 contains glucose, galactose, and mannose; *P. acnes* serotype II differs in containing only glucose and mannose. Rhamnose (along with galactose and mannose) is characteristic of *P. freundenreichii*.[2] Many of these differences were known before the potential of quantitative carbohydrate profiling by GC and definitive identification by MS was fully realized. A good example of the value of GC-MS for reinvestigating previously established features was recently shown by the identification of an unusual sugar, 2,3-diamino dideoxyglucuronic acid, as a constituent of *P. acnes* cell walls.[9]

Carbohydrate profiling by capillary GC is useful in differentiating streptococci.[10-12] Sugars derived from neutral polysaccharides and lipoteichoic acids can be released from whole bacterial cells by methanolysis and analyzed as trifluoroacetate derivatives using electron capture detection (Chapter 2). Species from groups A, B, C, D, F and G all were found to contain rhamnose, glucose and glucosamine. The major β-hemolytic pathogens (groups A and B) can be readily distinguished-- glucitol is present only in group B streptococci. Group C streptococci contain large amounts of galactosamine, presumably derived from carbohydrate side chains of the group antigen. The classic Lancefield serological grouping scheme for differentiating streptococci recognizes these same carbohydrate antigens. GC profiling offers a more quantitative and precise alternative. Although whole bacterial cells can be analyzed by GC or GC-MS without the need to isolate group antigens, the presence of glucosamine in GC profiles of whole cells may be difficult to interpret. Group A streptococci contain glucosamine in the side chains of the group antigen. Glucosamine derived from this source can not be differentiated from glucosamine derived from peptidoglycan by GC analysis of whole cells. This uncertainty complicates the use of glucosamine for identification of streptococci.

Analysis of the sugar content of isolated teichoic acids has been used to differentiate staphylococci. Most staphylococcci contain ribitol or glycerol teichoic acids; a minority of staphylococcci contain teichoic acids with N-acetyl glucosamine backbones. The major pathogen among the staphylococci, *Staphylococcus aureus*, can be readily differentiated from *Staphylococcus epidermidis* (a common contaminant of clinical samples) by their respective ribitol and glycerol contents. On extensive heating in acid, sugar alcohols are converted into their anhydrosugars. In one study, ribitol was partially converted into anhydroribitol and analyzed in this form. However, chromatograms of mixtures of the alditol and anhydrosugar may be difficult to interpret.[8] Shorter heating times may result in less quantitative release of alditols, but may produce simpler chromatograms containing only the alditol.

Gram Negative Bacteria

Members of the family *Legionellaceae*, a family of Gram negative bacteria, contain minute amounts of carbohydrates, but these sugars are useful for differentiation. Despite this fact, the major pathogens in the group, *Legionella pneumophila* and *Tatlockia micdadiae*, can be readily differentiated. *T. micdadiae* contains relatively large amounts of rhamnose and fucose (approximately 2% dry weight each); *L. pneumophila* contains only a small amount of rhamnose (about 0.3% dry weight). Using a modified alditol acetate procedure with low background noise, the two species may be differentiated using GC with a flame ionization detector.[7] However, comparison of Figures 2A and 2B shows that more easily interpretable chromatograms are obtained using selected ion monitoring (SIM) MS. GC-MS also enables identification of unusual sugars such as the aminodideoxysugars found in certain legionellae.[13,14] These sugars have subsequently been shown to originate in the LPS of *Legionella pneumophila*.[15]

The major surface antigens of *E. coli* are the LPS and capsular polysaccharides. In *E. coli* only one lipid A structure is present but there are five core structures and greater than 150 O-antigen types. In contrast to *Klebsiella* and *Salmonella*, *E. coli* can have acidic sugar components in their O-antigen. The capsular polysaccharides of *E. coli* are acidic polysaccharides including KDO, N-acetyl neuraminic acid, and N-acetyl mannosaminuronic acid.[1] Chapter 10 discusses the analysis of the lipid A portion of LPS by soft ionization MS methods.

Neisseria meningitidis groups are normally differentiated using specific antisera. Alternatively, analysis of phenol extracts of whole cells which contain nucleic acids, lipopolysaccharides, and capsular material were analyzed after methanolysis and trifluoroacetylation.[16,17] Most groups could be readily differentiated. However, in some instances it was difficult to determine whether sugars were derived from LPS or from capsular material and sub-fractionation by ultracentrifugation was performed. The separation of LPS from capsule increases the complexity of the procedure.

DETECTION OF BACTERIAL SUGARS IN MAMMALIAN BODY FLUIDS AND TISSUES

The detection of chemical markers for bacteria in body fluids and tissues is a more difficult task than profiling of isolated bacterial cells for several reasons. In a tissue analysis the bacteria are only one component of the sample and bacterial sugars are present at low concentrations. Potential chemical markers for bacteria may be present either as a free sugar or a component of macromolecules. It may also be difficult to resolve by chromatography the bacterial marker from natural tissue components present in greater abundance. However, the prior fractionation of tissues and body fluids (e.g., using monoclonal antibodies) to isolate or concentrate bacterial structures before analysis has not been explored extensively and may have utility.

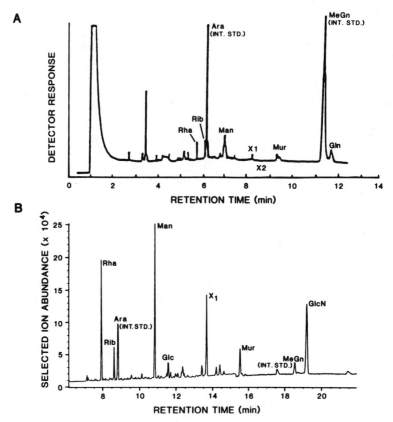

Figure 2. A comparison of chromatograms of whole cell
hydrolysates of *Legionella pneumophila* (as alditol
acetates) analyzed (A) using a flame ionization
detector and (B) using selected ion monitoring GC-MS.
Peak identification: dRib = deoxyribose, Rha =
rhamnose, Fuc = fucose, Rib = ribose, Ara = arabinose
(internal standard, Int. Std.), Xyl = xylose, Man =
mannose, Gal = galactose, Glc = glucose, D,D-Hep =
D-glycero-D-mannoheptose, L,D-Hep = L-glycero-D-
mannoheptose, D,L-Hep = D-glycero-L-mannoheptose,
Mur = muramic acid, MeGN = methylglucamine (Int. Std.),
GlcN = glucosamine, ManN = mannosamine, GalN =
galactosamine. Chromatographic conditions:
(A) borosilicate glass capillary column coated with
SP-2330, 30 s hold then temperature programmed from
100°C to 245°C at 30°C/min; (B) 30 m x 0.32 mm id fused
silica capillary column coated with SP-2330, 45 s hold
then temperature programmed from 100°C to 230°C at
30°C/min, 4°C/min to 250°C and 3°C/min to 265°C.
Reprinted with permission from references 7 and 40.

Normal mammalian tissues and body fluids have been shown to contain a number of sugar alditols and aldoses. These include alditols (ribitol, xylitol, arabinitol, mannitol, glucitol, galactitol, myo-inositol[18-20], and aldoses (fucose, mannose, galactose, glucose, glucosamine, galactosamine).[21-23] Levels for some of the alditols in serum are low (about 100 ng/mL of serum[19]) in comparison to glucose (present at mg/mL concentrations[24]). Mannosamine is not a major component of normal mammalian tissues. Rhamnose has not been found in uninfected mammalian tissues, although its isomer fucose does occur. Muramic acid, if present, is at exceedingly low levels (pg/mg dry weight of tissue) in mammalian tissues.[25]

The use of muramic acid as a chemical marker for detection of bacteria or bacterial debris in rheumatoid arthritis has been proposed.[26] Rhamnose and muramic acid were also used as chemical markers to quantitate levels of bacterial cell walls in tissues of arthritic rats. These sugars can be detected at levels as low as 1 ng/mg of wet tissue.[21-23] Figure 3 illustrates the GC-MS detection of bacteria or their constituents in mammalian tissues. A number of other unique sugars found in bacteria and fungi (including heptoses and ketodeoxyoctonic acid as markers for Gram negative bacteria) may be of use for detecting bacteria in body fluids and tissues. The metabolite, arabinitol, has been successfully used as a marker for *Candida* and other fungi.[19,20] This subject is explored in Chapter 15.

ANALYSIS OF CARBOHYDRATES BY GC and GC-MS

Comparison of Derivatives

GC methods for carbohydrates are suitable for neutral and amino aldoses since both amino and hydroxyl moieties will readily react with common derivatizing reagents such as acetic or trifluoroacetic anhydrides or trimethylsilyl reagents.[10,27-32] Acidic sugars are more complex to analyze because of the carboxyl group.[33] Analysis of aldoses fall in two categories: methods that eliminate the anomeric center and which can produce single peaks (the aldononitrile acetate and alditol acetates) or two peaks (O-methyl oxime acetates), and other methods that give multiple peaks for each single sugar (trimethylsilyl and trifluoroacetyl procedures). Unlike aldoses, alditols produce only single peaks on derivatization since they lack an anomeric center.

Aldoses exist in aqueous solution in dynamic equilibrium between ring and straight chain forms. In the straight chain form, an aldehyde group is accessible; in the ring form, this moiety is part of a hemiacetal ring (pyranose or furanose form). Two anomers can be formed on cyclization (α and β) in which the newly formed hydroxyl group is above or below the plane of the ring (Figure 4). If the anomeric center is not destroyed, acylation of the aldose ring freezes the structure in the α or β anomeric form producing multiple peaks (anomers). As shown in Figure 1, certain sugars can exist only in six-membered (pyranose) or 5-membered (furanose) ring forms, and for other sugars both forms can occur. This implies that either 2 or 4 multiple chromatographic peaks are produced from each sugar (2 from pyranose anomers and 2 from furanose anomers) on acylation. Since these anomers are usually resolved by capillary GC, complicated chromatograms can result. Even though high resolution capillary columns can adequately resolve complex mixtures, multiple peaks may confound both qualitative identification and quantitative measurement.

Production of multiple peaks for each single sugar of interest also may adversely affect the limit of detection. In the case of the aldononitrile acetate procedure the aldose is reacted with hydroxylamine and a nitrile is produced after dehydration. Although the anomeric center is eliminated, an

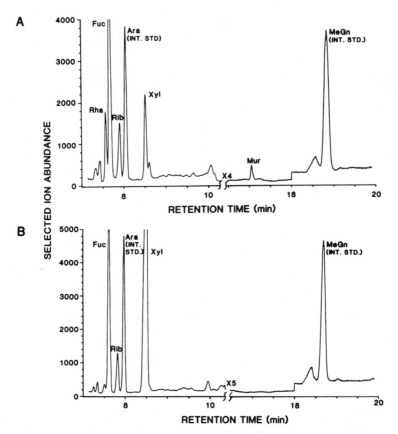

Figure 3.   Selected ion monitoring chromatograms of alditol
            acetates prepared from hydrolysates of 100 mg (wet
            weight) of joint samples from (A) a rat 4 days after
            intraperitoneal injection of streptococcal cell walls
            and (B) a control rat.   Peak identification and chromato-
            graphic conditions as in Figure 2B.   Modified from
            reference 23 with permission.

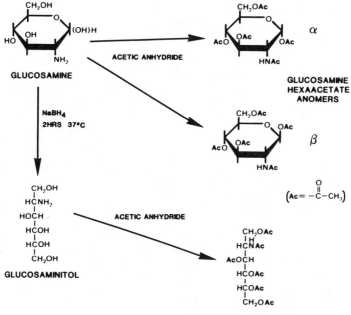

Figure 4. Acetylation of glucosamine with or without prior
reduction showing some possible reaction products.

additional active nitrile site is generated which may explain the difficulty
experienced by some workers in the analysis of amino sugars by this
technique. Contaminating peaks can also be produced by reaction of
hydroxylamine with the acylating agent. The O-methyl oxime acetate procedure
involves reaction of the aldose with O-methyl hydroxylamine. Anti and syn
isomers are produced which on certain capillary columns result in two peaks
for each sugar.[32] The alditol acetate procedure has the advantages of
derivatizing the reduced sugar and producing single peaks without production
of an active site on the anomeric center.

The instability of some carbohydrate derivatives in the presence of
moisture from the air or certain reagents can cause problems in GC analysis.
Trifluoracetyl derivatives of aldoses and alditols, often used for high
sensitivity electron capture GC or negative ion MS detection, are unstable
and not readily amenable to post-derivatization clean-up procedures.
Trimethylsilyl derivatives are also unstable. Acetate derivatives, including
alditol acetates, are extremely stable.[34,35]

In summary, the alditol acetate method has advantages over other
methods for sugar analysis: derivatives are relatively inert (unlike
aldononitrile acetate derivatives), one peak is produced for each sugar
(unlike trimethylsilyl, trifluoro, O-methyloxime acetates), and derivatives
are stable (unlike trimethylsilyl or trifluoroacetates). Some versions of
the alditol acetate method do, however, require extensive manual processing
steps.

The Alditol Acetate Procedure

Determining the carbohydrate composition of a polysaccharide requires
initial hydrolysis of the polymer to release its monomeric constituents.

79

Sugars are unstable when heated to high temperatures in strong acid solutions. Any such degradation of the sample is undesirable because it results in erroneously low values for the monomeric sugars and may introduce additional chromatographic peaks. Two commonly used acids for release of sugars from polymers are hydrochloric acid and sulfuric acid. Hydrochloric acid introduces extraneous peaks in the chromatogram, either due to side reactions or sugar degradation. The use of sulfuric acid and the careful exclusion of air during hydrolysis can minimize these effects. Steps taken to remove the acid after hydrolysis may cause losses of some sugars. Hydrochloric acid is usually removed by lyophilization which is simple but takes an extensive time period (usually about 24 h). Additional degradation of the sugar may occur during this evaporation period. Sulfuric acid is not volatile, but can be easily removed by using a water immiscible N,N-dioctylmethylamine in chloroform solution for neutralization. Internal standards are usually added to the sample after the hydrolysis step and carried through the remaining sample handling steps.

Following hydrolysis, many researchers perform derivatization without using any cleanup procedures. We employ disposable hydrophobic C_{18} columns to remove lipids released by the hydrolysis reaction. Cation exchange resins have also been used at this point to remove amino acids and other charged compounds. Amino sugars are unfortunately separated from the neutral sugars by ion exchange and require separate analysis when this approach is adopted.

The first step in the chemical derivatization of a monomer is a sodium borohydride reduction of the aldose to an alditol, preventing anomerization on acylation. This step is illustrated in Figure 4 using the derivatization of glucosamine as an example. If glucosamine is acylated without reduction, two pyranose anomers can be produced. This sugar can also form furanose rings resulting in two further anomers (not shown). If the sugar is first reduced with sodium borohydride to glucosaminitol, derivatization with acetic anhydride produces the single product, glucosaminitol hexaacetate. Most workers have added sodium borohydride in aqueous solution to reduce the aldose to the sugar alditol. The reduction can be accomplished in under 2 hours at room temperature or at $37°C$; or more conveniently, overnight at $4°C$. Sodium borohydride can be dissolved in methylimidazole to form a stable solution. However, as noted below, N-methylimidazole can be difficult to remove from the sample and can be a source of extraneous peaks.[35-37]

The next step in the alditol acetate derivatization involves multiple evaporations, generally with methanol-acetic acid, to remove borate (as tetramethyl borate gas) and to avoid inhibition of the acetylation (the next step in the method). Borate forms covalent complexes with diols which inhibit acylation at alkaline or neutral pH. Sodium borohydride decomposes to sodium borate in the presence of acetic acid and, after evaporation with methanol-acetic acid, sodium acetate is left behind. Sodium acetate serves as a catalyst for the acylation reaction. The removal of borate is the most time consuming part of the procedure.

Acetylation of the hydroxyl and amino groups on the alditol to form the corresponding esters and amides is the next step in the alditol acetate method. Acetylation eliminates hydrogen bonding and the resulting alditol acetate is sufficiently volatile for GC analysis. Acetylation of free hydroxyl groups in the alditol also decreases interactions with other components of the sample and the GC system.

Pyridine, methylimidazole, and sodium acetate have been employed as catalysts for acetylation. The sample is thoroughly dried before proceeding to the acetylation step if pyridine or sodium acetate are used. Drying is particularly critical if muramic acid is to be analyzed. Using methylimidazole as a catalyst, acetylation can proceed without removal of water or borate generated during the reduction step, allowing for rapid and simple derivatization. Methylimidazole is an excellent catalyst for the acetylation of many neutral and amino sugars, however it is not suitable for the analysis of muramic acid.[35]

When pyridine or methylimidazole are used as catalysts for acetylation, browning of the sample occurs and interfering peaks appear in the chromatogram. Tarry reaction products are also deposited in the GC injection port. These products result from the reaction between the organic base (pyridine or methylimidazole) and acetic anhydride. Performing acylation in the presence of water also reduces sensitivity for certain sugars.

Methylimidazole has utility as a catalyst when sugars are in high abundance (above 10 μg per mg of starting material) and if the sample matrix is of limited complexity.[35] Acetylation using sodium acetate catalysis may be preferred in trace analysis work because of the low chromatographic background and excellent yields of all neutral and amino sugars, including muramic acid. Sodium acetate does not react with acetic anhydride, the acetylating reagent, and more extreme acetylation conditions may be used to give higher yields.

Following the acetylation, many researchers dry the sample and inject aliquots into the gas chromatograph without clean-up. Sample clean-up, however, reduces contamination of the GC system by salts and organic acids or bases initially present in the samples or resulting from the derivatization. Clean-up may also remove extraneous peaks from the chromatogram. A simple method is performed by the following steps. Acetic acid is generated by the addition of water (with chloroform) to the sample in the acylation solution. Basic compounds are discarded in the aqueous phase. The sugar derivative remains in the chloroform phase. Acidic contaminants are removed by extraction with base (ammonium hydroxide) and by applying the mixture to a hydrophilic (magnesium sulfate) clean-up column. The samples are also dried at this point. The time required for the use of these columns is short and no appreciable sample loss occurs. Clean-up considerably reduces discoloration in derivatized samples and decreases background peaks. The final step in the analysis of alditol acetates is the GC and MS analysis.

Use of the Alditol Acetate Procedure

The reduction of aldoses to alditols and their conversion to alditol acetates simplifies chromatograms by producing only one peak for each aldose. A significant loss of information may be incurred in the use of this reduction step for certain mixtures of sugars. Mixtures of aldoses and their corresponding alditols would yield only alditols, because they can not be reduced further by sodium borohydride (e.g., glucose and glucitol both produce glucitol). Certain pairs of sugars will produce the same alditol (e.g., gulose and glucose). Aldoses and the related ketoses may also produce the same alditols on reduction and cannot be differentiated (e.g., glucose and fructose both produce glucitol). The presence of ketoses can confound the analysis by producing two sugar alcohols, each of which produces a separate chromatographic peak (e.g., fructose produces glucitol and mannitol). Alditols can be analyzed without prior reduction. Aldoses or ketoses can be differentiated from alditols using sodium borodeuteride in place of sodium borohydride for reduction. Newly formed alditols (produced from aldoses or ketoses on reduction) are deuterated while pre-existing alditols in the sample are not. After acetylation, one deuterium remains on the alditol (derived from aldoses or ketoses) and permits differentiation from the alditols originally present (which do not contain deuterium) by the mass/charge differences observed by MS[13].

Some researchers have reported difficulties in the GC analysis of amino sugar alditol acetates, primarily poor sensitivity and long retention times. The reasons for these problems are now understood and solutions primarily relate to using GC systems with minimal active sites and suitable chromatographic conditions.[38] The alditol acetate procedure is suitable for the analysis of amino sugars whether alone or in mixtures with neutral sugars. Nevertheless, lower response factors for amino sugars compared to neutral sugars may occur.

Acidic sugars are not amenable to analysis by the alditol acetate procedure without prior derivatization of the carboxyl group.[33] The analysis of muramic acid is one exception. Muramic acid (3-O-lactyl-D-glucosamine) contains lactic acid in an ether linkage on the number three carbon of the sugar ring and a free carboxyl group in the lactyl moiety. Under dehydrating conditions, the elimination of water between the amino and carboxyl group of muramicitol creates a lactam ring, thus avoiding the problems of a free carboxyl group.[21]

An internal standard for GC of alditol acetates should be as similar as possible to the components of interest, should not naturally be present in the sample, and should not co-elute either with the components of interest or with background peaks. For the analysis of neutral sugars, any neutral sugar not present in the analytical sample could be employed. Arabinose and xylose are often used rather than the more common sugars such as ribose, or glucose. Inositol is sometimes used as an internal standard but, because it is already in the reduced form, does not control for the reduction step. The internal standard could be added to the sample prior to the hydrolysis step or immediately after this step. Adding the internal standard after the hydrolysis step does not properly allow for correction of any losses in sugar components that might have occurred during the hydrolysis. Adding the internal standard prior to the hydrolysis may cause variability due to differing rates of breakdown of the polymer sugar components compared to the added internal standard monomer. Ideally, an internal standard whose degradation matches the degradation of the polymer sugar components should be employed. The practical use of internal standards does not achieve this ideal.

Many workers use a single internal standard for measuring amounts of both neutral and amino sugars. Dissimilarities in the structures of neutral and amino sugars, however, can lead to variability in the yields of derivatization reactions. Amino sugars also elute much later in the chromatogram than neutral sugars and may present different peak integration problems. Losses of amino sugars relative to neutral sugars during chromatography are also greater. These sources of variability will result in poor reproducibility if a single internal standard is used for such mixtures. Most accurate quantitative analysis of amino sugars requires a second internal standard such as methylglucamine, which is effective for quantitation and does not occur in most biological samples.[34]

Significant improvement in the GC separation of alditol acetates resulted from fused silica capillary column technology. Coupled with the use of helium or hydrogen as a carrier gas, chromatography of complex mixtures of sugars is rapid. The separation of neutral and amino sugar alditol acetates on a fused silica capillary column coated with SP-2330 is illustrated in Figure 5. Monomeric sugar standards in the sample included deoxyribose, methylpentoses (rhamnose, fucose), pentoses (arabinose, ribose, xylose), hexoses (mannose, galactose, glucose), heptoses (D-glycero-D-mannoheptose, L-glycero-D-mannoheptose, D-glycero-L-mannoheptose), methyl glucamine (internal standard), muramic acid, and amino hexoses (glucosamine, galactosamine, mannosamine).

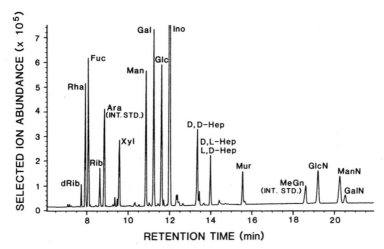

Figure 5.   SIM chromatograms of a standard mixture of alditol
acetates of neutral and amino sugars.   Peak
identification and chromatographic conditions as in
Figure 2B.   Modified from reference 29 with permission.

Alditol acetates do not generally give a molecular ion in electron
impact MS at 70 eV and the spectra of isomers tend to be similar.   Structural
analysis of carbohydrates by MS is well established, although the majority of
applications have used derivatives other than alditol acetates.   The base
peak in the EI spectra of alditol acetates is the acetylinium ion ($CH_3C\overset{\bullet}{\equiv}O$,
m/z 43).   Primary fragments are produced by elimination of the acetoxyl group
($CH_3CO_2^{\bullet}$, m/z 59), acetic acid (m/z 60), ketene (m/z 42), or by cleavage of
the alditol chain.   The mass spectra of a number of alditol acetates have
been published.[29]

ANALYTICAL METHODOLOGY

Pre-derivatization Clean-up

Depending on the type of material to be analyzed, different sample pre-
processing steps may be needed to manipulate the sample into a form ready for
hydrolysis and further derivatization.   In tissue analysis, we usually
dialyze the sample after homogenization to remove free monomeric sugars and
amino acids.   For other samples, this step is not needed or may require
modification.   Tissue homogenates in 0.05% Tween 80 in phosphate buffered
saline are dialyzed three times against distilled water.

Hydrolysis and Neutralization

The pH of the aqueous samples is adjusted by adding sulfuric acid to
reach a concentration of 2 N in a volume of 1 mL in a hydrolysis tube.   These
hydrolysis tubes are commercially available (Pierce, Rockford, IL) or can be
made by sealing high vacuum stopcocks at one end (Chemglass, Inc., Vineland,
NJ).   We employ a custom-made apparatus permitting 20 hydrolysis tubes to be
alternately evacuated and flushed with nitrogen to remove oxygen.   The
evacuation apparatus consists of a nitrogen tank and vacuum pump, each
connected to a copper welded manifold and isolated by needle valves.   Samples
are frozen in the hydrolysis tubes prior to evacuation to reduce bumping,
then unfrozen under vacuum and flushed several times with nitrogen.   The PTFE
valves on the hydrolysis tubes are closed under vacuum and hydrolysis

performed in a Pierce Reacti-Therm heating module at 100°C for 3 h. After the tubes are cooled to room temperature, 2 μg of arabinose and methylglucamine (as internal standard for neutral and aminosugars) in 200 μL water and 2.5 mL of 40% N, N-dioctylmethylamine in chloroform are added. The mixtures are vigorously mixed on a Vortex mixer and allowed to settle into two layers on standing. A 1 mL C$_{18}$ column (Analytichem, Harbor City, CA) is prepared for each sample by rinsing with 2 mL methanol followed by 2 mL of distilled water. The upper aqueous layer of each hydrolysis mixture is applied to a column. Each mixture is then pulled through the column by vacuum into a reaction vial, which is subsequently fitted with a screw cap and a PTFE silicone liner (Pierce) and the column is washed with 1.0 mL of distilled water. The vacuum-elution system used permits up to 21 samples to be run simultaneously.

Derivatization

Fifty μL of sodium borohydride (100 mg/mL in water) are added to the sample. The reduction is allowed to proceed overnight in a refrigerator or for 90 min at 37°C. Excess sodium borohydride is destroyed by adding 2 mL of acetic acid: methanol (1:200 v/v) to the sample which is then evaporated to dryness in a Vortex evaporator (Buchler, Fort Lee, NJ) at 60°C under vacuum. Up to 36 samples can be evaporated to dryness simultaneously. The evaporation step is repeated four additional times to ensure complete removal of borate. The samples are allowed to dry for 3 h after the last evaporation. After cooling to room temperature, 300 μL of acetic anhydride are added to each vial and the sample heated at 100°C for 13-16 h in the Reacti-Therm heating module.

Post-derivatization Sample Clean-up

Samples are cooled in an ice bath, and 0.75 mL of water is added to each and left for 30 min in order to decompose excess acetic anhydride to acetic acid. Following the addition of one mL of chloroform the sample is mixed by vortexing and the aqueous phase is removed and discarded. To the chloroform phase, 0.8 mL of cold concentrated ammonium hydroxide (diluted 80:20 with water) is added. The mixture is poured onto a magnesium sulfate (Chem Elut) column and eluted with 2 mL chloroform. The chloroform solution is evaporated to dryness under vacuum and redissolved in about 40 μL of chloroform before analysis.

GC-MS Analysis Conditions

GC/MS analyses are routinely performed in our laboratories using a HP-5890 gas chromatograph equipped with a capillary column interfaced via a capillary direct inlet to a 5970 Mass Selective Detector (Hewlett-Packard, Palo Alto, CA). Other low cost MS instruments including the "Ion Trap" (Finnigan, Cincinatti, OH), as well as more complex quadrupole or magnetic sector systems are equally suitable. Fused silica capillary columns, 25 m x 0.22 mm, and coated with either BP10 (an OV-1701 bonded phase from SGE, Austin, TX) or with SP-2330 (Supelco, Bellefonte, PA) are used. Typical GC analyses are in the splitless mode.

MS data my be accumulated in the total ion mode with complete scanning of the mass spectra of eluting components, or in a selected ion mode that focuses for greater sensitivity on only a few of the prominent ions present in the sugars of interest. Variations in the SIM parameters can be employed depending on the target carbohydrate and the purpose of the analysis. Generally, ions are selected which are of higher abundance (for increased sensitivity) and of higher mass (for greater selectivity). For example, in the mass spectrum of the alditol acetate of muramic acid m/z 168 is a lower molecular weight ion present in high abundance, while m/z 403 (M - 42) and m/z 445 (the molecular ion) are of high molecular weight, but low abundance. The mass spectra of a number of common sugars from which ions suitable for SIM may be selected are available,[29] and ions appropriate for the analysis of different classes of sugars have been published.[39]

CONCLUSION

Carbohydrates are present in many different bacterial structures. We have described the fundamentals of carbohydrate chemistry, the basis of sugar analysis by GC-MS, and presented practical examples of analytical microbiological applications. Profiling of carbohydrates for bacterial identification and trace detection of carbohydrate chemical markers for bacteria are both readily achievable. With improvements in ease of sample preparation, faster analyses, increased sensitivity, and automated data handling, widespread use of these techniques is anticipated.

ACKNOWLEDGEMENTS

This work was supported by grants from the U. S. Army Office of Research and by the National Institutes of Health.

REFERENCES

1. K. Jann and B. Jann, Polysaccharide antigens of *Esherichia coli*, Rev. Inf. Dis. 9:S517 (1987).
2. J. L. Johnson and C. S. Cummins, Cell wall composition and deoxyribonucleic acid similarities among the anaerobic coryneforms, classical propionibacteria and strains of *Arachnia propionica*, J. Bact. 109:1047 (1972).
3. A. Fox and S. L. Morgan, The chemotaxonomic characterization of microorganisms by capillary gas chromatography and gas chromatography-mass spectrometry, in: "Instrumental Methods for Rapid Microbiological Analysis," W. H. Nelson, ed., Verlag Chemie, Deerfield Beach (1985), p. 135.
4. K. H. Schleifer, Analysis of the chemical composition and primary structure of murein, Meth. in Microbiol. 18:123 (1985).
5. K. H. Schleifer and O. Kandler, Peptidoglycan types of bacterial cell walls and their taxonomic implications, Bacteriol. Rev. 36:407 (1972)
6. K. Bryn and E. Jantzen, Analysis of lipopolysaccharides by methanolysis, trifluoroacetylation and gas chromatography on a fused silica capillary column, J. Chromatog. 240:405 (1982).
7. A. Fox, P. Y. Lau, A. Brown, S. L. Morgan, Z.-T. Zhu, M. Lema, and M. D. Walla, Capillary gas chromatographic analysis of carbohydrates of *Legionella pneumophila* and other members of the family *Legionellaceae*, J. Clin. Micro. 19:326, (1984).
8. J. Endl, H. P. Seidl, F. Fiedler, K. H. Schleifer, Chemical composition and structure of cell wall teichoic acids of staphylococci, Arch. Microbiol. 135:215 (1983).
9. C. S. Cummins and R. H. White, Isolation, identification and synthesis of 2,3-diamino-2,3-dideoxyglucuronic acid: a component of *Propionibacterium acnes* cell wall polysaccharide, J. Bact., 153:1388 (1983).
10. D. J. Pritchard, J. E. Coligan, S. E. Speed, and B. M. Gray, Carbohydrate fingerprints of streptococcal cells, J. Clin. Microbiol. 13:89 (1981).
11. D. G. Prichard, G. B. Brown, B. M. Gray, and J. E. Coligan, Glucitol is present in the group-specific polysaccharide of group B streptococcus, Current Microbiol. 5:283 (1981).

12. C. S. Smith, S. L. Morgan, C. D. Parks, A. Fox, and D. G. Pritchard, Chemical marker for the differentiation of group A and group B streptococci by pyrolysis gas chromatography-mass spectrometry, Anal. Chem. 59:1410 (1987).

13. M. D. Walla, P. Y. Lau, S. L. Morgan, A. Fox, and A. Brown, Capillary gas chromatography-mass spectrometry of carbohydrate components of legionellae and other bacteria, J. Chromatogr. 288:399 (1984).

14. J. Gilbart, A. Fox, and S. L. Morgan, Carbohydrate profiling of bacteria by gas chromatography-mass spectrometry: chemical derivatization and analytical pyrolysis, Eur. J. Clin. Micro. 6:715 (1987).

15. S. Otten, S. Iyer, W. Johnson, and R. Montgomery, Serospecific antigens of Legionella pneumophila, J. Bact., 167:893 (1986).

16. K. Bryn, L. O. Froholm, E. Holten, and K. Bovre, Gas-chromatographic screening of capsular polysaccharides of Neisseria meningitidis. NIPH Ann. 6:91 (1983).

17. K. Bryn and E. Jantzen, Quantification of 2-Keto-3-Deoxyoctonate in (lipo)polysaccharides by methanolytic release, trifluoroacetylation and capillary gas chromatography, J. Chromat. 370:103 (1986).

18. C. Jakobs, T. G. Warner, L. Sweetman, and W. L. Nyhan, Stable isotope dilution analysis of galactitol in amniotic fluid: an accurate approach to the prenatal diagnosis of galactosemia, Pedia. Res., 18:714 (1984).

19. J. Roboz, R. Suzuki, and J. F. Holland, Quantitation of arabinitol in serum by selected ion monitoring as a diagnostic technique in invasive candidiasis, J. Clin. Micro. 12:594 (1980).

20. J. Roboz, D. C. Kappatos, and J. F. Holland, Role of individual serum pentitol concentrations in the diagnosis of disseminated visceral candidiasis, Eur J. Clin. Microbiol. 6:708 (1987).

21. A. Fox, J. H. Schwab and T. Cochran, Muramic acid detection in mammalian tissues by gas-liquid chromatography-mass spectrometry, Infect. Immun. 29:526 (1980).

22. J. Gilbart, A. Fox, R. S. Whiton, and S. L. Morgan, Rhamnose and muramic acid: chemical markers for bacterial cell walls in mammalian tissues, J. Micro. Methods 5:271 (1986).

23. J. Gilbart and A. Fox, Elimination of group A streptococcal cell walls from mammalian tissues, Infect. Immun. 55:1526 (1987)

24. O. Pelletier and S. Cadieux, Quantitative determination of glucose in serum by isotope dilution mass spectrometry following gas liquid chromatography with fused silica column, Biomed. Mass Spectrom. 10:130 (1983).

25. Z. Sen and M. L. Karnovsky, Qualitative detection of muramic acid in normal mammalian tissues, Infect Immun 43:937 (1984).

26. D. G. Pritchard, S. L. Settine, and J. L. Bennett, Sensitive mass spectrometric procedure for the detection of bacterial cell wall components in rheumatoid joints, Arth. Rheum. 23:608 (1980).

27. S. W. Gunner, J. K. N. Jones, and M. B. Perry, Analysis of sugar mixtures by gas-liquid partition chromatography, Chem. Ind. (London) 255 (1961).

28. J. S. Sawardeker, J. H. Sloneker, and A. Jeanes, Quantitative determination of monosaccharides as their alditol acetates, Anal. Chem. 37:1602 (1965).

29. A. Fox, S. L. Morgan, and J. Gilbart, Preparation of alditol acetates and their analysis by gas chromatography and mass spectrometry, Chapter 5 in: "Analysis of Carbohydrates by GLC and MS," C. J. Bierman and G. McGinnis, eds., CRC Press, Boca Raton (1989), pp. 87-117.

30. C. C. Sweeley, R. Bentley, M. Makita, and W. W. Wells, Gas liquid chromatography of trimethylsilyl derivatives of sugars and related substances, J. Am. Chem. Soc. 85:2497 (1963).

31. R. Varma and R. S. Varma, Simultaneous determination of neutral sugars and hexosamines in glycoproteins and acid mucopolysaccharides (glycosaminoglycans) by gas-liquid chromatography, J. Chromatogr. 128:45 (1976).

32. T. P. Mawhinney, M. S. Feather, G. J. Barbero, and J. R. Martinez, The rapid, quantitative determination of neutral sugars (as aldononitrile acetates) and amino sugars as (O-methyloxime acetates) in glycoproteins by gas-liquid chromatography, Anal. Biochem., 101:112 (1980).

33. J. Lehrfeld, GLC determination of aldonic acids as acetylated aldonamides, Carb. Res. 135:179 (1985).

34. A. Fox, S. L. Morgan, J. R. Hudson, Z.-T. Zhu, and P. Y. Lau, Capillary gas chromatographic analysis of alditol acetates of neutral and amino sugars in bacterial cell walls, J. Chromatogr. 256:429 (1983).

35. R. S. Whiton, P. Y. Lau, S. L. Morgan, J. Gilbart, and A. Fox, Modifications in the alditol acetate method for analysis of muramic acid and other neutral and amino sugars by gas chromatography-mass spectrometry with selected ion monitoring, J. Chromatogr. 347:109 (1985).

36. A. B. Blackeney, P. J. Harris, R. J. Henry, and B. A. Stone, A simple and rapid preparation of alditol acetates for monosaccharide analysis, Carbohydr. Res. 113:291 (1983).

37. G. D. McGinnis, Preparation of aldononitrile acetates using N-methylimidazole as catalyst and solvent, Carb. Res. 108:284 (1982).

38. J. R. Hudson, S. L. Morgan, and A. Fox, High-resolution glass capillary columns for the gas chromatographic analysis of alditol acetates of neutral and amino sugars, HRC & CC 5:285 (1982).

39. J. Gilbart, J. Harrison, C. Parks, and A. Fox, Analysis of the amino acid and sugar composition of streptococcal cell walls by gas chromatography-mass spectrometry, J. Chromatogr. 441:323 (1988)

40. A. Fox, J. Gilbart, B. Christensson, and S. L. Morgan, Analysis of carbohydrates for profiling and detection of microorganisms, in: "Rapid Methods and Automation in Microbiology and Immunology," A. Balows, R. C. Tilton, and A. Turano, eds., Brixia Academic Press, Brescia (1989), pp. 379-388.

CHAPTER 6

ANALYSIS OF BACTERIAL AMINO ACIDS

Alvin Fox

Department of Microbiology & Immunology
School of Medicine
University of South Carolina
Columbia, SC 29208

Kimio Ueda and Stephen L. Morgan

Department of Chemistry
University of South Carolina
Columbia, SC 29208

INTRODUCTION

Two major classes of amino acid containing polymers in bacteria are proteins and cell wall peptidoglycan (PG). Prokaryotic proteins contain various proportions of 20 common L-amino acids. These proteins have regulatory or catalytic functions and are involved in various binding and transport functions in bacterial membranes. Bacterial exotoxins are also proteins. The polypeptide sidechains and cross-links of peptidoglycan contain both L- and D-amino acids, including several unusual amino acids not found in proteins. The variety of amino acids present in microorganisms makes their detection and identification valuable for differentiating and identifying isolated bacteria.[1-3] Amino acids are also useful as chemical markers for detecting, without prior culture, the presence of bacteria in complex matrices such as mammalian body fluids and tissues.[4-6]

The most common approach for analysis of amino acid content is the amino acid analyzer. Manual hydrolysis of polypeptides or proteins is followed by liquid chromatography (LC) and on-line post-column derivatization with a color or fluorescence-enhancing derivatization reagent. Although acid hydrolysis can be slow, this chromatographic analysis is rapid and works well for protein amino acids. Unusual non-protein amino acids present in bacteria are not often easily identified. Sensitivity in LC-based methods, however, is often a problem.

Amino acids are easily identified by their mass spectra and GC-MS using selected ion monitoring (SIM) offers selectivity and sensitivity. GC-MS is notably suited for the detection of bacteria in complex matrices such as

Analytical Microbiology Methods
Edited by A. Fox *et al.*
Plenum Press, New York, 1990

.vironmental samples or infected body fluids and tissues. Amino acid derivatives for GC generally involve esterification of carboxyl groups with alcohols (such as butanol) and acylation of amino groups with halogenated anhydrides (simultaneously with reaction of hydroxyl and sulfhydryl groups). Butyl or isobutyl heptafluorobutyryl derivatives of amino acids are commonly employed for GC analysis. The excellent electron capturing characteristics of such derivatives makes them suitable for trace analysis using chemical ionization with negative ion detection.[7] MS can also be used to correct for racemization occurring during hydrolysis. Analysis of D- and L-amino acid mixtures is then achieved by separation on a capillary column coated with a chiral stationary phase.[8]

Extensive reviews of amino acid analysis by GC have appeared, including an excellent chapter in the previous volume to this book.[9,10] More recent improvements in simplified sample preparation and cleanup of butyl heptafluorobutyryl (BHFB) derivatives of amino acids are described here. A variety of microbiological applications emphasizing the analysis of amino acids found in bacterial PG are also presented.[6,11]

AMINO ACIDS PRESENT IN BACTERIA

Besides protein, a major source of amino acids in bacteria is the bacterial cell wall. The amino acid content and differences between the peptide composition of isolated peptidoglycans have been used extensively for chemotaxonomic differentiation of bacteria.[1-3] Bacterial cell walls often contain less common amino acid constituents (including ornithine, D-D, L-L, and *meso* diaminopimelic acid, D-alanine, and D-glutamic acid) that may be used as chemical markers to detect the presence of bacteria.[1,2]

PG consists of a backbone of alternating units of muramic acid and glucosamine with attached peptide side-chains consisting of pentapeptides or tetrapeptides. These sidechains are often cross-linked by other peptides. The side chains consist of alternating L- and D- amino acids. The first position is generally occupied by L-alanine, position 2 by D-glutamic acid (or one of its derivatives, glutamine or isoglutamine), position 3 almost always by L-lysine or *meso* diaminopimelic acid, and position 4 (and position 5 if present) is occupied by D-alanine. The cross-bridge and its linkage is quite variable among bacterial species; it can consist of the same peptide sequence as the side-chain or, alternatively, an entirely new peptide. For example, both *Streptococcus pyogenes* and *Staphylococcus aureus* have very similar pentapeptide sidechains (L-alanine-D-isoglutamine-L-lysine-D-alanine-D-alanine). The cross-bridge in *S. pyogenes* is an L-alanine dipeptide; in *S. aureus* it is a glycine pentapeptide.[12,13] Protein contaminants in isolated cell wall can be readily observed by the presence of aromatic amino acids which are rarely found in cell walls (e.g., tryptophan or histidine). The structure of PG is shown in Chapter 1 (Figure 5).

Proteins of eukaryotes (including yeast) and prokaryotes contain almost exclusively L-amino acids. In the Gram positive bacterial cell envelopes, the PG layer is multilamellar, whereas in Gram negative bacteria it is a monolayer.[1,2] Gram positive bacteria have a higher absolute content of D-alanine and D-glutamic acid, as well as higher D-alanine/L-alanine and D-glutamic acid/L-glutamic acid ratios when compared to Gram negative bacteria. Both Gram positive and negative bacteria can be readily differentiated from yeasts by differences in D/L amino acid ratios.[4] D-amino acids are also rarely found in plants or mammals, thus D-amino acid markers are also useful for detecting the presence of bacterial peptidoglycan. D-alanine has been employed to detect bacteria in environmental samples and mammalian tissues.[4-6]

THE BUTYL HEPTAFLUOROBUTYRYL PROCEDURE-- A GENERAL DISCUSSION

The classic BHFB method produces one peak for each protein amino acid (except histidine for which two peaks occur) and entails several processing steps prior to chromatographic analysis.[14,15] These steps include hydrolysis in hydrochloric acid, removal of the acid followed by butyl esterification of carboxyl groups with butanol/acetyl chloride, and heptafluorobutyrylation of amino, hydroxyl and sulfhydryl groups with heptafluorobutyric anhydride (HFBA). Long hydrolysis times at low temperature (24 h at 100°C) or shorter hydrolysis times at higher temperature (6 hours at 150°C) can also be used. The latter conditions are useful in routine analysis, but can introduce high levels of racemization making analysis of racemic mixtures of amino acids more difficult.[6] Ion exchange resins for pre-derivatization clean-up are most commonly employed. However, the columns must be used in the H^+ form which requires preconditioning of columns immediately prior to use with dilute acid. Amino acids bind to the cationic resin and non-binding contaminants are removed with water washes. Finally the amino acid is eluted with strong acid. The acid must be removed by evaporation. These steps are rather tedious to perform.[4,6] A simple clean-up based on the pre-derivatization use of hydrophobic C_{18} columns and a post-derivatization extraction using aqueous phosphate buffer is sufficient in many cases.[11] The first clean-up step removes fatty acids and other hydrophobic materials that might be present. The second clean-up step removes heptafluorobutyric acid generated from the acylating reagent (heptafluorobutyric anhydride), eliminates many extraneous peaks, and decreases baseline drift. The presence of highly acidic reagents such as HFBA in the analytical sample should be avoided since they can rapidly damage the stationary phase and reduce column life.

During acid hydrolysis to release monomeric amino acids from peptidoglycan, racemization can confound accurate measurement of the original amounts of individual D- and L-amino acids. Liardon suggested hydrolysis in deuterated HCl to differentiate by MS D-amino acids formed by racemization during hydrolysis from those naturally present in protein samples.[8] During hydrolysis, any racemization that occurs will be accompanied by hydrogen exchange. If the hydrolysis is performed using deuterated HCl, the enantiomer undergoing racemization is labelled. The original quantities of amino acid enantiomers can be calculated by examining ratios of specific ion masses. At low D/L ratios, interference of ions from the labelled D-amino acids reduces the possibility of detecting trace amounts of unlabelled (original) D-amino acids.[4-6]

Two primary approaches to the analysis of mixtures of D-and L-amino acids have been proposed: separation as diastereoisomers on a conventional column, or separation as racemers on a chiral column. The diastereomers are produced by esterification with D- (or L-) isobutanol to produce D-isobutyl-L-amino acid or D-isobutanol-D-amino acid).[7] Totally pure L- or D-isobutanol that is not contaminated with the other racemer is uncommon. Other diastereoisomers (L-isobutyl-L-amino acid and L-isobutyl- D-amino acid) are produced as contaminants. These contaminants co-elute with their isomers and their interfering background makes trace analysis of D-amino acids in the presence of L-amino acids difficult. Alternatively, reaction with non-optically active butanol and separation of the D-and L-isomers (after acylation) on a chiral column avoids this problem.[4,6]

Samples can be analyzed using flame ionization detectors or conventional electron impact (EI) or chemical ionization (CI) MS. Alternatively since the derivatives have excellent electron capture characteristics, electron capture detection (ECD) or CI with negative ion detection by MS can be employed for increased sensitivity.[5,6]

ANALYTICAL METHODOLOGY

The method described here is based on the work of MacKenzie and D. Tenaschuk[14-15] as modified for microbiological applications by Odham and coworkers.[4-7] Samples are hydrolyzed in 0.5 mL 6 N HCl or DCl (if racemers are to analyzed) for 6 h at 150°C or for 16 h at 105°C under vacuum. Samples are often evaporated to dryness in a lyophilizer in a flask equipped with a sodium hydroxide trap to prevent acid contamination of the vacuum pump and refrigeration unit. Evaporation is tedious and acid can be removed more rapidly using a vortex evaporator (Buchler) also equipped with a suitable trap at 60°C.

Samples can be redissolved in 1.0 mL of water and extracted on a C_{18} column (Analytichem, Harbor City, CA). These clean-up columns are pre-conditioned by washing with methanol and water. The samples are then dried under nitrogen, 1 mL of dichloromethane is added, and the samples are evaporated to dryness. Alternatively, prior to use, 1 mL SCX cation exchange resin columns (Analytichem, Harbor City, CA) are washed with two column volumes (2 mL) of methanol and 3 volumes of water rapidly, followed with 3 volumes of 4 N HCl at a flow rate of 0.5 mL/min and then washed rapidly with 6 volumes of water. After applying hydrolysates to the columns, sample tubes are rinsed with 0.5 mL of 0.1 N HCl which is also added to the columns. The resins are washed with 2 volumes of water, after which amino acids are eluted with 2 volumes of 4 N HCl. The eluent is dried by evaporation as described above.[16]

Acetyl chloride (Aldrich, Gold Label) is added to N-butanol (Burdick and Jackson, Muskegon, WI, HPLC grade) in a ratio of 1:4, and 50-100 μL of this reagent is added to each vial which is then gently flushed with nitrogen. Esterification is performed at 100°C for 20 min. The samples are then cooled and excess reagent removed in a stream of nitrogen. Dichloromethane (0.2 mL) is added and the samples are evaporated to dryness. Acylation is performed with 100 μL of heptafluorobutyric anhydride (HFBA) (Fluka, Ronkonkoma, NY) at 150°C for 12 min.

After cooling at room temperature, 0.5 mL chloroform and 2 mL 1 M phosphate buffer are added to each sample; tubes are shaken and centrifuged at 1000 g. The chloroform phases are removed and samples evaporated to dryness under nitrogen.

GC/MS analyses can be carried out using a benchtop GC-MS system. We employ a a HP-5890 GC interfaced to a 5970 Mass Selective Detector (Hewlett-Packard, Palo Alto, CA). Programmed temperature GC separations are done on a 25 m x 0.22 mm CPSil-19 (OV-1701 bonded phase) fused silica column (Chrompack, Raritan, NJ.) or a 25 m x 0.22 mm fused silica capillary column coated with Chirasil-L-Val (Chrompack, Raritan, NJ). In the EI MS mode a single ion (usually the base peak) is selected for SIM, although in some instances two prominent ions can be selected. Samples can also be analyzed using negative ion detection following CI with methane as a reagent gas at 1 torr pressure. A prominent M - 20 ion is produced (e.g., m/z 321 for D-and L-alanine).

APPLICATIONS OF THE METHOD

Excellent separations of mixtures of BHFB derivatives can be obtained on an OV-1701 fused silica capillary column (Figure 1). In this instance, alanine, valine, glycine, threonine, leucine, isoleucine, serine, proline, cysteine, methionine, aspartic acid, phenylalanine, glutamic acid, tyrosine, lysine, arginine, tryptophan, histidine, cystine, and internal standards (α-amino isobutyric acid, norleucine, and ornithine) were separated. Three

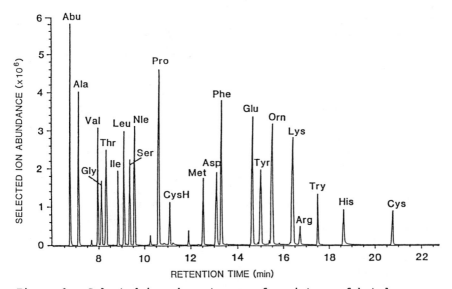

Figure 1. Selected ion chromatogram of a mixture of butyl
heptafluorobutyryl derivatives: Abu = alpha amino
isobutyric acid (internal standard), Ala = alanine,
Val = valine, Gly = glycine, Thr = threonine,
Ile = isoleucine, Leu = leucine, Ser = serine,
Nle= norleucine (internal standard), Pro = proline,
CysH = cysteine, Met = methionine, Asp = aspartic acid,
Phe = phenylalanine, Glu = glutamic acid,
Tyr = tyrosine, Orn = ornithine (internal standard),
Lys = lysine, Arg = arginine, Try = trytophan,
His = histidine, Cys = cystine. Separations were
performed on a OV-1701 fused silica capillary column
with detection by EI MS. Reprinted with permission
from reference 11.

internal standards are used because better reproducibility is obtained if the
standard elutes close to the peak of interest. A good internal standard for
alanine is α-amino isobutyric acid; norleucine is suitable for valine through
phenylalanine; and ornithine is suitable for glutamic acid through histidine.
These internal standards do not often occur in proteins or cell walls.

During derivatization, glutamine is converted to glutamic acid and
asparagine to aspartic acid; each pair of amino acids produce one
chromatographic peak. Certain amino acids with nitrogen containing moieties
are known to be difficult to derivatize; arginine contains a guanidino group
and histidine has an imidazole group in the chain.[15] Arginine was found to
produce a small peak relative to other amino acids. Arginine requires
vigorous acylation conditions and the absence of water. In the presence of
even small amounts of water, heptafluorobutyric acid is formed which reduces
the yield of the HFB derivative of arginine.[17] We further note that this
peak disappears rapidly on storage in ethyl acetate. As noted previously,
histidine produces two peaks which represent the mono and diacyl derivatives.
In the case of the diacyl derivative, the imidazole nitrogen is
heptafluorbutyrylated. The second peak tails, a property noted by McKenzie
for the monoacyl derivative.[9,10] In the SIM chromatogram (Figure 1), only the
second histidine peak is seen since the ion monitored, m/z 407 (the molecular
ion for butyl monoacyl histidine), is considerably less abundant in the first
histidine peak.

Chromatograms can be simplified by using SIM for major ions present at the correct retention time for all amino acids of interest. Butyl HFB derivatives generally produce an abundant ion (EI mode) between m/z 200-300 derived from the parent structure. The use of high mass ions provides excellent sensitivity and selectivity. Background in the EI mode is primarily produced by fragmentation of the heptafluorobutyryl group (mainly ions m/z 119 and 147). Figure 2 shows a selected ion chromatogram of BHFB derivatives of a hydrolysate of a streptococcal cell wall preparation separated on an OV-1701 fused silica capillary column.[11] The major peaks (alanine, lysine and glutamic acid) are readily visualized, but also low level protein contamination of the preparation can be assessed by the levels of other non-cell wall amino acids.

When amino acids are heated in strong acid, some racemization of L-amino acids to their D-isomers occurs. If the L-amino acids are in large excess, as in whole bacterial cells, background racemization can obscure the detection of small amounts of D-amino acids derived from the peptidoglycan. By hydrolyzing with deuterium hyrdrochloride, newly formed molecules of D-amino acids are labelled with deuterium; pre-existing "natural" D-alanine molecules remain unlabelled. D-alanine formed by racemization is labelled with deuterium and gains an additional unit of mass (m/z 322), while unlabelled D-alanine has a mass of m/z 321. Racemized D-alanine and natural D-alanine can be differentiated by MS using these different characteristic ions.[6]

Because yeasts have not been reported to contain D-amino acids in their cell walls, background racemization levels can be established by analysis of whole yeast cells. The D-alanine/L-alanine ratios for *Candida albicans* and *Cryptococcus neoformans* were about 0.3 to 0.4% using both EI or CI MS with negative ion detection. Under these same conditions, bovine serum albumin contains a D-alanine/L-alanine ratio of 0.3%.[6] Levels of D-alanine in Gram negative bacteria (*E. coli* and *Pseudomonas aeruginosa*) were about 10 times

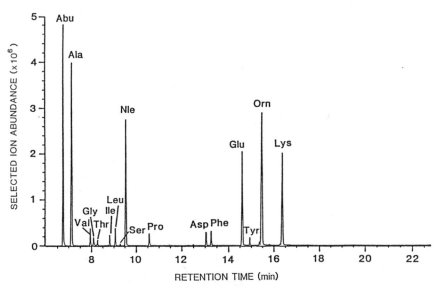

Figure 2.   Selected ion chromatogram of butyl heptafluorobutyryl derivatives of a hydrolysate of a streptococcal cell wall preparation separated on an OV 17-01 fused silica capillary column. Peak identification and chromatographic conditions as in Figure 1. Reprinted with permission from reference 11.

Table 1. D-alanine and D-glutamic acid content of
         microorganisms determined by GC-MS with chemical
         ionization and negative ion detection

Organism	D-alanine as % of L-alanine	D-glutamic acid as % of L-glutamic acid
Bacillus anthracis	26.8	32.7
Staphylococcus aureus	55.6	32.6
Escherichia coli	2.8	5.0
Pseudomonas aeruginosa	3.5	3.2
Candida albicans	0.4	0.7
Cryptococcus neoformans	0.3	0.7

higher. Gram positive bacteria (*Bacillus anthracis* and *Staphylococcus aureus*) had D-alanine/L-alanine ratios an order of magnitude higher than for Gram negative bacteria (Table 1). D-glutamic acid/L-glutamic acid ratios in *Cryptococcus neoformans* were about 1.0%, representing background racemization. The D-glutamic acid/L-glutamic acid ratios in Gram negative bacteria (*E. coli* and *Pseudomonas aeruginosa*) were over four fold higher. Gram positive bacteria (*Bacillus anthracis* and *Staphylococcus aureus* contained considerably higher amounts of D-glutamic acid than Gram negative bacteria (Table 1). Figure 3 illustrates the differences in the D-/L-amino acid ratios of Gram positive bacteria and fungi. This work and the nature of difficulties in trace analysis of D-glutamic acid are discussed in greater detail elsewhere.[18]

D-amino acids can be also be used for detection of bacteria or bacterial cell walls in complex samples such as mammalian tissues. Negative ion CI MS may be helpful in such work because trace detection in a complex matrix requires the highest sensitivity and selectivity. BHFB derivatives have high efficiency electron capturing characteristics. Molecular radical ions of BHFB derivatives in common with other halogenated derivatives commonly lose HF, and give a base peak of M - 20.[19] SIM is employed here rather than total ion abundance MS; focussing on selected prominent ions in the mass spectra of the target compounds yields both greater selectivity and sensitivity.

Figure 4 compares SIM chromatograms of hydrolysates of liver samples from rats injected one day or 13 days previously with a streptococcal cell wall preparation.[6] Figure 4C shows a SIM chromatogram of the control rat liver. Two ions (m/z 335 for the internal standard, and m/z 321 for D-/L-alanine) were monitored in each chromatogram. Clear differences in the amount of D-alanine between the experimental and control samples are seen.

Diaminopimelic acid may have advantages over D-amino acids as an amino acid chemical marker for bacteria because background interference from racemization does not occur. Using two-dimensional GC with ECD, diaminopimelic acid has been recently detected at trace levels in infected but not normal human urine samples.[20]

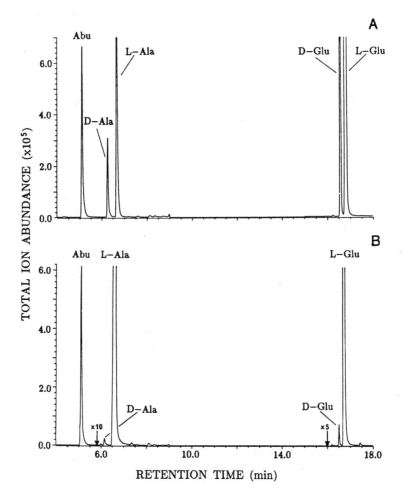

Figure 3. Selected ion chromatograms of butyl
heptafluorobutyryl derivatives of hydrolysates
of (A) *Bacillus anthracis*; and (B) *Candida albicans*.
Peak identification: Abu = Alpha amino isobutyric
acid (internal standard), D-ala = D-alanine,
L-ala = L-alanine, D-glu = D-glutamic acid,
L-glu = L-glutamic acid. Separations were
performed using a Chirasil-L-Val fused silica
capillary column and CI MS with negative ion
detection.

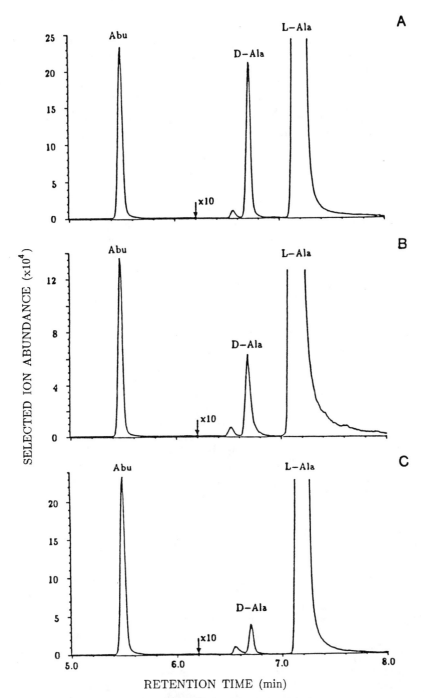

Figure 4. Selected ion chromatograms of butyl heptafluorobutyryl derivatives of hydrolysates of rat livers from (A) an animal 1 day after injection of bacterial cell walls; (B) an animal 13 days after injection of bacterial cell walls; (C) a control animal. For peak identification and chromatographic conditions see Figure 3. Reprinted with permission from reference 6.

CONCLUSION

The modified butyl heptafluorobutyryl procedures described here allow batches of samples to be rapidly analyzed. High temperature hydrolysis requires only a few hours to completely release amino acid monomers. The acid can be removed rapidly by heating in a vacuum evaporator. Pre-derivatization clean-up with C_{18} columns and post-derivatization clean-up by aqueous extraction are simple to perform. In some instances, ion exchange resins may provide additional clean-up. The derivatization itself only takes a few hours. Excellent amino acid separations can be obtained using conventional fused silica capillary columns and D-/L-amino acid mixtures can be analyzed on chiral columns. For many applications, EI with a bench top MS system is adequate. For greatest sensitivity, CI MS and negative ion detection is desirable.

The method is suitable for characterization of simple samples such as isolated bacterial cell walls, whole bacteria, or more complex environmental or clinical samples. D-amino acids and diaminopimelic acid may be valuable chemical markers for a variety of analytical microbiological applications.

ACKNOWLEDGEMENTS

This work was supported by grants from the U. S. Army Office of Research and by the National Institutes of Health.

REFERENCES

1. K. H. Schleifer, Analysis of the chemical composition and primary structure of murein, Meth. in Microbiol. 18:123 (1985).
2. K. H. Schleifer and O. Kandler, Peptidoglycan types of bacterial cell walls and their taxonomic implications, Bacteriol. Rev. 36:407 (1972)
3. A. G. O'Donnell, D. E. Minnikin, M. Goodfellow, and J. H. Partlett, The analysis of actinomycete wall amino acids by gas chromatography, FEMS Micro. Lett. 15:75 (1982).
4. S. Sonesson, L. Larsson, A. Fox, G. Westerdahl, and G. Odham, Determination of environmental levels of peptidoglycan and lipopolysaccharide using gas chromatography with negative-ion chemical-ionization mass spectrometry utilizing bacterial amino acids and hydroxy fatty acids as biomarkers, J. Chromatogr. Biomed. Appl. 431:1 (1988).
5. A. Tunlid, and G. Odham, Diastereomeric determination of R-alanine in bacteria using positive/negative ion mass spectrometry, Biomed. Mass. Spectrom. 11:428 (1984).
6. K. Ueda, S. L. Morgan, A. Fox, J. Gilbart, A. Sonesson, L. Larsson, and G. Odham, D-alanine as a chemical marker for the determination of streptococcal cell wall levels in mammalian tissues by gas chromatography/negative ion chemical ionization mass spectrometry, Anal. Chem. 61:265 (1989).
7. A. Tunlid and G. Odham, Capillary gas chromatography using electron capture or selected ion monitoring detection for the detemination of muramic acid of muramic acid, diaminopimelic acid and the ratio of D/L alanine in bacteria, J. Microbiol. Methods 1:63 (1983).
8. R. Liardon, S. Ledermann, and U. Ott, Determination of D-amino acids by deuterium labelling and selected ion monitoring, J. Chromatogr. 203: 385 (1981).
9. S. L. Mackenzie, Amino acids and peptides, in: "Gas Chromatography Mass Spectrometry Applications in Microbiology," G. Odham, L. Larsson, and P.-A. Mardh, eds., Plenum Press, New York (1984).

10. S. L. Mackenzie, Recent developments in amino acid analysis by gas liquid chromatography, Methods of Biochemical Analysis 27:1 (1981).

11. J. Gilbart, J. Harrison, C. Parks, and A. Fox, Analysis of the amino acid and sugar composition of streptococcal cell walls by gas chromatography-mass spectrometry, J. Chromatogr. 441:323 (1988).

12. E. Munoz, J.-M. Ghuysen, and H. Heymann, Cell Walls of *Streptococcus pyogenes*, Type 14, C polysaccharide-peptidoglycan and G polysaccharide-peptidoglycan complexes, Biochemistry 6:3659 (1967).

13. W. W. Karakawa, H. Lackland, and R. M. Krause, An immunochemical analysis of bacterial mucopeptides, J. Immunol. 97:797 (1966).

14. S. L. MacKenzie and D. Tenaschuk. Quantitative formation of N(O,S)-heptafluorobutyryl isobutyl amino acids for gas chromatographic analysis. I. Esterification, J. Chromatogr. 171:195 (1979).

15. S. L. MacKenzie and D. Tenaschuk. Quantitative formation of N(O,S)-heptafluorobutyryl isobutyl amino acids for gas chromatographic analysis. II. Acylation, J. Chromatogr. 173:53 (1979).

16. I. M. Kapetanovic, W. D. Yonekawa, and H. J. Kupferberg, Determination of 4-aminobutyric acid, aspartate, glutamate, and glutamine and their ^{13}C stable isotopic enrichment in brain tissue by gas chromatography-mass spectrometry, J. Chromatogr. 414:265 (1987).

17. I. M. Moodie, J. A. Burger, G. S. Shephard, and D. Labadarios, Gas liquid chromatography of amino acids, J. Chromatogr. 347:179 (1985).

18. K. Ueda, S. L. Morgan, and A. Fox, Analysis of D-alanine and D-glutamic acid in microorganisms by capillary gas chromatography-mass spectrometry with electron impact and electron capture negative ion chemical ionization, manuscript in preparation (1990).

19. G. K. Low and A. M. Duffield, Positive and negative ion chemical ionization mass spectra of amino acid carboxy-n-butyl ester N-pentafluoropropionate derivatives, Biomed. Mass Spectrom. 11:223 (1984).

20. A. Sonesson, L. Larsson, and J. Jimenez, "Two-dimensional gas chromatography with electron-capture detection used in the determination of specific peptidoglycan and lipopolysaccharide constituents of Gram-negative bacteria in infected human urine, J. Chromatogr. 490:71 (1989)

CHAPTER 7

HEADSPACE ANALYSIS BY GAS CHROMATOGRAPHY-MASS SPECTROMETRY OF VOLATILE
ORGANIC COMPOUNDS ASSOCIATED WITH *CLOSTRIDIUM* CULTURES

Alain Rimbault

Laboratoire de Microbiologie
Faculte des Sciences Pharmaceutiques
et Biologiques de Paris
F-75270 Paris Cedex 06
France
(1) 43 29 12 08 ext. 237

INTRODUCTION

Clostridia form a heterogeneous group of bacteria that are widespread
in natural environments and have numerous metabolic properties (e.g.,
fermentation of carbohydrates, amino acids, purines, and pyrimidines,
cellulolysis, pectinolysis, acetogenesis).[1,2] Their various degradative
properties explain their involvement in the different aspects of degradation
of organic matter. Many of the metabolites of clostridia are gaseous or
volatile; since they are present in the headspace above culture media,
analysis can be performed directly by gas chromatography (GC). Headspace GC
is a rapid and sensitive analytical method for detection of trace volatile
compounds in complex gaseous mixtures. When headspace GC is combined with
mass spectrometry (MS), chemical identification of these volatiles is
possible.[3-6]

This chapter illustrates the application of static headspace analysis
by GC-MS for the detection and identification of trace gaseous compounds
associated with cultures of clostridia. Headspace GC-MS can also provide
information on the biochemical and/or chemical pathways leading to these
compounds.

HEADSPACE ANALYSIS

A headspace is a vapor phase in a closed system which is in
thermodynamic equilibrium with a solid or, more often in microbiology, a
liquid phase. Vapor-liquid equilibria in a multicomponent system are
complex; for dilute solutions, however, the equations can be simplified
(Figure 1).[7] For a dilute solution of a component i that is completely
soluble in the liquid phase, the partial pressure (p_i) is proportional to its
activity coefficient (γ_i), its molar fraction in the liquid phase (x_i), and
the saturation vapor pressure of pure component i (p_i°). The saturation
vapor pressure is greatly dependent on temperature (as a result of the
Clausius-Clapeyron's law) and, to a lesser extent, on the overall pressure of

Analytical Microbiology Methods
Edited by A. Fox *et al.*
Plenum Press, New York, 1990

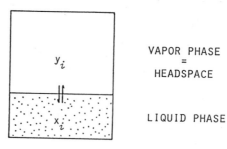

VAPOR PHASE
=
HEADSPACE

LIQUID PHASE

<u>Closed System in Thermodynamic Equilibrium</u>

$$P_i \;=\; \gamma_i \cdot x_i \cdot P_i^o \;=\; y_i \cdot P$$

DILUTE SOLUTION : γ_i = CONSTANT

IDEAL SOLUTION : γ_i = 1 (RAOULT'S LAW)

P_i^o = f (T, P)

FOR EXAMPLE, P = CONSTANT, P_i^o = f (T) :

$$\frac{\text{Ln } P_i^o}{dT} \;=\; \frac{\Delta H}{RT^2} \quad \text{(CLAUSIUS–CLAPEYRON'S LAW)}$$

γ_i : activity coefficient of component i,

x_i : molar fraction of component i in the liquid phase,

y_i : molar fraction of component i in the vapor phase,

P_i : partial pressure of component i dissolved in the liquid phase,

P_i^o : saturated vapor pressure of pure component i (at its own pressure),

P : overall pressure of the system,

T : temperature (°K) of the system,

R : ideal gas constant,

ΔH : latent heat of vaporization.

Figure 1. Static headspace: simplified theory of vapor-liquid equilibria.

the system. Consequently, samples must be compared under isothermal and isobaric conditions, and temperature must be well controlled.

The partial pressure of an analyte can be increased by raising the temperature (and thus the saturation vapor pressure); alternately, the activity coefficient can be increased by adding a mineral salt, causing the "salting-out effect".[3] However, high temperature can cause thermal decomposition of labile compounds and the addition of a mineral salt can introduce organic contaminants into the system. For these reasons, such measures should be avoided if possible when analyzing for trace or labile volatile compounds.

There are two classical approaches to the GC of headspace vapors: static analysis and dynamic analysis. This chapter is primarily concerned with static analysis, in which a fraction of the headspace is withdrawn with a gas-tight syringe or by a gas sampling valve and injected into the GC column. Derivatization is not required and, since there is no solvent, early eluting peaks are not masked by a solvent front. Dynamic headspace analysis is beyond the scope of this chapter and will not be discussed here.

HEADSPACE ANALYSIS BY GC OR BY GC-MS OF VOLATILE COMPOUNDS ASSOCIATED WITH CULTURES OF CLOSTRIDIA

Volatile compounds associated with cultures of clostridia and analyzed by headspace GC or headspace GC-MS include fatty acids, alcohols, carbonyl compounds, amines, and hydrogen. They are summarized in Table 1.

Volatile fatty acid profiling by headspace analysis has been used to identify anaerobic bacteria, particularly clostridia[8-14] and is an alternative to the analysis of volatile fatty acids found in liquid culture media.[15,16] Identification of clostridia and other bacteria of clinical interest by headspace analysis is discussed in Chapter 8.

Headspace GC has also been used for detecting growth of clostridia in blood cultures[17,18] and for on-line monitoring of the acetone-butanol fermentation carried out by *C. acetobutylicum*.[19-21]

Hydrogen is classically detected by headspace GC in headspace vapors of cultures of clostridia.[15,16] Trace amounts of methane have been found for cell extracts of *C. pasteurianum* strain w.[22] Methane and hydrogen have been detected for growing cultures of clinical isolates of clostridia.[23] The occurence of trace ethylene was reported for a soil isolate identified as *C. butyricum*.[24]

A number of volatile non-acidic compounds have been identified by headspace GC-MS in the headspace vapors from clostridia that cause gas gangrene.[25-28] However, headspace collection over mercury (using a modified van Slyke device filled with mercury) required large sample volumes and mercury-reactive compounds (such as thiols) had to be removed by alkalinization.

Our attention has been focused on the detection of these less studied volatile compounds and, using reference strains as much as possible, we have made the following three modifications. First, the headspace fraction is sampled with a gas sampling valve to avoid losses of volatile compounds during transfer steps. Second, a removal reagent was not used, thus minimizing artifacts and decomposition of unstable compounds and enlarging the scope of the volatile compounds. Third, the method was adapted to allow small volumes of culture medium to be used; this modification permits stable isotope labeling of volatile compounds.

Table 1. Headspace GC or GC-MS analysis of volatile compounds detected from *Clostridium* cultures

Compound	Microorganism	Column	Reference
Volatile fatty acids	Various clinical isolates	Packed	8
C_2-C_5 alcohols C_2 and C_4 fatty acids	*C. perfringens* (clinical isolate)	Packed	9
C_2-C_5 alcohols C_2-C_6 fatty acids C_2-C_6 amines	*C. septicum* (clinical isolate)	Packed	10
C_2-C_6 fatty acids	*C. sporogenes* and *C. difficile* (clinical isolates)	Capillary	11
C_2-C_5 alcohols C_2-C_6 fatty acids	Various clinical isolates	Capillary	12
Ethanol C_2 and C_4 fatty acids	*C. perfringens* (blood culture)	Capillary	17
C_2-C_6 fatty acids	Various isolates	Capillary	13
C_1-C_5 alcohols C_2-C_8 fatty acids	Various isolates	Capillary	14
Ethanol, butanol, Propanone	*C. acetobutylicum* ATCC 824[T]	Packed	19-21
Ethanol	*C. perfringens* (simulated blood culture)	Packed	18
C_1-C_4 alcohols, trimethylamine, propanone, butanone, dimethyl disulfide	Various isolates causing gas gangrene	Packed	25-28

OPTIMIZATION OF A HEADSPACE PROCEDURE

Bacterial Strains

Clostridium strains were obtained from the American Type Culture Collection (ATCC), the "Deutsche Sammlung von Mikroorganismen" (DSM), the National Culture of Industrial Bacteria (NCIB), the "Unite des Anaerobies" (AIP, Institut Pasteur, Paris, France), and the Anaerobe Laboratory (VPI, Virginia Polytechnic Institute, Blacksburg, VA). Whenever possible, type ([T]) strains were included.

Organisms studied included *Clostridium* sp. DSM 1786 and ATCC 25772, *C. bifermentans* (ATCC 638[T] and ten other strains), *C. cadaveris* ATCC 25783[T], *C. difficile* ATCC 9689[T], *C. ghonii* ATCC 25757[T], *C. glycolicum* ATCC 14880[T], *C. hastiforme* (ATCC 33268[T] and three other strains), *C. histolyticum* (ATCC 19401[T] and four other strains), *C. lituseburense* ATCC 25759[T], *C. mangenotii* ATCC 25761[T], *C. perfringens* (one strain), *C. septicum* (one strain), *C. sordellii* (NCIB 10717[T] and seventeen other strains), *C. sporogenes* (two strains), and *C. subterminale* (ATCC 25774[T] and four other strains). The strain DSM 1786 (ATCC 25772) is no longer the type strain of *C. hastiforme* and is now assigned to a *Clostridium* species.[29] "*C. aerofoetidum*" strain WS and "*Plectridium putrificum*" strain Bienstock were from the collection of Prevot (Institut Pasteur, Paris, France). For the former, the nomenspecies does not appear in the "approved lists of bacterial names".[30]

Growth Conditions and Media

Cultures were grown at 37°C, usually for seven days, under reduced pressure (residual pressure in a water-free vial of *ca.* 0.3 kPa) in 330 mL[31-35] culture flasks or 28 mL[36-38] culture tubes containing 50 mL or 5 mL of culture medium, respectively. Usually, clostridia were grown in basal thioglycolate-Trypcase-yeast extract (TTY) medium consisting (weight/volume) of 1.5% Trypcase (BioMerieux, Lyon, France), 0.5% yeast extract (BioMerieux), 0.25% sodium chloride (Merck, Darmstadt, Federal Republic of Germany), and 0.05% sodium thioglycolate (Merck). In some cases the TTY medium was modified (e.g., the medium was supplemented with D-glucose or L-amino acids, or 2-mercaptoethanol was used instead of sodium thioglycolate).[32] To determine the origin of some volatile compounds, deuterium-labeled substrates [L-(*methyl*-$^2H_3$) methionine, L-(2',3',4',5',6'-$^2H_5$) phenyl (2,3-$^2H_3$) alanine (Service des Molecules Marquees, Centre d'Etudes Nucleaires de Saclay, Gif-sur-Yvette, France), or (1-$^2H_3$) methanethiol[38]] were added to the basal TTY medium; the fate of the deuterated group was followed by headspace GC-MS. For the study of volatile amines[35], meat liver infusion (30 g per liter; Institut Pasteur Production, Marnes-la-Coquette, France) was preferred.

Culture Processing and Headspace Sampling

During the incubation period, gases (mainly hydrogen and carbon dioxide) and various volatile compounds are formed. At the end of the incubation period, the internal pressure remains in the sub-atmospheric range. When analyzing non-acidic compounds, either 6 M[31-33] or 1 M[39] sodium hydroxide solution was added to the culture (1 volume to 2.5 volumes of culture) before headspace sampling to trap acidic metabolites (e.g., carbon dioxide, fatty acids, and thiols). Light hydrocarbons and volatile organosulfur compounds[36-38] in the gas phase were directly injected into the GC column without any pretreatment of the culture. In some experiments, internal pressures were equalized to atmospheric pressure by adding the carrier gas, helium, through the culture vial stopcock before headspace sampling so as to produce isobaric conditions.

The culture vial was thermostated for at least 30 min at a low temperature (25°C) to avoid sample modification. The vial was then connected to a sampling device consisting of a six-way gas sampling valve fitted with a loop (5 mL or 15 mL) and tubing connections. The sampling device (Figure 2) was made of stainless steel for the analysis of volatile non-acidic compounds or was PTFE-lined for the analysis of light hydrocarbons or organosulfur compounds. After air evacuation from the sampling device, the headspace was allowed to expand in the whole system and a headspace fraction was injected by switching the gas sampling valve. When a small culture tube (28 mL) and consequently small culture medium volume (5 mL) were used, the volume of the sample loop was as great as 15 mL without injection problems.

The analysis of volatile aliphatic amines requires concentration of the sample.[35] After alkalinization with sodium carbonate, the culture was distilled at 37°C under reduced pressure. Volatile aliphatic amines were trapped as their hydrochloride salts, which were then distilled to dryness under reduced pressure. After transfer in a special headspace vial which was subsequently air-evacuated, these hydrochloride salts were decomposed by alkalinization and volatile amines were released into the headspace. The sample pressure was adjusted to atmospheric pressure with helium, and a fraction of the headspace sampled by a syringe.

GC AND GC-MS CONDITIONS

The several GC procedures used here are listed in Table 2. Usually, volatile compounds were separated on packed columns with helium as the carrier gas (flow rate *ca.* 17 mL/min) using a GC (Girdel 300 and 30 series, Delsi-Nermag, Argenteuil, France) fitted with flame ionization detectors. The oven temperature was usually programmed (procedures A, B, E) and high sensitivity (in the pA full scale range) was needed for the detection of a wide range of volatile compounds, particularly those present at trace levels. Except for procedure F (methane analysis), packed columns were used in the dual mode to compensate for baseline drift due to temperature programming or the gradual increase in ammonia levels in the carrier gas for the analysis of volatile aliphatic amines (procedures C and D). In that case, two packing materials were used.

Figure 2.   General view of a headspace sampling device with (a) the culture tube (volume, *ca.* 28 mL) thermostated in a water bath, (b) connections (PTFE tubing, see procedure E in Table 2), (c) the gas sampling valve, (d) the gas sampling loop (volume, *ca.* 15 mL, PTFE tubing), (e) the bellows valve, and (f) the vacuum line.

Table 2. Gas chromatographic conditions for analysis of volatile
compounds associated with *Clostridium* cultures

Procedure A: Analysis of volatile non-acidic compounds in the headspace from
alkalinized cultures using a stainless steel column (3 m x 1/8 inch o.d.) packed
with 10% Carbowax 600, 2% potassium hydroxide on 60-80 mesh chromosorb W AW,
dual mode; helium (17 mL/min); temperature programming (5°C for 1 min, then
2.5°C/min up to 90°C); injector and detector temperature, 100°C; gas sampling
valve (140°C), stainless steel loop (5 mL); culture flask (330 mL) with 50 mL of
medium. References: 31-34, 39.

Procedure B: Analysis of methane, other light hydrocarbons, and volatile
organosulfur compounds in the headspace from alkalinized cultures using a
stainless steel column (1.50 m x 1/8 inch o.d.) packed with 100-120 mesh n-
octane-Porasil C (Alltech Associates, Deerfield, IL), dual mode; helium (17
mL/min); temperature programming (5°C for 1 min, then 5°C/min up to 70°C);
injector and detector temperature, 100°C; gas sampling valve (140°C), stainless
steel loop (5 mL); culture flask (330 mL) with 50 mL of medium. The initial
temperature was achieved by a home-made cryogenic device with carbon dioxide
coolant.

Procedure C: Analysis of volatile aliphatic amines after reduced pressure
distillation of alkalinized cultures using a borosilicate glass column (2.10 m x
2 mm i.d.) packed with 38% Pennwalt 223, 4% potassium hydroxide on 80-100 mesh
Gas Chrom R (Alltech Associates), dual mode; helium (17 mL/min) plus ammonia;
oven temperature, 80°C; injector and detector temperature, 220°C; syringe
sampling (5 mL). Reference: 35.

Procedure D: Analysis of volatile aliphatic amines after reduced pressure
distillation of alkalinized cultures using a borosilicate glass column (2.10 m x
2 mm i.d.) packed with 4.8% Carbowax 20M, 0.3% potassium hydroxide on 100-120
mesh Carbopack B (provided by Dr. Di Corcia, Istituto di Chimica Analitica,
Rome, Italy; reference: 40), dual mode; helium (17 mL/min) plus ammonia; oven
temperature, 50°C; injector and detector temperature, 220°C; syringe sampling (5
mL). Reference: 35.

Procedure E: Analysis of light hydrocarbons and volatile organosulfur compounds
in the headspace from cultures without pretreatment using a borosilicate glass
column (1.40 m x 2 mm i.d.) packed with 100-120 mesh n-octane-Porasil C (Alltech
Associates), dual mode; helium (17 mL/min); temperature programming (5°C for 1
min, then 5°C/min up to 100°C); injector and detector temperature, 100°C; gas
sampling valve (140°C), PTFE loop (15 mL); culture tube (28 mL) with 5 mL of
medium. The initial temperature was achieved by a home-made cryogenic device
with carbon dioxide coolant. References: 36-38.

Procedure F: Analysis of methane in the headspace from cultures without
pretreatment using a borosilicate glass column (3.10 m x 2 mm i.d.) packed with
40-60 mesh molecular sieve 5 Å (Alltech Associates), single column mode; helium
(17 mL/min); oven temperature, 40°C; injector and detector temperature, 100°C;
gas sampling valve (140°C), stainless steel loop (5 mL); culture tube (28 mL)
with 5 mL of medium. References: 36-38.

Procedure G: Analysis of methane in the headspace from cultures without
pretreatment using a soft glass capillary column (100 m x 0.25 mm i.d.,
Chrompack International B.V., Middelburg, The Netherlands) conditioned with
water vapors; helium plus 20% (vol/vol) nitrogen; oven temperature, -196°C;
syringe sampling (50 μL). References: 38, 42.

The commercially available Carbowax 20M stationary phase was found to produce unsuitable variability in retention times. A packing material provided by Di Corcia et al.[40] was preferred. To minimize adsorption and ghosting phenomena associated with polar amines, ammonia was added to the carrier gas.[41] Capillary GC was employed for the separation of ($^2H_3$) methane from methane (procedure G).[42]

Identification of peaks was based on both GC-MS and co-chromatography with authentic standards (with the exception of S-($^2H_3$)methyl volatile compounds or compounds obtained by synthesis such as ($^2H_3$) methane prepared from ($^2H_3$) methyl iodide and magnesium). A Nermag R 10-10 quadrupole mass spectrometer (Delsi-Nermag, Argenteuil, France) coupled to a PDP/8M computer (Digital Equipment Corporation, Maynard, MA) and a Girdel 30S gas chromatograph was operated in the electron-impact (EI) mode at 70 eV and, for volatile aliphatic amines, in the positive ion chemical ionization mode with ammonia as the reactant gas (0.1 kPa). Mass spectra of ($^2H_3$) volatile organosulfur compounds were interpreted on the basis of the fragmentation pattern of the corresponding non-deuterated compounds.

RESULTS AND DISCUSSION

Volatile Non-acidic Compounds

Various volatile non-acidic compounds were identified, some of them having been previously described.[25-28] Several patterns were observed after clostridia were grown in TTY medium containing D-glucose (5 g per liter) with addition of sodium hydroxide at the end of the incubation period.[31-33] Some GC profiles are given in Figure 3.

Compounds identified included C_2-C_8 primary alcohols with straight or branched chains (iso and anteiso isomers). For C. ghonii ATCC 25757^T, they correspond to the C_2-C_6 volatile fatty acids previously reported.[15] Carbonyl compounds (such as propanone, 2-methylbutanal, and 3-methylbutanal) were identified for C. glycolicum ATCC 14880^T as shown in Figure 4. Dimethyl disulfide and ethylene sulfide were detected for C. sporogenes AIP G01 and C. ghonii ATCC 25757^T. Toluene was identified as a volatile compound over cultures of "C. aerofoetidum" strain WS. Trimethylamine and methane (eluting as a non-retained solute in an unresolved peak) were also detected in several clostridia.

The presence of methane was confirmed by GC on other packing materials (molecular sieve 5 Å, n-octane-Porasil C, phenylisocyanate-Porasil C). For example, significant amounts of methane were detected in cultures of C. histolyticum AIP Tro2E, C. ghonii ATCC 25757^T, and Clostridium sp. ATCC 25772, whereas only a small peak of methane was observed for C. cadaveris ATCC 25783^T, C. perfringens AIP Lechien, and for the uninoculated medium (Figure 5).

An interesting application of headspace analysis is the differentiation of two closely-related species, C. sordellii and C. bifermentans.[39] Among seventeen strains labeled C. sordellii, only three were toxigenic, and one strain had a phenotypic pattern characteristic of C. bifermentans (lecithinase positive, toxin and urease negative[43]). When grown in TTY medium supplemented with D-glucose, the GC patterns were quite similar for seventeen strains of C. sordellii and ten strains of C. bifermentans. When D-glucose was omitted, all strains of C. sordellii formed the two isopentanals (3-methylbutanal and 2-methylbutanal) in amounts ranging from 1 to 20 μmole/liter. In the strains of C. bifermentans, isopentanals were undetectable or in low concentrations (below 0.4 μmole/liter). The

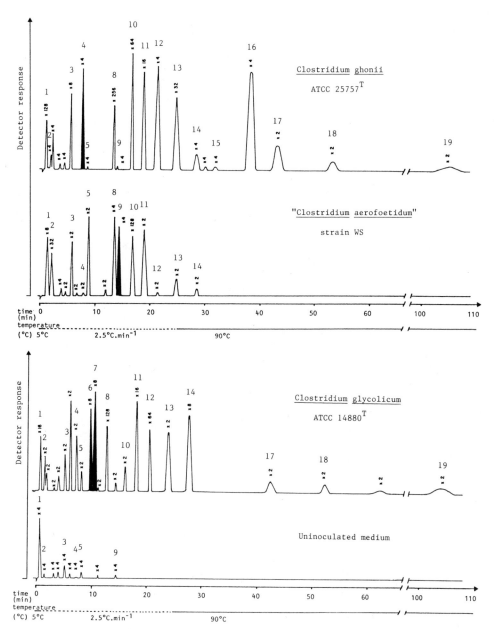

Figure 3. Gas chromatographic patterns (redrawn) obtained for
reduced-pressure headspace samples of cultures of
Clostridium ghonii ATCC 25757[T], "*C. aerofoetidum*" strain
WS, *C. glycolicum* ATCC 14880[T] in TTY medium supplemented
with D-glucose and uninoculated medium. For conditions
used see procedure A, Table 2. Sensitivity at
attenuation x 2, 50 pA f.s. Peak identification:
1 = complex peak containing methane; 2 = trimethylamine;
3 = propanone; 4 = ethylene sulfide; 5 = butanone;
6 = 2-methylbutanal; 7 = 3-methylbutanal; 8 = ethanol;
9 = toluene; 10 = dimethyl disulfide; 11 = 1-propanol;
12 = 2-methyl-1-propanol; 13 = 1-butanol; 14 = 3-methyl-
1-butanol and/or 2-methyl-1-butanol; 15 = 1-pentanol;
16 = 4-methyl-1-pentanol; 17 = 1-hexanol; 18 = 5-methyl-
1-hexanol; 19 = 1-octanol.

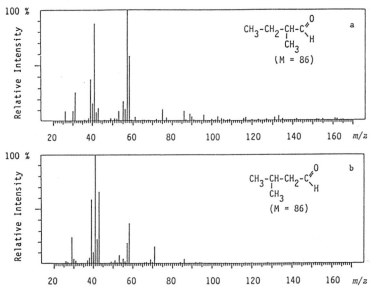

Figure 4. Mass spectra (EI, 70 eV) of (a) 2-methylbutanal and
(b) 3-methylbutanal obtained from a gas sample of a culture of
C. glycolicum ATCC 14880[T] grown in TTY medium supplemented with
D-glucose (5 g per liter) which was alkalinized before analysis.

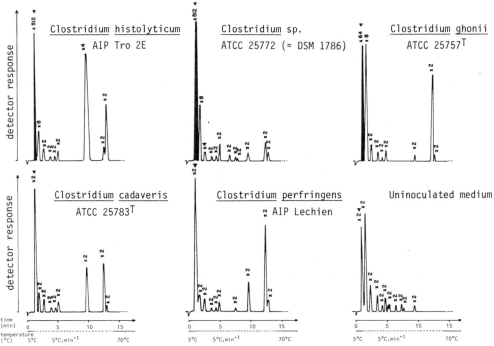

Figure 5. Gas chromatographic patterns (redrawn) obtained for
reduced-pressure headspace samples of cultures of several
clostridia grown in TTY medium supplemented with D-glucose
(5 g per liter) and a control. For analytical conditions
see Procedure B, Table 2. Sensitivity at attenuation x 2,
50 pA f.s. The methane peak is marked with a "v", other
peaks are light hydrocarbons.

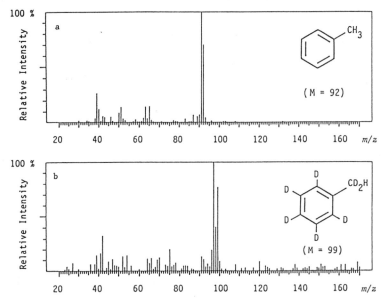

Figure 6. Mass spectra (EI, 70 eV) of (a) toluene and (b) ($^2H_2$) methyl (2,3,4,5,6-$^2H_5$) benzene from a gas sample of a culture of "*C. aerofoetidum*" strain WS (redrawn with permission from reference 34).

formation of these isopentanals may be useful in differentiating *C. bifermentans* from *C. sordellii*.

Deuterium-labeled substrates are useful in elucidating the origin of toluene in "*C. aerofoetidum*" strain WS. This hydrocarbon most likely originates from aromatic L-amino acids or from their metabolites. The formation of toluene was enhanced after addition to TTY medium of 100 mM sodium phenylacetate (2.4-fold) or 100 mM L-phenylalanine plus 100 mM L-methionine (4.9-fold). Addition of L-phenylalanine alone or in combination with another L-amino acid caused smaller increases in toluene production. To test the hypothesis that L-phenylalanine yields toluene, 60 mM L-(2',3',4',5',6'-$^2H_5$) phenyl (2,3-$^2H_2$) alanine was added with 60 mM L-(methyl-$^2H_3$) methionine to TTY medium.[34] The detection of ($^2H_2$) methyl (2,3,4,5,6-$^2H_5$) benzene confirmed this hypothesis. The deuterated compound represented about 85% of the toluene peak (based on the abundance of tropylium ions at m/z 97 and 91). Mass spectra of toluene and ($^2H_2$) methyl (2,3,4,5,6-$^2H_5$) benzene are shown in Figure 6.

Volatile Aliphatic Amines

Table 3 lists the several C_1-C_5 aliphatic amines that were detected after preconcentration, including methylamine, dimethylamine, trimethylamine, 2-methylbutylamine, and 3-methylbutylamine.[35] Ethylamine was previously reported to occur in a clinical isolate of *C. septicum*.[10] Since we detected only the tertiary amine, trimethylamine, for several clostridia, headspace analysis may not be suitable for trace detection of amines if a preconcentration step is not included in sample processing. Indeed, various amines have been reported in cultures of clostridia after distillation and separation by techniques other than GC[44-48,57] or after suitable derivatization and separation by GC.[49-54,58] Trimethylamine has also been detected by thin layer chromatography when clostridia were grown in the presence of choline or betaine.[55] Table 3 summarizes these results.

111

Table 3. Volatile aliphatic amines detected in cultures of clostridia. Polyamines (e.g., putrescine and cadaverine) and non-volatile amines (e.g., tryptamine and histamine) are not included.

Compounds	Paper Chromatography Electrophoresis Chromatoionophoresis [a]	Thin Layer Chromatography [b]	Gas Chromatography Derivatization [c]	Gas Chromatography Derivatization [d]	Headspace [e]	Headspace [f]	Headspace [g]
methylamine	•		•				•
dimethylamine							•
trimethylamine	•	•					•
ethylamine	•				•	•	•
2-methylpropylamine	•		•				
butylamine			•	•			
dibutylamine			•				
2-methylbutylamine				•			
3-methylbutylamine ("isoamylamine")	•			•			•
pentylamine ("amylamine")							•
hexylamine			•	•			
decylamine			•	•			
2-phenylethylamine	•		•				

a References: 44-48, 57.
b Reference: 55.
c N-heptafluorobutyryl and N-trifluoroacetyl derivatives; references: 49-51, 58.
d N-trifluoroacetyl derivatives; references: 52-54.
e References: 25-28, 31-33.
f Reference: 10.
g Reference: 35.

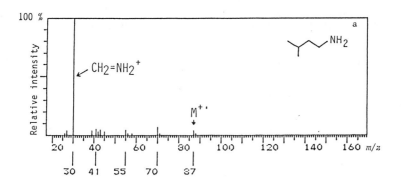

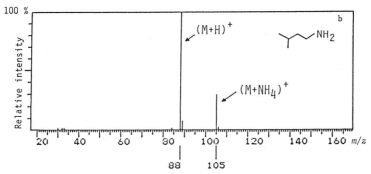

Figure 7.   Mass spectra of 3-methylbutylamine (M = 87)
(a) electron impact ionization (70 eV) of peak from
a culture of *C. mangenotii* ATCC 25761[T] and (b) positive
ion chemical ionization (ammonia, 0.1 kPa) of a peak
from a culture of *C. sordellii* AIP 82. Reproduced from
reference 35 with permission.

The usefulness of positive ion chemical ionization MS for the
identification of 3-methylbutylamine is illustrated in Figure 7. After EI
MS, the mass spectrum shows a base peak at m/z 30 for the ion $CH_2NH_2^+$, and
there is almost no molecular peak at m/z 87 (Figure 7a). Positive ion
chemical ionization with ammonia yields two prominent peaks, one at m/z 88
for the adduct $(M + H)^+$ and another at m/z 105 for the adduct $(M + NH_4)^+$
(Figure 7b). Selected ion monitoring of m/z 30 made possible the detection
of both 2-methylbutylamine and 3-methylbutylamine in cultures of *C.
mangenotii* ATCC 25761[T]. In the total ion mode, only very small peaks were
observed for these components (Figure 8).

Light Hydrocarbons and Volatile Organosulfur Compounds

The formation of several S-methyl volatile compounds (i.e.,
methanethiol, dimethyl mono-, di-, and trisulfides, and a thioester, S-methyl
thioacetate) and methane was also observed.[36] The headspace GC profile for
Clostridium sp. DSM 1786 is shown in Figure 9. Methane is detected for
Clostridium sp. DSM 1786 (isolated from a human post-appendicular peritonitis
by H. Beerens), for *C. hastiforme* ATCC 33268[T], or for telluric isolates
(e.g., *C. ghonii* ATCC 25757[T], *C. histolyticum* AIP Tro2E, *C. histolyticum* DSM
1126). The amount of methane for *Clostridium* sp. DSM 1786 after culture for
seven days in 5 mL TTY medium is about 60 nmoles,[37] which is far below the
values observed with methanogens *sensu stricto*. However, these small amounts
of methane observed for growing cultures of type strains must be considered.

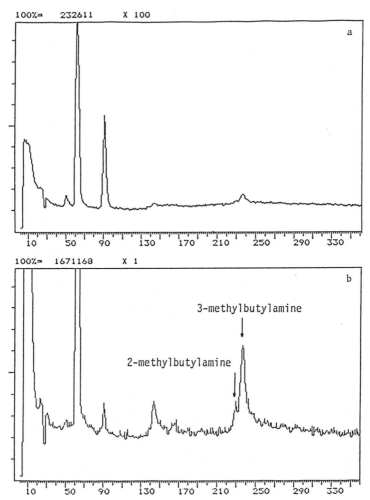

Figure 8. Gas chromatographic-mass spectrometric detection
of 2-methylbutylamine and 3-methylbutylamine in
cultures of *C. mangenotii* ATCC 25761[T]. (a) Total
ionic current (electron-impact ionization, 70 eV) and
(b) single ion monitoring for m/z 30. GC conditions
as given in procedure D, Table 2.

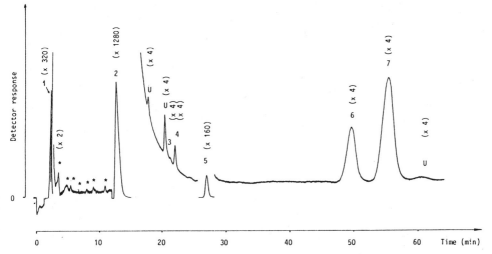

Figure 9.   Gas chromatographic pattern obtained for a
reduced-pressure headspace sample from a culture of
Clostridium sp. DSM 1786 grown in TYY medium. GC
conditions are given in Procedure E, Table 2.
Sensitivity at attenuation x2, 50 pA f.s.   Peak
identification: 1 = methane; 2 = methanethiol;
3 = ethylene sulfide; 4 = dimethyl sulfide;
5 = dimethyl disulfide; 6 = dimethyl trisulfide;
7 = S-methyl thioacetate; * = light hydrocarbons;
U, unidentified.   Reproduced with permission from
reference 36.

Although the five strains of *C. histolyticum* studied were from
different sources (ATCC, DSM, AIP) and from different origins (human
myositis, soil, aquatic sediments), the same hydrocarbons and volatile
organosulfur compounds were detected.   Methane and a small amount of
methanethiol were also among the compounds identified.   The production of 3-
methyl-1-butene and 2-methyl-2-butene differentiated *C. histolyticum* from all
other strains of clostridia studied so far.

Ethylene sulfide was identified for *C. sporogenes* AIP GO1 after culture
in TTY medium supplemented with both D-glucose and sodium thioglycolate (up
to 5 g per liter).   This rather unusual compound is the sulfur analog of two
other hazardous chemicals, ethylene oxide and ethyleneimine, and is found in
the headspace, with or without alkalinization of the culture.[32,36]   Ethylene
sulfide has not been previously reported as a microbial metabolite although
propylene sulfide (a homolog of ethylene sulfide) has been identified by GC-
MS from cultures of *Pseudomonas putida*.[56]   Only 30 pmoles of ethylene sulfide
were detected for *C. sporogenes* AIP GO1 when grown in TTY medium, whereas
about 80 nmoles of ethylene sulfide were detected in a medium containing D-
glucose and 2-mercaptoethanol instead of sodium thioglycolate (Table 4).   It
is not known whether sodium thioglycolate and/or 2-mercaptoethanol are direct
precursors of ethylene sulfide and whether this compound is formed in complex
ecosystems such as the human bowel.

The influence of several compounds, added to TTY medium, on methane
formation by *Clostridium* sp. DSM 1786 was also studied.[37]   The mini-
methanogenesis was increased after addition of 100 mM L-methionine (2.3-fold)
and, to a lesser extent, after addition of 100 mM sodium pyruvate (1.25-
fold), but was depressed after addition of 100 mM sodium acetate (0.65-fold).
No change was observed after addition of 100 mM L-alanine or 1 μM sodium 2-
bromoethanesulfonate.   To determine the role of L-methionine in this mini-

115

Table 4. Variations in amounts of ethylene sulfide in the headspace from cultures of *C. sporogenes* AIP G01 after changes in composition of the growth medium. GC procedure E was used

Medium	Amount of ethylene sulfide[a] (nmoles per 5 mL of medium)	
	After bacterial growth	Uninoculated medium
TTY medium without sodium thioglycolate	Not detected	Not detected
TTY medium	0.03 ± 0.01	Not detected
TTY medium supplemented with sodium thioglycolate up to 5 g per liter (43.8 mM)	0.20 ± 0.01	< 0.01
TTY medium supplemented with D-glucose (5 g per liter)	6.90 ± 0.30	< 0.01
TTY medium supplemented with D-glucose (5 g per liter) and 2-mercaptoethanol (43.8 mM)	77.80 ± 3.70	1.14

[a]mean and standard deviation calculated from three experiments

methanogenesis, 100 mM L-(*methyl*-$^2H_3$) methionine was added to growing cultures of *Clostridium* sp. DSM 1786 in TTY medium. The labeled methyl group in methane and other S-methyl volatile compounds (methanethiol, dimethyl di- and trisulfides and S-methyl thioacetate) was found unchanged. Mass spectra of non-deuterated and deuterated methane and S-methyl volatile compounds are given in Figure 10 and Figure 11, respectively. The proportion of ($^2H_3$) methane in the methane peak, determined by both GC-MS and capillary GC, was found to be about 90%. The role of (1-$^2H_3$) methanethiol as a precursor of ($^2H_3$)methane was also demonstrated.[38]

It was also of interest to know whether L-ethionine, the S-ethyl homolog of L-methionine, is metabolized with the formation of ethane and S-ethyl volatile compounds. Ethanethiol, ethyl methyl sulfide, ethyl methyl disulfide, diethyl di- and trisulfides, and two thioesters, S-ethyl thioacetate and S-ethyl thiopropionate, were identified in the headspace from *Clostridium* sp. DSM 1786 (Figure 12) and a non-quantifiable increase for ethane and/or ethylene was observed.[37] From these observations, it might be assumed that the thiol released from the S-alkyl group of the corresponding L-S-alkyl homocysteine is dimerized into the symmetric disulfide and reacts with volatile fatty acid activated derivatives, yielding S-alkyl thioalkanoates. Methanethiol can be reduced into methane, which was unambiguously demonstrated with the use of L-(*methyl*-$^2H_3$) methionine and (1-$^2H_3$) methanethiol.

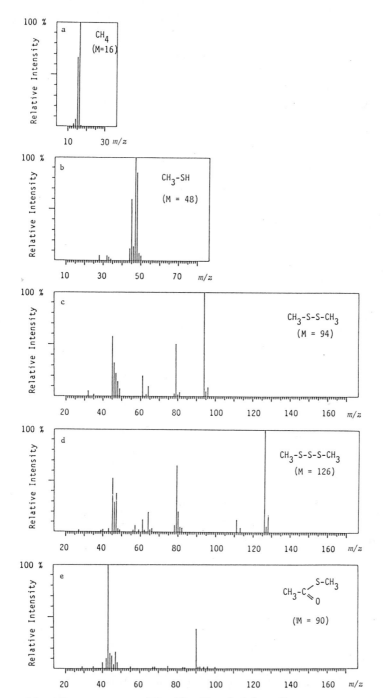

Figure 10. Mass spectra (EI, 70 eV) of (a) methane,
(b) methanethiol, (c) dimethyl disulfide,
(d) dimethyl trisulfide, and (e) S-methyl thioacetate,
from the headspace from a culture of *Clostridium* sp.
DSM 1786 grown in TTY medium (b, c) or in TTY medium
amended with 100 mM L-methionine (a, d, and e). (d)
and (e) reproduced with permission from reference 36.

117

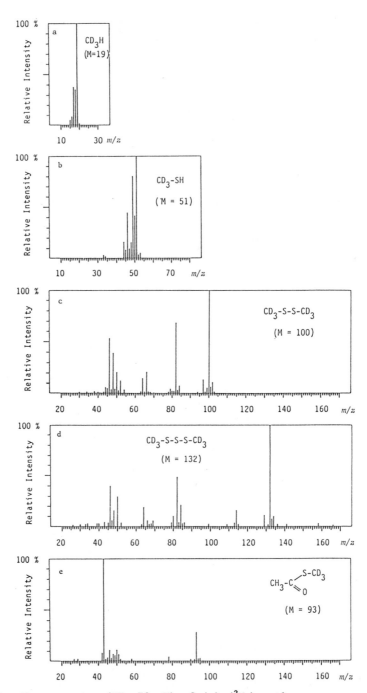

Figure 11. Mass spectra (EI, 70 eV) of (a) $(^2H_3)$ methane, (b) $(1-^2H_3)$ methanethiol, (c) $(^2H_6)$ dimethyl disulfide, (d) $(^2H_6)$ dimethyl trisulfide, and (e) $S-(^2H_3)$ methyl thioacetate, obtained from a gas sample of a culture of *Clostridium* sp. DSM 1786 in TTY medium supplemented with 100 mM L-(*methyl*-$^2H_3$) methionine. Reproduced with permission from reference 38.

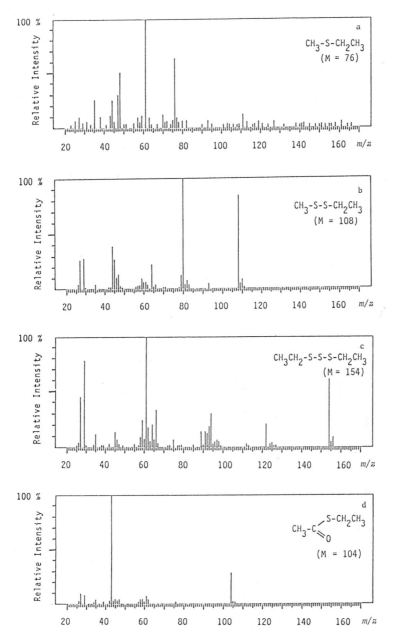

Figure 12. Mass spectra (EI, 70 eV) of (a) ethyl methyl sulfide, (b) ethyl methyl disulfide, (c) diethyl trisulfide, and (d) S-ethyl thioacetate, obtained from a gas sample of a culture of *Clostridium* sp. DSM 1786 in TTY medium amended with 100 mM L-ethionine. Reproduced with permission from reference 37.

CONCLUSION

Headspace analysis by GC-MS is a powerful approach to the study of volatile compounds associated with cultures of clostridia. Information from headspace analysis can lead to better knowledge of clostridial metabolism, as demonstrated here for amino acid metabolism by using deuterated substrates to determine the origin of volatiles. Headspace profiling of volatile compounds in addition to the classical profiling of volatile fatty acids is also a promising method for the chemotaxonomic characterization of clostridia.

Clostridia are widespread in natural environments and are able, *in vitro*, to form trace levels of various volatile organic compounds (e.g., methane and other one-carbon unit compounds, isopentenes, isopentanals, ethylene sulfide and toluene). Headspace analysis by GC-MS can be a useful tool to unravel the complex interrelationships between clostridia and other microorganisms involved in degradation of organic matter. This technique may also be used to evaluate the ecological significance of metabolic pathways that generate volatiles in complex ecosystems such as aquatic sediments and the digestive tracts of humans and animals.

REFERENCES

1. G. Gottschalk, J. R. Andreesen, and H. Hippe, The genus *Clostridium* (nonmedical aspects), in: "The Prokaryotes, A Handbook on Habitats, Isolation, and Identification of Bacteria," M. P. Starr, H. Stolp, H. G. Truper, A. Balows, and H. G. Schlegel, eds., Springer-Verlag, Berlin, pp. 1767 (1981).
2. E. P. Cato, W. L. George, and S. M. Finegold, Genus *Clostridium*, in "Bergey's Manual of Systematic Bacteriology," P. H. A. Sneath, N. S. Nair, M. E. Sharpe, and J. G. Holt, eds., Williams and Wilkins, Baltimore, 2:1141 (1986).
3. H. Hachenberg and A. P. Schmidt, "Gas Chromatographic Headspace Analysis," Heyden and Son Ltd., London (1977).
4. B. Kolb, "Applied Headspace Gas Chromatography," Heyden and Son Ltd., London (1980).
5. B. V. Ioffe and A. G. Vitenberg, "Head-Space Analysis and Related Methods in Gas Chromatography," John Wiley and Sons, Inc., New York (1984).
6. L. Larsson, P. -A. Mardh, and G. Odham, Analysis of volatile metabolites in identification of microbes and diagnosis of infectious diseases, in: "Gas Chromatography/Mass Spectrometry Applications in Microbiology, G. Odham, L. Larsson, and P. -A. Mardh, eds., Plenum Press, New York, p. 207 (1984).
7. J. Drozd and J. Novak, Headspace gas analysis by gas chromatography, J. Chromatogr. 165:141 (1979).
8. B. S. Drasar, P. Goddard, S. Heaton, S. Peach, and B. West, Clostridia isolated from faeces, J. Med. Microbiol. 9:63 (1976).
9. L. Larsson, P. -A. Mardh, and G. Odham, Detection of alcohols and volatile fatty acids by head-space gas chromatography in identification of anaerobic bacteria, J. Clin. Microbiol. 7:23 (1978).
10. L. Larsson, P. -A. Mardh, and G. Odham, Analysis of amines and other bacterial products by head-space gas chromatography, Acta Pathol. Microbiol. Scand., Sect. B 86:207 (1978).
11. L. Larsson, E. Holst, C. G. Gemmell, and P. -A. Mardh, Characterization of *Clostridium difficile* and its differentiation from *Clostridium sporogenes* by automatic head-space gas chromatography, Scand. J. Infect. Dis., Suppl. 22:37 (1980).
12. L. Larsson and E. Holst, Feasibility of automated head-space gas chromatography in identification of anaerobic bacteria, Acta Pathol. Microbiol. Immunol. Scand., Sect. B, 90:125 (1982).

13. H. S. H. Seifert, D. Hoffmann, and H. Bohnel, Differenzierung pathogener Clostridia mit Hilfe der Dampfraum-Gaschromatographie und statistischen Auswertung ihrer metabolisch gebildeten Fettsauremuster, Fortschr. Vet. Med. 37:214 (1983).

14. H. S. H. Seifert, H. Bohnel, S. Giercke, A. Heine, D. Hoffmann, U. Sukop, and D. H. Boege, Routine identification of clostridia using headspace GC and integral biometric analysis, Int. Lab. 17:46 (1986).

15. L. V. Holdeman, E. P. Cato, and W. E. C. Moore, "Anaerobe Laboratory Manual," 4th ed., Virginia Polytechnic Institute and State University, Blacksburg (1977).

16. V. L. Sutter, D. M. Citron, and S. M. Finegold, "Wadsworth Anaerobic Bacteriology Manual," 3rd ed., C. V. Mosby Co., St. Louis (1980).

17. L. Larsson, P. -A. Mardh, G. Odham, and M. -L. Carlsson, Diagnosis of bacteraemia by automated head-space capillary gas chromatography, J. Clin. Pathol. 35:715 (1982).

18. M. B. Huysmans and W. J. Spicer, Assessment of head-space gas-liquid chromatography for the rapid detection of growth in blood cultures, J. Chromatogr. (Biomed. Appl.) 337:223 (1985).

19. D. M. Comberbach, J. M. Scharer, and M. Moo-Young, An improved chromatographic procedure for the rapid determination of volatile solvents via the headspace gas, Biotechnol. Letters 6:91 (1984).

20. J. K. McLaughlin, C. L. Meyer, and E. T. Papoutsakis, Gas chromatography and gateway sensors for on-line state estimation of complex fermentations (butanol-acetone fermentation), Biotechnol. Bioengng. 27:1246 (1985).

21. C. L. Meyer, J. K. McLaughlin, and E. T. Papoutsakis, On-line chromatographic analysis and fermenter state characterization of butanol/acetone fermentations, Ann. N. Y. Acad. Sci. 469:350 (1986).

22. J. R. Postgate, Methane as a minor product of pyruvate metabolism by sulphate-reducing and other bacteria, J. Gen. Microbiol. 57:293 (1969).

23. L. F. McKay, W. P. Holbrook, and M. A. Eastwood, Methane and hydrogen production by human intestinal anaerobic bacteria, Acta Pathol. Microbiol. Immunol. Scand., Sect. B, 90:257 (1982).

24. J. Pazout, M. Wurst, and V. Vancura, Effect of aeration on ethylene production by soil bacteria and soil samples cultivated in a closed system, Plant Soil 62:431 (1981).

25. J. Bory, G. Leluan, and L. Benichou, Composition des gaz degages par Clostridium septicum, Clostridium chauvoei et Welchia perfringens cultives en milieu au thioglycolate de sodium sous vide, apres alcalinisation du milieu en fin de culture, C. R. Seances Acad. Sci. 274(D):2382 (1972).

26. L. Benichou, G. Leluan, and J. Bory, Composition des gaz degages par Clostridium sporogenes, Clostridium bifermentans, Clostridium sordellii, Clostridium oedematiens et Plectridium putrificum cultives en milieu au thioglycolate de sodium sous vide, apres alcalinisation du milieu en fin de culture, C. R. Seances Acad. Sci. 277(D):2825 (1973).

27. J. Bory, G. Leluan, and L. Benichou, Composition des gaz degages par Clostridium histolyticum cultive en milieu au thioglycolate de sodium sous vide, apres alcalinisation du milieu en fin de culture, C. R. Seances Acad. Sci. 279(D):611 (1974).

28. G. Leluan, L. Benichou, and J. Bory, Composition des gaz degages par Clostridium aerofoetidum et Clostridium fallax cultives en milieu au thioglycolate de sodium sous vide, apres alcalinisation du milieu en fin de culture, C. R. Seances Acad. Sci. 278(D):1971 (1974).

29. E. P. Cato, D. E. Hash, L. V. Holdeman, and W. E. C. Moore, Electrophoretic study of Clostridium species, J. Clin. Microbiol. 15:688 (1982).

30. V. B. D. Skerman, V. McGowan, and P. H. A. Sneath, Approved lists of bacterial names, Int. J. Syst. Bacteriol. 30:225 (1980).

31. A. Rimbault and G. Leluan, Composes neutres et basiques presents dans les gaz produits par *Clostridium histolyticum, Clostridium hastiforme* et *Clostridium ghoni* cultives sous vide en milieu glucose au thioglycolate de sodium, C. R. Seances Acad. Sci. 295(III):219 (1982).

32. A. Rimbault and G. Leluan, Composes neutres et basiques presents dans les gaz produits par *Clostridium sporogenes, Plectridium putrificum* et *Plectridium glycolicum* cultives sous vide en milieu glucose au thioglycolate de sodium, C. R. Seances Acad. Sci. 295(III):299 (1982).

33. A. Rimbault and G. Leluan, Etude des composes organiques volatils produits par des bacteries anaerobies sporulees appartenant au genre *Clostridium*, Colloque de la Societe Francaise de Microbiologie, p. 263 (1982).

34. J. -L. Pons, A. Rimbault, J. C. Darbord, and G. Leluan, Biosynthese de toluene chez *Clostridium aerofoetidum* souche WS, Ann. Microbiol. (Inst. Pasteur) 135B:219 (1984).

35. J. -L. Pons, A. Rimbault, J. C. Darbord, and G. Leluan, Gas chromatographic-mass spectrometric analysis of volatile amines produced by several strains of *Clostridium*, J. Chromatogr. (Biomed. Appl.) 337:213 (1985).

36. A. Rimbault, P. Niel, J. C. Darbord, and G. Leluan, Headspace gas chromatographic-mass spectrometric analysis of light hydrocarbons and volatile organosulphur compounds in reduced-pressure cultures of *Clostridium*, J. Chromatogr. (Biomed. Appl.) 375:11 (1986).

37. A. Rimbault, P. Niel, J. -L. Pons, and G. Leluan, Gas chromatographic studies on the formation of methane and volatile organosulphur compounds by *Clostridium* sp. DSM 1786 in "Biology of Anaerobic Bacteria," H. C. Dubourguier, G. Albagnac, J. Montreuil, C. Romond, P. Sautiere, and J. Guillaume, eds., Elsevier, Amsterdam, p. 86 (1986).

38. A. Rimbault, P. Niel, H. Virelizier, J. C. Darbord, and G. Leluan, L-methionine, a precursor of trace methane in some proteolytic clostridia, Appl. Environ. Microbiol. 54:1581 (1988).

39. G. Leluan, M. Leluan, and A. Rimbault, The contribution of headspace analysis to the differentiation between *Clostridium sordellii* and *Clostridium bifermentans*, in "Biology of Anaerobic Bacteria," H. C. Dubourguier, G. Albagnac, J. Montreuil, C. Romond, P. Sautiere, and J. Guillaume, eds., Elsevier, Amsterdam, p. 80 (1986).

40. A. Di Corcia, R. Samperi, and C. Severini, Improvements in the gas chromatographic determination of trace amounts of aliphatic amines in aqueous solutions, J. Chromatogr. 170:325 (1979).

41. S. R. Dunn, M. L. Simenhoff, and L. G. Wesson, Gas chromatographic determination of free mono-, di-, and trimethylamines in biological fluids, Anal. Chem. 48:41 (1976).

42. G. Berger, C. Prenant, J. Sastre, and D. Comar, Separation of isotopic methanes by capillary gas chromatography. Application to the improvement of $^{11}CH_4$ specific radioactivity, Int. J. Appl. Radiat. Isot. 34:1525 (1983).

43. M. R. Popoff, J. -P. Guillou, and J. -P Carlier, Taxonomic position of lecithinase-negative strains of *Clostridium sordellii*, J. Gen. Microbiol. 131:1697 (1985).

44. A. -R. Prevot and A. Sarraf, Recherches sur les odeurs degagees par les anaerobies. I. *Inflabilis lacustris*, Ann. Inst. Pasteur, 5:629 (1960).

45. C. Billy and A. -R. Prevot, Recherches sur les odeurs degagees par les anaerobies. II. *Clostridium corallinum*, Ann. Inst. Pasteur 100:475 (1961).

46. A. Frangopoulos and C. Billy, Recherches sur les odeurs degagees par les anaerobies. III. *Clostridium histolyticum*, Ann. Inst. Pasteur 101:136 (1961).

47. A. K. Kundu, P. Kaiser, and C. Billy, Recherches sur les odeurs degagees par les anaerobies. IV. Les pectinolytiques rouisseurs, Ann. Inst. Pasteur 102:69 (1962).

48. A. -R. Prevot and H. Thouvenot, Recherches sur les odeurs degagees par les anaerobies. VIII. *Cl. botulinum* et *Cl. sordellii*, Ann. Inst. Pasteur, 103:925 (1962).

49. J. B. Brooks and W. E. C. Moore, Gas chromatographic analysis of amines and other compounds produced by several species of *Clostridium*, Can. J. Microbiol. 15:1433 (1969).

50. J. B. Brooks, C. W. Moss, and V. R. Dowell, Differentiation between *Clostridium sordellii* and *Clostridium bifermentans* by gas chromatography, J. Bacteriol. 100:528 (1969).

51. J. B. Brooks, C. C. Alley, J. W. Weaver, V. E. Green, and A. M. Harkness, Practical methods for derivatizing and analyzing bacterial metabolites with a modified automatic injector and gas chromatograph, Anal. Chem 45:2083 (1973).

52. A. Tavakkol and D. B. Drucker, Qualitative GC analysis of bacterial amines as their acylated derivatives, J. Chromatogr. Sci. 22:12 (1984).

53. D. B. Drucker and A. Tavakkol, Chromatographic analysis of volatile end products of Gram-positive bacteria, in "Chemical Methods in Bacterial Systematics," M. Goodfellow and D. E. Minnikin, eds., Academic Press, London, p. 301 (1985).

54. A. Tavakkol, D. B. Drucker, and J. M. Wilson, Gas chromatography/mass spectrometry of bacterial amines, Biomed. Mass Spectrom. 12:359 (1985).

55. B. Moller, H. Hippe, and G. Gottschalk, Degradation of various amine compounds by mesophilic clostridia, Arch. Microbiol. 145:85 (1986).

56. E. V. Bowman, L. R. Freeman, D. W. Later, and M. L. Lee, Comparison of volatiles produced by selected pseudomonads on chicken skin, J. Food Sci. 48:1358 (1983).

57. C. Billy, Recherches sur les odeurs degagees par les anaerobies. VII. *Welchia perfringens* et *Welchia agni*, Ann. Inst. Pasteur 103:464 (1962).

58. J. B. Brooks, V. R. Dowell, D. C. Farshy, and A. Y. Armfield, Further studies on the differentiation of *Clostridium sordellii* from *Clostridium bifermentans* by gas chromatography, Can. J. Microbiol. 16: 1071 (1970).

CHAPTER 8

IDENTIFICATION OF PATHOGENIC BACTERIA BY HEADSPACE GAS CHROMATOGRAPHY

Horst S. H. Seifert, Sabine Giercke-Sygusch,
and Helge Boehnel

Georg-August Universitaet
Goettingen, D3400
West Germany

INTRODUCTION

Gas chromatographic (GC) analysis of metabolically produced alcohols, short and long chain fatty acids, and cell components provide precise criteria for the identification of bacteria.[1,2] Generally, prior to GC analysis, volatile compounds are extracted from the growth broth and/or the washed bacteria. However, this extraction process often leads to the loss of some of the alcohols and short chain fatty acids because of their high volatility.[3]

In comparison, headspace analysis provides better reproducibility, and more reliable identification of bacteria since the extraction step is eliminated.[2,4-8] Furthermore, analysis is faster since sample preparation is minimized. Headspace analysis is typically performed as follows: 10 μl of the bacterial culture is mixed with 10 μL sodium sulfate, stored in a sealed vial and thermostated. Generally the sample is heated at 125°C for two h. At this temperature, volatiles are fully vaporized and non-volatile bacterial components are exposed to a low temperature carbonization. When equilibrium between the liquid phase of the sample and the gas phase is reached, the test tube is automatically pressurized with N_2 and a small amount of the vapor phase introduced into the GC. Headspace sampling extracts only the gaseous volatile constituents and the non-vaporizable residues remain in the sample vial. Volatiles are released from metabolites in the media, the bacterial cell envelope, and cytoplasm. Headspace analysis, unlike extraction methods, allows analysis of a number of alcohols in addition to short chain fatty acids.

Head-space analysis of clostridia provide an excellent illustration of the technique. When grown under standardized conditions these organisms can produce at least sixteen different compounds including alcohols (methanol, ethanol, isopropanol, propanol, butanol, and pentanol) and short chained fatty acids (acetic acid, propanoic acid, 2-methyl propanoic acid, butanoic acid, 1-methyl butanoic acid, pentanoic acid, 4-methyl pentanoic acid, hexanoic acid, heptanoic acid, and octanoic acid) that are useful for identification. This number of relevant criteria is sufficient to allow for statistically reliable differentiation and identification.[4,5]

Analytical Microbiology Methods
Edited by A. Fox *et al.*
Plenum Press, New York, 1990

125

PREPARATION OF THE BACTERIOLOGICAL SAMPLE

Usually, the bacterial samples are stored in a freezer at -80°C. For analysis, two test tubes containing 5.0 ml each of culture medium (Reinforced Clostridial Medium/RCM) are inoculated and incubated for 48 h at a vacuum of 30 torr and a temperature of 37°C. From the tube containing the best growth, three new tubes of culture medium are inoculated. The sample is also streaked at this time onto nutrient agar using the inoculation pipet and incubated under aerobic conditions. This test serves as a check for contamination. Parallel to these efforts, the bacteria are also streaked on blood agar and incubated under anaerobic conditions for 48 h.

For the differentiation of pathogens, each strain is tested for motility and ability to produce phosphatase. Acidic phosphatase is an indicator of the presence of *Clostridium perfringens*. To reduce undesirable variability in metabolite production during growth, each of the three prepared samples is inoculated into 7 test tubes containing culture medium (either Peptone-Yeast-Glucose/PYG or RCM) and incubated for 24 h. Prior to the headspace GC tests, the samples are pooled, and then analyzed by two replicate GC runs[3,4,5] (Figure 1). With improved hardware, the reproducibility of the method has increased and standard deviations for replicate analyses are lower than 1%. For routine identification, samples may be analyzed directly without repetition and pooling.

ANALYTICAL METHODOLOGY AND INSTRUMENTATION

Headspace GC applications have been extended to microbiological applications only recently. Larsson et al.[8,9] described the analysis of several anaerobic species. We have confirmed these results and made some improvements by modifying the GC hardware. The Perkin Elmer system (Norwalk, CT) that we employ consists of a HS-100 Headspace Sampling System, Sigma 2000 GC, LCI-100 integrator, and an IBM PC/AT 2 for statistical evaluation of

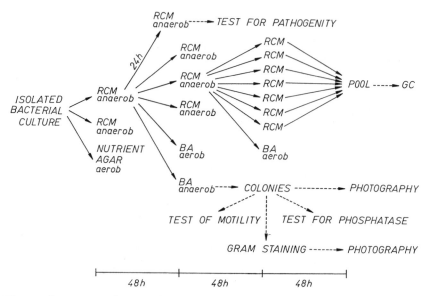

Figure 1. Steps in sample preparation for headspace analysis of bacteria by GC. RCM refers to the media used for anaerobic growth. BA refers to blood agar plates used for aerobic incubation.

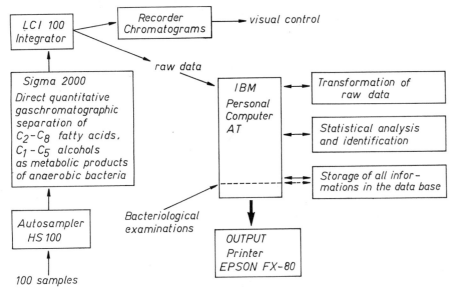

Figure 2. Steps in handling and processing of data obtained
after headspace analysis by GC.

data. The HS-100 is an automated headspace sampling system which allows
batch processing of large numbers of samples. Up to 100 samples may be
stored at room temperature within the magazine. With the present system,
sample vials are transported individually into the heating carousel where
each is thermostated precisely at 125°C for 2 hr. Prior to sampling, each
vial is pressurized with N_2 and an exact predetermined amount of sample is
injected into the analyzing system.

The Sigma 2000 GC GC is equipped with a capillary column and flame
ionization detector (FID). All temperatures and timed events for both the GC
and the headspace sampler are controlled by a microprocessor. The LCI-100
integrator with an internal printer/plotter serves as a user interface and is
interfaced to the computer for downloading of reduced data. The IBM Personal
Computer AT 2 receives raw data from the integrator and stores them on a hard
disc. Statistical evaluation of the chromatogram is performed by the
computer, supported by a data bank containing details of about 100 known
clostridia and bacteroides (Figure 2).[3-5]

DATA-FLOW

Electrical signals produced by the FID are transferred to the LCI-100
integrator and stored as a raw retention time and peak area data. The data
analyzing system of the integrator is standardized with defined standard
solutions. Since acetic acid is contained within the broth and appears in
each chromatogram, it is used as a reference peak but not further considered
as a diagnostic criteria. The sixteen selected peak areas are transformed
into g/L amounts of the respective chemical markers and transferred to the
IBM AT 2 as percentage of total alcohol content and total fatty acid content
respectively..

Due to the amount of possible data when establishing the data bank from
defined reference strains, random samples are analyzed from each strain and
the concentration of each chemical component processed as a percentage of the

127

total alcohol response or fatty acid concentration. The number of data points to processed for one reference strain is 336 (16 variables, 3 series with 7 replicates each). This step is valid because the percentage of metabolites remains the same under different growth conditions.[10] The mean of all random samples is evaluated for each component and used for the calculation within the biometric analysis. All data are stored on the PC hard disk (storage capacity 20 megabytes). Results are printed on a dot matrix printer. The data flow is illustrated in Figure 2.[4,5]

BIOMETRIC EVALUATION-- CLUSTER ANALYSIS

The aim of the statistical evaluation is to differentiate between known reference strains and to identify unknown field strains by grouping them as close as possible to the reference strains. The data from each bacterial sample is averaged as described above to produce a mean chromatogram which can be represented as a data point in a sixteen dimensional space. The familiar concept of a distance in two dimensions may be generalized in higher dimensions as a measure of similarity (actually, dissimilarity) of samples to one another. Chromatograms (and thus the bacterial samples they represent) that are similar to one another will be located close to one another (small distances) in this multidimensional space; chromatograms that are different will be well separated from one another (larger distances). The principle of cluster analysis is to group unknown individuals hierarchically in such a way that similar groups are constructed on different levels of distance and similarity. The "single linkage " (nearest neighbor) clustering algorithm is used. Samples ar put together by rough common differences in larger groups ("agglomerative"). These larger groups are then split into smaller groups ("divisive") if more distinctive characteristics are used.

The result of this process is a hierarchical system of clusters of samples. Two clusters at different levels of distance or similarity are either alien or one is contained within the other. Grouping may be done either through gradual refining or by increasing the coarseness of the partition. Through gradual division or agglomeration, clusters are positioned on different distance and similarity levels and the relationships among clusters displayed with a tree-structured clustering display or dendrogram (Figure 3).[2,3]

IDENTIFICATION OF A FIELD STRAIN

As an example of the identification process, the bacteriological characteristics and the biometric identification of the malagasy field strain 335[11] in comparison to the USDA *C. chauvoei* vaccine challenge strain IRP 206 are presented here.

For the data sheets of these two strains, the chromatograms with identified components are presented in Figures 4-5 respectively and Table 1. The cluster analysis output lists reference strains (Figure 6A) and plots a dendrogram using multivariate distances between each sample (Figure 6B). Then reference strains for the identification procedure are those *Clostridium* species which are international type strains, phosphatase-negative, and which are available in our data bank. Sample nr. 50 is classified in one group together with samples 2-46. The nearest neighbor is sample nr. 46 with a distance of 0.009 (the distances could be printed out separately). This means that sample nr. 46 and sample nr. 50 are not only similar but identical, and that the field strain 335 isolated in the province of Tulear/Madagascar is a typical black-leg causing pathogen like sample nr. 46, the US strain IRP 206.

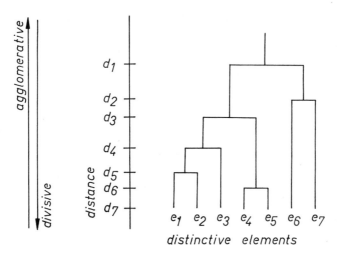

Figure 3.   Principle of construction of a hierarchical
            clustering dendrogram (single linkage algorithm).
            d refers to common level of distance.   e refers
            to distinctive elements.

ADVANTAGES OF THE ANALYZING SYSTEM

        Identification of bacteria by headspace GC has the following
advantages: (1) the headspace sample contains volatiles produced by
metabolism and as well as volatile constituents of the cells themselves; (2)
alcohols (critical for the identification of aerobes) and fatty acids are
determined in one run; (3) the results are reproducible with low relative
standard deviations; (4) the system can be accurately calibrated by
quantitative analysis of standard mixtures of free fatty acids and alcohols;,
and (5) the system runs automatically with a GC analysis cycle time of 30
min.

        During one week (168 working hours), about 300 analyses are possible.
The large throughput is an important requirement for routine applications.
Direct headspace analysis is not labor intensive.   The headspace GC procedure
described here does not require additional wet chemistry (e.g., extraction or
esterification).   In manual extraction procedures for determination of fatty
acid metabolites from pathogens, partial loss of the higher free fatty acids
occurs, or an esterification step must be included.   The extraction and
transfer of the volatile fatty acids is critical and results in poor relative
standard deviations (above 15%).[12]   The mechanization of the critical steps
of sample preparation and injection greatly improves the reliability of the
analysis.   Less skilled operators can routinely run the simplified procedure.
Automated classification and identification of field samples are possible by
computer-assisted cluster analysis using reference data from previously
identified strains.

        These headspace GC methods can be applied to the routine analysis of
anaerobic bacteria isolated from man, animals, or food.   Rapid automated
identification of many anaerobes in the clinical microbiology laboratory by
headspace GC can be readily accomplished.   For the identification of certain
exotic strains, it is almost mandatory to apply this procedure.   For example,
a tropical pathogen *C. haemolyticum* was identified in Mexico and atypical
field strains related to the gas edema pathogens were characterized.[13]
Headspace identification of previously uncharacterized pathogens may thus be
useful in development of vaccines.   Fast identification of anaerobic wound
infections can be of vital importance in medical hospitals.[14]

```
Collection  number          :  335
Entry number           : 77480
```

ORIGIN

```
Entry date               : 7/77
Identification number    : 220/77
Material                 : muscle, dried
Source                   : bovine
Origin identification    : contaminant of B.A.
Sender                   : Madagascar, Mahabo
```

BACTERIOLOGICAL EXAMINATION

Colonies on blood agar

```
Growth                   : localized
Edge                     : entire
Surface                  : convex
```

Morphology of cell

```
Gram stain               : positive
Width                    : short stout
Arrangement              : occurring singly
Spore                    : central / terminal

Motility                 : -
Hemolysis                : alpha
Phosphatase              : -
Indole                   : -
```

PATHOGENICITY

```
Guinea pig     (i.m)     : n.t.
Leth. mouse    (i.v)     : +++
```

CLASSIFICATION

```
Biometric                : identical to 1271 C. chauvoei
Final                    : Clostridium chauvoei
```

Figure 4.   (a) Data sheets for strains of *C. chauvoei* 335
            Madagascar.

Collection number   : 1271
Entry number   : 87112

ORIGIN

Entry date   : 04/07/87
Identification number   : IRP 206
Material   :
Source   :
Origin identification   : C. chauvoei
Sender   : NVSL, Ames USA

BACTERIOLOGICAL EXAMINATION

Colonies on blood agar

Growth   : localized
Edge   : entire
Surface   : smooth

Morphology of cell

Gram stain   : positive
Width   : polymorph
Arrangement   : occurring singly
Spore   : central / terminal

Motility   : +
Hemolysis   : beta
Phosphatase   : -
Indole   : -

PATHOGENICITY

Guinea pig   (i.m)   : ++
Leth. mouse   (i.v)   : +++

CLASSIFICATION

Biometric   : C. chauvoei
Final   : Clostridium chauvoei

Figure 4. (b) Data sheets for strains of *C. chauvoei* IRP 206/1271.

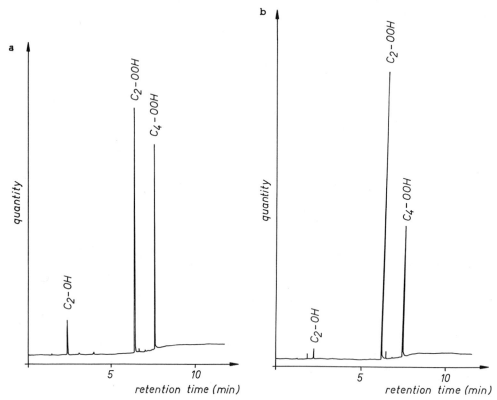

Figure 5. Headspace GC profiles for strains of *C. chauvoei* (a) 335 Madagascar (b) IRP 206/1271.

```
I.D. collection              identification
nr.        nr.


001        1018        Clostridium paraperfringens
002        1019        Clostridium fallax
003        1020        Clostridium difficile
004        1021        Clostridium clostridioforme
005        1022        Clostridium absonum
006        1023        Clostridium chauvoei
007        1024        Clostridium chauvoei
008        1025        Clostridium novyi A
009        1026        Clostridium septicum
010        1027        Clostridium bifermentans
011        1028        Clostridium botulinum A
012        1029        Clostridium botulinum B
013        1030        Clostridium botulinum C
014        1031        Clostridium botulinum D
015        1032        Clostridium botulinum E
016        1033        Clostridium botulinum F
017        1034        Clostridium butyricum
018        1035        Clostridium coccoides
019        1036        Clostridium histolyticum
020        1045        Clostridium putrefaciens
021        1046        Clostridium sphenoides
022        1047        Clostridium spiroforme
023        1048        Clostridium tertium
024        1049        Clostridium tetani type 1
025        1050        Clostridium tetanomorphum
026        1051        Clostridium villosum
027        1069        Clostridium haemolyticum
028        1070        Clostridium sordellii
029        1072        Clostridium beijerinckii
030        1073        Clostridium botulinum G
031        1074        Clostridium cadaveris
032        1075        Clostridium carnis
033        1076        Clostridium chauvoei
034        1077        Clostridium cochlearium
035        1078        Clostridium hastiforme
036        1079        Clostridium innocuum
037        1080        Clostridium lentoputrescens
038        1082        Clostridium novyi
039        1083        Clostridium paraputrificum
040        1084        Clostridium barati
041        1085        Clostridium parabotulinum
042        1086        Clostridium ramosum
043        1087        Clostridium sporogenes
044        1088        Clostridium subterminale
045        1089        Clostridium tetani
046        1271        Clostridium chauvoei
047        1272        Clostridium sordellii
048        1273        Clostridium novyi
049        1274        Clostridium haemolyticum

050         335        fieldstrain Madagascar
```

Figure 6.   (A) List of reference strains to be compared with
the field strain (example Madagascar field strain
nr. 335).

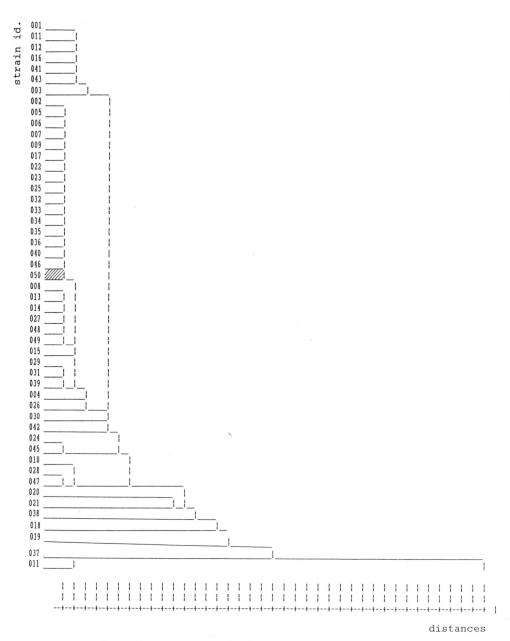

Figure 6.   (B) Dendrogram showing the grouping of the field
            strain Madagascar 335 in relation to the different
            distance levels of the reference strains listed in
            Figure 6A.

Table 1. Calculation of percentage of total alcohols and total fatty acids for two strains of *C. chauvoei*

	Percentage	
Component	335 Madagascar	IRP 206/1271
methanol	0.000	0.000
ethanol	94.093	100.000
isopropanol	5.907	0.000
butanol	0.000	0.000
pentanol	0.000	0.000
acetic acid[a]	-	-
propanoic acid	0.000	0.000
2-methyl propanoic acid	0.000	0.000
butanoic acid	100.000	95.263
1-methyl butanoic acid	0.000	0.000
pentanoic acid	0.000	4.737
4-methyl pentanoic acid	0.000	0.000
hexanoic acid	0.000	0.000
heptanoic acid	0.000	0.000
octanoic acid	0.000	0.000

[a]not calculated

Headspace profiles have been accumulated for many aerobic bacteria (including *Bacillus*). For example, *Bacillus anthracis* is an important apathogen, causing infectious diseases of cattle in the Third World. *B. anthracis* can be readily distinguished from *B. cereus* using headspace GC. Identification of other aerobic bacteria by this approach will require optimization of growth conditions for production of alcohols and short chain fatty acids useful in identification. With further development of this operational information and the continued accumulation of data banks for comparative purposes, headspace GC shows great promise as an analytical tool for the identification of microorganisms. Further improvements in headspace GC should allow the detection of up to C_{20} long chain fatty acids after simple esterification within the thermostated vial.[15]

REFERENCES

1. R. S. Breed, E. G. D. Murray, and N. R. Smith, "Bergey's Manual of Determinative Bacteriology," 7th ed., Williams and Wilkins, Baltimore (1957).
2. L. V. Holdemann, E. Cato, and W. E. C. Moore, "Anaerobe Laboratory Manual," 4th ed., Virginia Polytechnic Institute and State University, Blacksburg (1977).
3. H. S. H. Seifert, D. Hoffmann, and H. Boehnel, Differenzierung pathogener Clostridia mit Hilfe der Dampfraum-Gaschromatographie und statistischen Auswertung ihrer metabolisch gebildeten Fettsauremuster, Fortschr. Vet. Med. 37:214 (1983).
4. S. Giercke-Sygusch, Untersuchungen zur Erstellung eines Atlas anaerober Bakterien, Diss. sc. agr. Gottingen (1987).
5. S. Giercke, H. Boehnel, H. S. H. Seifert, Atlas der Clostridia und Bacteroiides-Arten (in Vorbereitung), Universitaet Goettingen (1986).

6. D. Hoffmann, Differenzierung von Clostridien mit Hilfe der Dampfraum-Gaschromatographie und der statistischen Auswertung ihrer Fettsaeuremuster, Diss. sc. agr. Goettingen (1984).

7. B. Kolb, P. Pospisil, M. Jaklin, and D. Boege, Analytic methods for head-space analysis, Chromatogr. Newsletter 7:1 (1979).

8. L. Larsson, P. A. Mardh, and G. Odham, Detection of alcohols and volatile fatty acids by head-space gas chromatography in identification of anaerobic bacteria, J. Clin. Microbiol. 7:23 (1978).

9. L. Larsson, P. A. Mardh, G. Odham, and M. L. Carlsson, Diagnosis of bacteriaemia by automated head-space capillary gas chromatography, J. Clin. Path. 35:715 (1982).

10. H. S. H. Seifert, H. Boehnel, and A. Ranaivoson, Verhuetung von Anaerobeninfektionen bei Wiederkaeuern in Madagaskar durch intradermale Applikation von ultrafiltrierten Toxoiden standortspezifischer Clostridia, Dtsch. tierarztl. Wschr. 90:274 (1983).

11. L. J. Turton, D. B. Drucker, L. A. Ganguli, and V. F. Hillier, Statistical significance of effect of growth medium on anaerobe fermentation products, J. Appl. Bact. 53:30 (1982).

12. K. Rieke, Differenzierung von Clostridia mit Hilfe der Gaschromatographischen Analyse metabolisch gebildeter Fettsaeuren, Diss. sc. agr. Goettingen (1981).

13. P. Gonzalez-Salinas, Aetiologie und Inzidenz von Bodenseuchen im Nord-Osten Mexikos, Diss. sc. agr. Goettingen (in press).

14. J. Wuest, Schnelldiagnose anaerober Infektionen mit der direkten gaschromatographischen Untersuchung von klinischem Material, Schweiz. med. Wschr. 110:362 (1980).

15. S. Heitefuss, to be published (1990).

CHAPTER 9

HIGH PERFORMANCE CHROMATOGRAPHIC AND SPECTROSCOPIC ANALYSES OF
MYCOBACTERIAL WAXES AND GLYCOLIPIDS

David E. Minnikin, Gary Dobson, Robert C. Bolton,
and James H. Parlett

Department of Organic Chemistry
The University
Newcastle upon Tyne NE1 7RU, United Kingdom

and

Anthony I. Mallet

The Institute of Dermatology
St. Thomas's Hospital
Lambeth Palace Road
London SE1 7EH, United Kingdom

INTRODUCTION

The cell envelopes of mycobacteria are composed of a wide range of unusual lipids which are probably of importance in the integrity and virulence of important pathogens such as *Mycobacterium leprae* and *M. tuberculosis*. The covalently-bound mycolic acids are long-chain 2-alkyl-branched, 3-hydroxy fatty acids which are considered to provide a basal lipid monolayer providing an anchorage for a range of very special free lipids.[1,2] Mycobacterial free lipids range in polarity from apolar waxes to highly polar glycolipids, many of the latter being surface lipid antigens.[3-7] Analyses of these characteristic free lipids is of value in several respects, the most important being classification, identification, and lipid function.

The dimycocerosates of the phthiocerol family (Table 1) are characteristic waxes produced by a limited range of mycobacterial species.[8] These waxes are closely related to the so-called phenolic glycolipids[3,7,10-15] which are glycosylphenolphthiocerol dimycocerosates (Table 2). The latter class of lipids have been extensively studied, particularly for the composition and antigenicity of the sugar units. Studies of the characteristic long-chain components of both these lipid classes are of value in understanding their function in the mycobacterial envelope. This chapter will illustrate how combinations of chromatographic and mass spectrometric techniques can be employed for the efficient analysis of these waxes and glycolipids and their long-chain components.

Analytical Microbiology Methods
Edited by A. Fox *et al.*
Plenum Press, New York, 1990

Table 1. Mycobacterial dimycocerosates of the phthiocerol family

Taxon	Mycocerosate	Phthiocerol family
M. africanum	$C_{27}-C_{32}$	C_{34}, C_{36}
M. bovis	$C_{27}-C_{32}$	C_{34}, C_{36}
M. microti	$C_{27}-C_{32}$	C_{34}, C_{36}
M. tuberculosis	$C_{27}-C_{32}$	C_{34}, C_{36}
M. kansasii[1]	$C_{27}-C_{32}$	C_{27}
M. leprae	$C_{30}-C_{34}$	C_{30}, C_{32}
M. marinum	$C_{27}-C_{30}$	C_{28}, C_{30}
M. ulcerans[1,2]	$C_{27}-C_{30}$	C_{28}, C_{30}

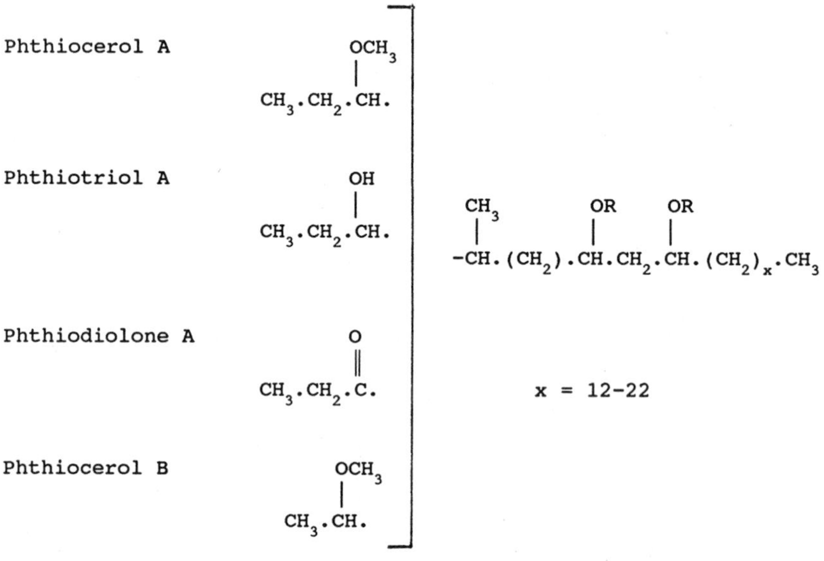

Phthiocerol A

$$CH_3 . CH_2 . \overset{\overset{\textstyle OCH_3}{|}}{CH} .$$

Phthiotriol A

$$CH_3 . CH_2 . \overset{\overset{\textstyle OH}{|}}{CH} .$$

Phthiodiolone A

$$CH_3 . CH_2 . \overset{\overset{\textstyle O}{\|}}{C} .$$

Phthiocerol B

$$CH_3 . \overset{\overset{\textstyle OCH_3}{|}}{CH} .$$

$$-\overset{\overset{\textstyle CH_3}{|}}{CH} . (CH_2) . CH . CH_2 . \overset{\overset{\textstyle OR}{|}}{CH} . (CH_2)_x . CH_3$$

$$x = 12-22$$

$$R = CH_3 . (CH_2)_y . (CH_2 . \overset{\overset{\textstyle CH_3}{|}}{CH} .)_z . CO-$$

$$y = 16-20, \quad z = 2-5$$

Mycocerosate

---

[1] Diesters of phthiodiolone only.

[2] The multimethyl branched acids had the opposite stereochemistry to that of the mycocerosates.

Table 2. Mycobacterial glycosylphenolphthiocerol dimycocerosates

Taxon	Sugar unit (X) of major lipid
M. bovis, M. africanum, M. microti	2-*O*-Me-rhamnose
M. marinum	3-*O*-Me-rhamnose
M. kansasii	2,6-dideoxy-4-*O*-Me-arabinohexose, 2-*O*-Me-fucose, 2-*O*-rhamnose, 2,4-*O*-Me$_2$-rhamnose
M. leprae	3,6-*O*-Me$_2$-glucose, 2,3-*O*-Me$_2$-rhamnose, 3-*O*-Me-rhamnose
M. tuberculosis (Canetti)	2,3,4-*O*-Me$_3$-fucose, rhamnose, 2-*O*-Me-rhamnose

Phenolphthiocerol A   OCH$_3$
$$CH_3 . CH_2 . CH.$$

Phenolphthiotriol A   OH
$$CH_3 . CH_2 . CH.$$

Phenolphthiodiolone A   O
$$CH_3 . CH_2 . C.$$

Phenolphthiocerol B   OCH$_3$
$$CH_3 . CH.$$

$$CH . (CH_2) . CH . CH_2 . CH . (CH_2)_x - C_6H_4 - O . X$$

with CH$_3$, OR, OR substituents

$$x = 16\text{-}22$$

$$R = CH_3 . (CH_2)_y . (CH_2 . \overset{CH_3}{CH} .)_z . CO-$$

$$y = 16\text{-}20, \quad z = 2\text{-}5$$

Mycocerosate

It is important to employ simple and efficient extraction methods which can be applied both on a microanalytical and preparative scale. The relatively non-polar waxes and glycolipids under consideration here can be obtained easily by a simple procedure designed to separate non-polar lipids from the bulk of the polar lipids.[6,8,11,16,17] The essence of this procedure is to expose the biomass to a biphasic mixture of petroleum ether and aqueous methanol, the non-polar lipids being recovered in the hydrocarbon phase. The presence of the aqueous methanol is essential to release the lipids. This polar phase, including the residual biomass, can be further processed by the well-proven Bligh and Dyer protocol[18] to provide polar lipids.[17,19]

Positive identification of dimycocerosates of the phthiocerol and glycosylphenolphthiocerol families can be readily achieved by semi-quantitative two-dimensional thin-layer chromatography (TLC) systems.[11,17,19] Examples of such separations are shown in Figure 1. Milligram quantities, or less, can be separated by preparative TLC and larger amounts isolated by column chromatography with TLC monitoring. There is also potential for the separation of intact lipids by high performance liquid chromatography (HPLC); the use of a normal phase would allow the determination of the lipids based on different members of the phthiocerol and glycosylphenolphthiocerol families and reverse phase media would separate at least some of the homologous components of each lipid class.

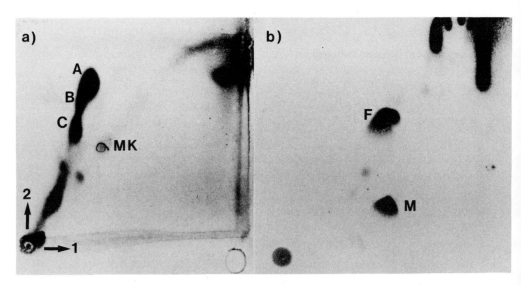

Figure 1.   TLC of mycobacterial free non-polar lipids.
(a) Phthiocerol dimycocerosates from *M. tuberculosis*.
First direction, petroleum ether (bp 60-80°C)-ethyl
acetate (95:5, thrice); second direction, petroleum
ether-acetone (98:2, once). (b) Glycosylphenolphthiocerol
dimycocerosates from *M. kansasii*. First direction,
chloroform-methanol (96:4); second direction,
toluene-acetone (80:20). Detection: charring with
5% ethanolic molybdophosphoric acid. Abbreviations:
A-C, dimycocerosates of phthiocerol A, B, and
phthiodiolone A; MK, menaquinone; F, free fatty
acid; M, glycosylphenolphthiocerol dimycocerosate.
Reproduced with permission from reference 19.

Preliminary characterization of intact lipids can be made with the usual organic chemical methods, but certain techniques are of particular value. High-field nuclear magnetic resonance (NMR) spectroscopy is a powerful non-destructive way to determine the essential structural details of an isolated lipid.[9,10,12-15] Examples of the high field NMR spectra of dimycocerosates of the phthiocerol and glycosylphenolphthiocerol families are shown in Figures 2 and 3, respectively. Mass spectrometry of intact lipids is also of value, a good example being the analysis by fast atom bombardment (FAB) of the phenolic glycolipids from *M. kansasii* which revealed the presence of a hitherto undetected sugar.[7,14]

LIPID DEGRADATION

The relative hydrophobicity of the waxes and glycolipids, under consideration, requires that special degradation methods are developed. Acid treatment is the method of choice for release of the sugar units in phenolic glycolipids, though care must be taken to avoid breakdown of acid-labile monosaccharides.[7,14] TLC patterns of methyl glycosides of sugars produced by acid methanolysis are of value in identifying the parent glycolipid.[11] The complete hydrolysis of the dimycocerosates of the phthiocerol and phenolphthiocerol families requires forcing conditions since simple boiling with aqueous alkali has practically no effect. An alternative to hydrolysis is reduction with lithium aluminium hydride,[8,17,20] but this has the disadvantage that ketonic components are reduced. Hydrolysis with KOH in aqueous 2-methoxyethanol (methyl cellosolve) at 110°C for 4 h has been recommended for hydrolysis of these molecules, without racemization. Experience in these laboratories has shown that, even using these conditions overnight, breakdown of the waxes was far from complete. Efficient alkaline degradation of this type of lipid can be achieved by heating at 100°C overnight with 30% KOH in methanol/toluene (1:1).[21,22] It is expected, however, that this latter method will produce some degree of racemization in the mycocerosate components.

Long chain components released from the lipids under consideration can be converted to various derivatives suitable for analysis. Free mycocerosic acids can be converted to methyl or pentafluorobenzyl (PFB) esters very efficiently using phase transfer catalysis[21,22] methyl esters can also be prepared by the traditional use of diazomethane.[23] Phase transfer catalyzed esterification with iodomethane is also suitable for placing a methyl group on the phenol unit of phenolphthiocerols.[11] Mycocerosic alcohols and long chain diols can be converted to t-butyldimethylsilyl (TBDMS) ethers for gas chromatographic (GC) and gas chromatographic-mass spectrometric (GC-MS) analyses.[8,24]

ANALYSIS OF LONG-CHAIN COMPONENTS

The mycocerosic acids occur as complex mixtures, varying both in chain length and in the numbers of methyl branches, so their complete analysis by a single chromatographic technique is difficult. The problem derives from the fact that a C_{30}-mycocerosate with four methyl branches, for example, will have retention values similar to those of a C_{29}-mycocerosate with only three methyl side chains. This relationship is clearly demonstrated by plotting retention times against numbers of carbons as shown in Figure 4 for TBDMS ethers of mycocerosic alcohols from the waxes of *M. tuberculosis*. The problem of overlapping mycocerosate peaks is overcome by GC-MS, using either alcohol TBDMS ethers[24] or PFB esters.[21] Reverse-phase HPLC does, however, provide clear separations of mycocerosate PFB esters as shown in Figure 5.

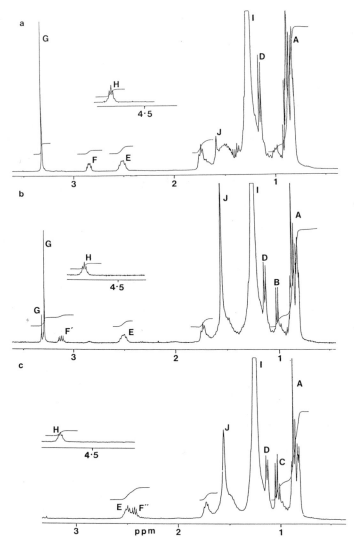

Figure 2. Proton NMR spectra (360 MHz) of phthiocerol-based
waxes. (a), (b) and (c) Dimycocerosates of phthiocerol
A, phthiocerol B and phthiodiolone A, respectively,
from *M. tuberculosis*. Peak assignments: A, methyl
groups; B, $\underline{CH}_3$.CH(OCH$_3$)-; C, $\underline{CH}_3$.CH$_2$.CO-; D,
-CH($\underline{CH}_3$).COO-; E, -C$\underline{H}$(CH$_3$).COO-; F, CH$_3$.CH$_2$.$\underline{CH}$(OCH$_3$)-;
F′,CH$_3$.$\underline{CH}$(OCH$_3$)-; F′′, CH$_3$.$\underline{CH}_2$.CO.$\underline{CH}$(CH$_3$)-;
G, CH$_3$.CH$_2$.CH(O$\underline{CH}_3$)-; G′, CH$_3$.CH(O$\underline{CH}_3$)-; H, >C$\underline{H}$.OOC-;
I, -C$\underline{H}_2$-; J, $\underline{H}_2$O.

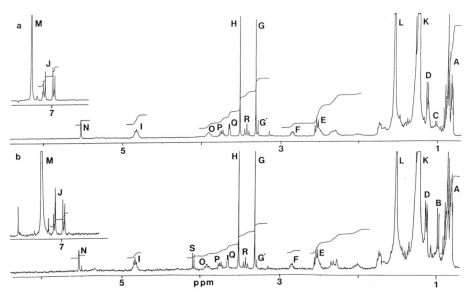

Figure 3. Proton NMR spectra (360 MHz) of phenolic glycolipids. (a) and (b) Dimycocerosates of 2-O-methylrhamnosyl phenolphthiocerols from *M. bovis* and *M. microti*, respectively. Peak assignments: A, methyl groups; B, unknown; C, $\underline{CH}_3$.CH(OCH$_3$)-; D, -CH($\underline{CH}_3$).COO-; E, -$\underline{CH}$(CH$_3$).COO-; F, CH$_3$.CH$_2$.$\underline{CH}$(OCH$_3$)-; G, CH$_3$.CH$_2$.CH(O$\underline{CH}_3$)-; G', CH$_3$.CH(O$\underline{CH}_3$)-; H, -O$\underline{CH}_3$ (sugar); I, >$\underline{CH}$.OOC-; J, aromatic ring; K, -$\underline{CH}_2$-; L, $\underline{H}_2$O; M, C$\underline{H}$Cl$_3$; N, O, P, Q, R, sugar ring protons 1, 3, 5, 2, 4, respectively; S, unknown.

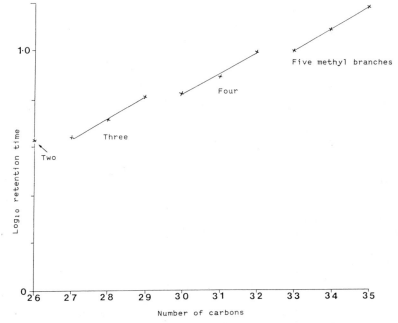

Figure 4. Gas chromatographic behavior of TBDMS ethers of mycocerosic alcohols from *M. tuberculosis*. Samples were analyzed on a Chrompak 25 m CP Sil-19 CB capillary column, isothermal run at 310°C.

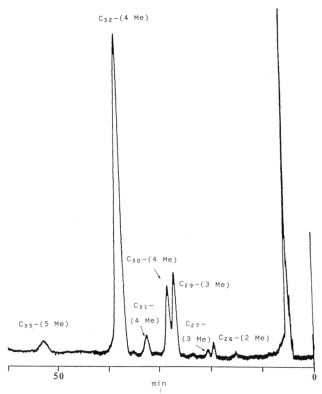

Figure 5.   Reverse phase HPLC of mycocerosic acid
            pentafluorobenzyl esters from *M. tuberculosis*. A
            Waters Radial Pak A cartridge was used with 1.5 mL/min
            tetrahydrofuran-acetonitrile (20:80) and detection at
            263 nm. Peak assignments exemplified by C_{32}-(4 Me)
            representing a 32 carbon mycocerosic acid derivative
            with 4 methyl branches.

    The characteristic profiles of mycocerosates (Table 1) allow the
analyses of these lipids to be used for sensitive detection of disease in
infected tissue, without prior cultivation of the causative agent. In early
studies, mycocerosic acid methyl esters were detected in five-day cultures by
selected ion monitoring mass spectrometry (SIM-MS).[25] Much lower detection
limits are provided by use of electron capture GC and chemical ionization
with negative ion detection MS, pentafluorobenzyl esters of mycocerosates
being ideal for both techniques. Electron capture GC of mycocerosate PFB
esters has been shown to be ideal for the detection of leprosy by direct
analysis of skin biopsies.[22] The potential of negative ion GC-MS for
detecting tuberculosis has been demonstrated.[21] Mycocerosate PFB esters give
spectra in which the only significant peak corresponded to the carboxylate
anion. This latter method has been applied to sputum samples from
tuberculosis patients and an example of a typical result is shown in Figure
6.

    The analysis of the long-chain diol components has been approached in
various ways. Early studies[26] showed that ethylidene acetals of the
phthiocerol family gave informative mass spectra but the complex breakdown
patterns are not ideal for sensitive detection using selected ion monitoring
MS. Di-TBDMS ethers of phthiocerols give excellent mass spectra since they

144

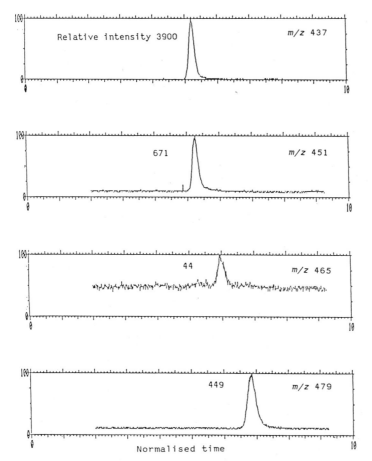

Figure 6. Selected ion monitoring negative ion chemical
ionization of mycocerosic acid pentafluorobenzyl
esters from a sputum sample collected from a
smear-positive tuberculosis patient (courtesy of
Dr. P. A. Jenkins, University Hospital of Wales,
Cardiff, UK). Sample preparation and analytical
conditions (see reference 21). The three most
abundant ions, monitored, correspond to carboxylate
anions of the major mycoserates from *M. tuberculosis*.

contain intense rearrangement ions formed by interaction of the adjacent
TBDMS ethers.[24] Selected ion monitoring of these rearrangement ions gives
good mass spectra, providing the overall size and structure of these
molecules. Members of the phenolphthiocerol family with the phenol group
methylated also produce good mass spectra as their di-TBDMS derivatives.[10]
To exploit sensitive methods for the detection of members of the phthiocerol
and phenolphthiocerol families, alternative derivatives are necessary.
Pentafluorobenzylidene acetals are attractive since they are readily prepared
by acid-catalyzed reaction with pentafluorobenzaldehyde and have potential
for electron capture GC and negative ion chemical ionization GC-MS. The GC
behavior is satisfactory but, on mass spectrometry, the negative charge is
concentrated in an ion at m/z 196, corresponding to loss of
pentafluorobenzaldehyde.[21] Nevertheless, this derivative can be used for
sensitive detection of phthiocerols but, since all components provide the
same diagnostic ion, the identification of characteristic patterns depends on
the GC retention data. A promising lead has been provided by preparing
acetals from 4-nitrobenzaldehyde.[27] These derivatives give good mass
spectra, as shown (Figure 7) for one of the homologues of phthiocerol A.
Nitro-derivatives are not ideal for GC, being relatively polar and subject to
variations in sensitivity with different injection systems. The search must
continue for better derivatives of members of the phthiocerol and
phenolphthiocerol families.

CONCLUSIONS

     The lipids described here are specific for the organisms that produce
them. Their analysis is important to be able to understand their role in the
cell envelope and they are also reliable chemotaxonomic markers, allowing
sensitive, reliable, detection of the parent bacterium. To exploit these
lipids, it is necessary to develop simple analytical methods suitable for
transfer to clinical laboratories. The special simple lipid extraction
procedures, outlined here for these lipids, allow multiple samples to be
processed easily. The semi-quantitative TLC systems (Figure 1) allow the
relative amounts of each lipid to be assessed. These methods can demonstrate
whether or not the molecules are reliable markers for a range of
representatives of a particular taxon. If it is necessary to degrade a
particular lipid, care should be taken to ensure that the method is simple
and efficient. The degradation of the lipids under study here are a good
example of this point. Again, time spent in choosing and developing good

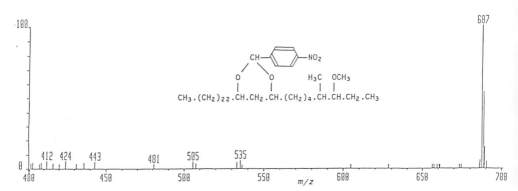

Figure 7. Negative ion chemical ionization mass spectrum of one
           of the two major components of the phthiocerol A
           4-nitrobenzylidene acetals from *M. tuberculosis*.
           The dominant peak corresponds to the molecular ion
           (m/z 687).

derivatization and analytical methods will pay dividends in producing efficient protocols. Exploiting modern sensitive methods such as electron-capture GC and negative ion chemical ionization GC-MS will provide increasingly important information on lipid structure and occurrence in microorganisms.

ACKNOWLEDGEMENTS

Support is acknowledged from the Medical Research Council (G8216538), the British Leprosy Relief Association (LEPRA), and the WHO Programme for Vaccine Development.

REFERENCES

1. D. E. Minnikin, Lipids: Complex lipids, their chemistry, biosynthesis and roles, in: "The Biology of the Mycobacteria," C. Ratledge and J. L. Stanford, eds., Academic Press, London, p. 95 (1982).
2. D. E. Minnikin, Chemical targets in the cell envelope of the leprosy bacillus and related mycobacteria, in: "Chemotherapy of Tropical Diseases," M. Hooper, ed., Wiley, Chichester, p. 19 (1987).
3. P. J. Brennan, New-found glycolipid antigens of mycobacteria, in: "Microbiology-1984," L. Lieve and D. Schlessinger, eds., American Society for Microbiology, Washington, D.C., p. 366 (1984).
4. M. Ridell, D. E. Minnikin, J. H. Parlett, and I. Mattsby-Baltzer, Detection of mycobacterial lipid antigens by a combination of thin-layer chromatography and immunostaining, Lett. Appl. Microbiol. 2:89 (1986).
5. D. E. Minnikin, M. Ridell, I. Mattsby-Baltzer, and J. H. Parlett, Thin-layer chromatography in combination with immunostaining for analysis of mycobacterial lipid antigens, in: "Topics in Lipid Research: from Structural Elucidation to Biological Function," R. Klein and B. Schmitz, eds., Royal Society of Chemistry, London, p. 144 (1986).
6. D. E. Minnikin, M. Ridell, J. H. Parlett and R. C. Bolton, Direct detection of *Mycobacterium tuberculosis* lipid antigens by thin-layer chromatography, FEMS Microbiol. Lett. 48:175 (1987).
7. J. J. Fournie, M. Riviere, and G. Puzo, Structural elucidation of the major phenolic glycolipid from *Mycobacterium kansasii*, J. Biol. Chem. 262:3174 (1987).
8. D. E. Minnikin, G. Dobson, M. Goodfellow, M. Magnusson, and Ridell, M., Distribution of some mycobacterial waxes based on the phthiocerol family, J. Gen. Microbiol. 131:1375 (1985).
9. S. W. Hunter, T. Fujiwara, and P. J. Brennan, Structure and antigenicity of the major specific glycolipid antigen of *Mycobacterium leprae*, J. Biol. Chem. 257:15072 (1982).
10. G. Dobson, D. E. Minnikin, G. S. Besra, A. I. Mallet, and M. Magnusson, Characterization of phenolic glycolipids from *Mycobacterium marinum*, Biochim. Biophys. Acta, submitted (1989).
11. D. E. Minnikin, G. Dobson, J. H. Parlett, M. Goodfellow, and M. Magnusson, Analysis of dimycocerosates of glycosyl phenolphthiocerols in the identification of some clinically significant mycobacteria, Eur. J. Clin. Microbiol. 6:703 (1987).
12. M. Daffe, C. Lacave, M. A. Laneelle, and G. Laneelle, Structure of the major triglycosyl phenolphthiocerol of *Mycobacterium tuberculosis* (strain *Canetti*), Eur. J. Biochem. 167:155 (1987).
13. M. Daffe, M. A. Laneelle, C. Lacave, and G. Laneelle, Monoglycosyldiacylphenol-phthiocerol of *Mycobacterium tuberculosis* and *Mycobacterium bovis*, Biochim. Biophys. Acta, 958:443 (1988).
14. J. J. Fournie, M. Riviere, and G. Puzo, G., Absolute configuration of the unique 2,6-dideoxy-4-O-methyl-arabino-hexapyranose of the major

phenolic glycolipid antigen from *Mycobacterium kansasii*, Eur. J. Biochem. 168:181 (1987).

15. M. Riviere, J. J. Fournie, and G. Puzo, A novel mannose containing phenolic glycolipid from *Mycobacterium kansasii*, J. Biol. Chem., 262:14879 (1987).

16. D. E. Minnikin, A. G. O'Donnell, M. Goodfellow, G. Alderson, G. Athalye, A. Schaal, and J. H. Parlett, An integrated procedure for the extraction of bacterial isoprenoid quinones and polar lipids, J. Microbiol. Meth. 2:233 (1984).

17. G. Dobson, D. E. Minnikin, S. M. Minnikin, J. H. Parlett, M. Goodfellow, M. Ridell, and M. Magnusson, Systematic analysis of complex mycobacterial lipids, in: "Chemical Methods in Bacterial Systematics," M. Goodfellow, and D. E. Minnikin, eds., Academic Press, London p. 237. (1985).

18. E. G. Bligh, and W. J. Dyer, A rapid method for total lipid extraction and purification, Can. J. Biochem. Physiol. 37:911 (1959).

19. D. E. Minnikin, Isolation and purification of mycobacterial wall lipids, in: "Bacterial Cell Surface Techniques," I. Hancock and I. Poxton, eds., Wiley, Chichester, p. 125 (1988).

20. P. Draper, S. N. Payne, G. Dobson, and D. E. Minnikin, Isolation of a characteristic phthiocerol dimycocerosate from *Mycobacterium leprae*, J. Gen. Microbiol. 129:859 (1983).

21. D. E. Minnikin, R. C. Bolton, G. Dobson, and A. I. Mallet, Mass spectrometric analysis of multimethyl branched fatty acids and phthiocerols from clinically-significant mycobacteria, Proc. Jap. Soc. Med. Mass Spectrom., 12:23 (1987).

22. D. E. Minnikin, G. Dobson, K. Venkatesan, A. K. Datta, J. H. Parlett, and R. C. Bolton, Lipid analysis in the detection of the leprosy bacillus in infected tissue, Quaderni di Cooperazione Sanitaria (Health Cooperation Papers), 7:211 (1988).

23. M. Daffe, M. A. Laneelle, J. Roussel, and C. Asselineau, Lipides specifique de *Mycobacterium ulcerans*, Ann. Microbiol. 135A:191 (1984).

24. A. I. Mallet, D. E. Minnikin, and G. Dobson, Gas chromatography mass spectrometry of tert-butyldimethylsilyl ethers of phthiocerols and mycocerosic alcohols from *Mycobacterium tuberculosis*, Biomed. Mass Spectrom. 11:79 (1984).

25. L. Larsson, P. A. Mardh, G. Odham, and G. Westerdahl, Use of selected ion monitoring for detection of tuberculostearic and C_{32} mycocerosic acid in mycobacteria and in five-day-old cultures of sputum specimens from patients with pulmonary tuberculosis, Acta Path. Microbiol, Scand. Sect. B 89:245 (1981).

26. D. E. Minnikin and N. Polgar, Studies relating to phthiocerol. Part V. Phthiocerol A and B, J. Chem. Soc. C:2107 (1966).

27. A. I. Mallet, Analysis of lyso-platelet activating factor by negative ion gas chromatography-mass spectrometry of the nitrobenzyl acetal derivatives, Biomed. Environ. Mass Spectrom., 16:207 (1988).

CHAPTER 10

STRUCTURAL ELUCIDATION OF LIPOPOLYSACCHARIDES AND THEIR LIPID A COMPONENT:
APPLICATION OF SOFT IONIZATION MASS SPECTROMETRY

Buko Lindner, Ulrich Zahringer, Ernst Th. Rietschel, and
Ulrich Seydel

Forschungsinstitut Borstel
Institut fur Experimentelle Biologie und Medizin
D-2061 Borstel, FRG

INTRODUCTION

Gram negative bacteria express at their surface various amphiphilic
macromolecules among which the lipopolysaccharides (LPS) are of special
biomedical significance. They represent the endotoxins of Gram negative
bacteria eliciting in higher organisms typical pathophysiological effects
such as fever, hypotension, dermal skin necrosis (local Shwartzman
phenomenon) and irreversible shock and they represent the O-antigens,
determining the serospecificity of LPS and of the bacteria containing them.[1]
Chemically, LPS consist of a polysaccharide or oligosaccharide portion
covalently linked to a lipid component, termed lipid A which anchors the
lipopolysaccharide in the outer leaflet of the outer cell membrane. The
polysaccharide part consists of an O-specific chain which expresses high
structural variability and a core oligo-saccharide which is less variable or
structurally identical for many different Gram negative bacteria. The
chemical structure of lipid A shows remarkable conformity for all members of
the Enterobacteriaceae family and many other Gram negative bacteria. It
represents the most conservative LPS region and consists of β-D-glucosaminyl-
(1-6)-α-D-glucosamine disaccharide which is phosphorylated in positions 4' of
the nonreducing glucosaminyl residue (GlcN II) and in position 1 of the
reducing glucosaminyl group (GlcN I). This hydrophilic backbone carries in
ester and amide linkage up to seven hydroxylated and nonhydroxylated
saturated fatty acid residues with chain lengths typically between ten and
sixteen carbon atoms[2,3]. Lipid A represents the endotoxic principle of LPS
being responsible for all pathophysiological effects of endotoxins. It was
and still is of particular interest to find out whether the biological
activity of lipid A's of various origin can be attributed to a common
chemical composition and whether a minimal endotoxically active structure can
be defined. Furthermore, a correlation between the chemical structure,
particularly the number and type of fatty acid residues and the degree of
phosphorylation on the one hand, and the physical properties of LPS membrane
systems on the other hand had been demonstrated. Thus, the thermotropic
phase behavior, i.e., the various three-dimensional supramolecular phase
states and the state of order adopted under physiological conditions was
shown to depend on these chemical parameters.[3]

Analytical Microbiology Methods
Edited by A. Fox *et al.*
Plenum Press, New York, 1990

In general, the determination of the chemical structure of the various components of LPS, and in particular of lipid A, is hampered by its intrinsic heterogeneity and the limitations of the available analytical methodology. Heterogeneity is a problem because purification of the amphiphatic LPS species without prior chemical modification is extremely difficult and the results may not be interpretable qualitatively. Conventional structural analysis involves chemical procedures as, e.g., methylation analysis in combination with gas-liquid chromatography/mass spectrometry (GLC-MS). This approach gives information on the molar ratios of and the linkage between the various structural components of LPS. In the case of fatty acid residues these techniques allow statements as to the nature and number of acyl groups, whether these are ester- or amide-linked and whether hydroxy fatty acids are substituted at their hydroxyl groups. With these methods, however, it can not be decided how fatty acids are distributed over the available hydroxyl and amino groups. Thus, it is not possible to definitely determine the number and type of fatty acids linked to GlcN I or GlcN II, nor to detect heterogeneity in the fatty acid distribution caused by non-stoichiometric substitution or statistically varying positions of the acyl chains. Two dimensional proton magnetic resonance spectroscopy (2D-NMR) notably yields information concerning the location of fatty acids, but not on the precise distribution of 3-acyloxyacyl groups on the lipid A backbone.

The introduction of soft-ionization mass-spectrometric techniques, such as fast atom bombardment (FAB) and laser desorption (LD) mass spectrometry to the analysis of LPS and lipid A offered a way to solve some of the problems unaccessible to the conventional analytical methods. Further, soft ionization mass spectra contained information on the different components in crude preparations of LPS, lipid A's, and oligosaccharides, particular on the molecular weights of the intact molecules and, thus, offered first indications on intrinsic biological heterogeneities.

In the present article a short description of the main principles of the soft ionization mass-spectrometric techniques FAB-MS and LD-MS is given and their potential for the structural elucidation of LPS and LPS subunits are discussed on the basis of some selected examples. Some further discussion on the basis of soft ionization MS techniques can be found in Chapter 2.

SOFT IONIZATION MASS SPECTROMETRY

The mass spectral analysis of large organic molecules, notably of highly polar compounds, involves a number of problems arising from thermolability and nonvolatility. Some of these problems could be overcome after the introduction of new, so called soft ionization techniques such as fast atom bombardment (FAB), secondary ion mass spectrometry (SIMS), californium plasma desorption (^{252}Cf-PD), field desorption (FD) and laser desorption (LD). By the bombardment with energetic particles (FAB, SIMS, ^{252}Cf-PD), the application of high electric fields (FD), the disruption of liquid surfaces (thermospray), or by irradiation with a laser beam (LD) the molecules are transferred directly (desorbed) from the condensed phase (solid or liquid phase) into the gas phase and, following this process, a certain percentage is ionized. The required energy is, in all these processes, applied in such a way that thermal stress of the compounds is largely avoided. Here, only those soft ionization techniques will be described which are most commonly utilized for the structural elucidation of LPS and LPS part structures, i.e., FAB and LD-MS. For more detailed information on these techniques the reader is referred to some recent reviews.[5,6] More recently, ^{252}Cf-PD-MS has also been applied for this purpose.[7]

Fast Atom Bombardment (FAB) or Liquid Secondary Ion Mass Spectrometry (liquid SIMS)

A differentiation between FAB and SIMS originates from the character of the primary particles: bombardment with KeV atoms (e.g., Ar) or with KeV ions (e.g., Cs^+). It appears that the charge of the incident particles is not of prime importance for biological molecules analyzed out of a liquid matrix, rather, the most important factor for biological samples seems to be the nature of the liquid matrix in which it is dispersed. Therefore, it was recommended to use the term "liquid SIMS" to describe both techniques.[8]

The liquid matrix provides several advantages: (1) high and long-lasting secondary molecular ion signals are obtained from the dispersed sample by the application of high incident particle fluxes because this sputtering process continuously leads to the renewal of the surface and thus to the supply of intact sample molecules; (2) abundant molecular ion signals are recorded even from thermolabile large molecules; (3) knowledge of solvent chemistry and ion chemistry in solution can be utilized to control the intensity and type of molecular ions formed, the type and nature of fragments, and the degree of fragmentation, to a large extent.[9] In particular, glycerol, thioglycerol and glycol have proven to be favorable matrix components because of their low vapor pressure and good solvent properties for biomolecules. Additives may serve to enhance molecular or (due to the attachment of cations) quasimolecular ion yields or to overcome charge effects arising from highly polar substances. Thus far, improvements of the matrices via its chemical composition have been made empirically, but it is to be expected that the increasing knowledge of the sputtering processes and the availability of physical constants of the matrix fluids will allow their predictive development for a particular problem.[10] Because the positively or negatively charged ions are produced continuously, registration of the ions by high resolution mass spectrometers is possible.

Laser Desorption

In LD-MS the sample is irradiated by a laser either from the opposite side of the detection system (transmission-type instruments) or from the same side (reflection-type instruments). In contrast to the situation encountered with liquid SIMS, a large variety of different ionization sources, i.e., laser systems, are utilized, varying as well in wavelength (from IR to UV) as in pulse duration (from continuous wave to picosecond pulses), in size of the irradiated sample area (from several mm^2 to μm^2), and in laser irradiance (from approximately 10^5 to 10^{12} W cm^{-2}).[11] This large range in laser irradiance and geometries leads to different physical processes during laser-sample interaction which, again, cause different fragmentation pathways. For applications relevant to LPS analysis exclusively pulsed ion sources have been used. This implies a combination with either Fourier-transform (FT) or time-of-flight (TOF) mass spectrometers, the latter offering only a limited mass resolution.

For the analysis of LPS and lipid A two rather different laser ionization sources have been used, a CO_2-pulse laser (pulse duration 40 ns, effective wavelength 10.6 μm, diameter of irradiated sample area 1 mm, irradiance 10^5 W cm^{-2}, reflection type geometry)[12] and a Nd YAG-pulse laser (pulse duration 10 ns, effective wavelength 265 nm, diameter of irradiated area 5 μm, irradiance typically 10^{11} W cm^{-2}, transmission type geometry).[13] The small laser focus diameter of the latter system is based on the fact that it is part of a laser microprobe mass analyzer (LAMMA 500, Leybold-Heraeus, Koln, W. Germany), which was used for organic mass spectrometry, an application other than originally intended for this instrument.

Mass spectrometric techniques contributed valuable information to the structural elucidation of different components of LPS from various species of *Enterobacteriacae* including free lipid A and partial structures (e.g., biosynthetic precursors, synthetic lipid A analogues), oligosaccharides of the O-specific chain and the core region, and complete (deep rough mutant) LPS. Problems which may arise from negative charges of the phosphate groups present in lipid A and, 3-deoxy-*D-manno*-octulosonic acid (KDO) containing inner core, or of sugar acids can be solved either by removal of phosphate groups (e.g., by hydrogen fluoride or hydrochloric acid) or by blocking negative charges via derivatization (e.g., by methylation). By applying different hydrolysis conditions different numbers of phosphate groups can be removed in a controlled fashion. It can not be excluded that under certain (strong) hydrolysis conditions the molecular structure is further changed, e.g., by cleavage of ester-bound fatty acids or even by breaking the glycosidic bond between GlcN II and GlcN I. These structural changes can, on the other hand, yield valuable additional structural informations.

Highly purified (HPLC or TLC) fractions of methylated mono- or bisphosphoryl and, in one case, also of nonmethylated bisphosphoryl lipid A's of various bacterial origin and of lipid A precursors have been analyzed using liquid SIMS in the positive and in the negative ion mode.[14-19] Both modes gave molecular ions allowing the exact determination of the molecular weights. By this procedure the chemical analyses of chromatographic fractions could easily be confirmed, showing that they differed in the number and/or type of the fatty acid substituents. Diagnostic cleavage occurred at the glycosidic bond yielding an oxonium ion in the positive ion mode which represents the distal glucosamine subunit (GlcN II). The fact that the oxonium ion exclusively derives from GlcN II allowed the determination of the distribution of fatty acids over the two glucosamine residues of the lipid A backbone. Further information on the nature of fatty acids could be obtained from additional fragment ion peaks which result from cleavage of acyl groups. A differentiation between ester- and amide-bound fatty acids is possible because amide linkages are cleaved very rarely. Similar observations have been made with LD-MS. The cleavage of ester-bound acyl chains may occur on either side of the connecting oxygen atom leading to a difference of 18 atom mass units (amu) between the corresponding fragment ions.[16] For details concerning the methodology and results the reader is particularly referred to references 16 and 19.

Some highly purified mono- or bisphosphoryl lipid A and LPS from Re mutant methyl esters have been investigated by Cotter and coworkers with a LD-MS instrument.[18,21] By this methodology information on the molecular weight and the fatty acid distribution was obtained but the oxonium ion was not observed as in liquid SIMS. Rather a prominent fragmentation within the sugar rings was described as previously reported.[23] This fragmentation provided direct information on the identity and positions of both ester- and amide-linked fatty acids. These authors proposed three major fragmentation pathways of the reducing sugar moiety: simultaneous cleavages of the C_1-O and C_4-C_5 bonds, cleavages of the C_1-C_2 and C_3-C_4 linkages and separation of the C_4-C_5 and C_1-C_2 bonds.[21]

With this technique lipid A's of *Salmonella thyphimurium*, *Neisseria gonnorhoeae*, *Rhodopseudomonas sphaeroides*, and *Escherichia coli* have been investigated and their acyl pattern elucidated.[16,20-22]

Our group started with the mass spectrometric analysis of nonpurified dephosphorylated free lipid A preparations using the LAMMA instrument.[24,25] Figure 1 shows the positive ion LD-mass spectrum of free lipid A of *E. coli* (strain F515) mixed with KCl. Free lipid A was obtained from LPS by hydrogen

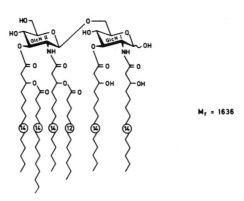

$M_r = 1636$

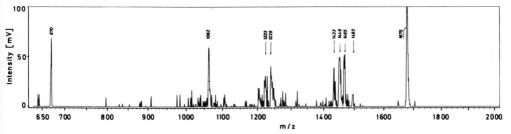

Figure 1. Chemical structure and positive ion LD-mass
spectrum of dephosphorylated free lipid A of
E. coli strain F 515 LPS mixed with KCl.
Reproduced with permission from reference 24.

fluoride hydrolysis (H_2F_2, 48%, 4°C, 48 h) and subsequent acid treatment (0.1
M HCl, 100°C, 1 h). The precipitate obtained was dialyzed (H_2O) and
lyophilized. The spectrum contains, beside the abundant quasi-molecular peak
$(M + K)^+$ at m/z 1675, two further main other peaks at m/z 670 and 1062, which
could be assigned to the GlcN I and GlcN II, respectively, each residue being
acylated. Additional peaks at m/z 1493, 1465, 1449, and 1239 obviously
correspond to chemical structures which lack one or two fatty acid residues
with respect to the quasimolecular ion (m/z 1675). The decision as to
whether an ion signal originates from fragmentation or from a priori present
molecular ion species is difficult because it is not possible to
unequivocally differentiate between, e.g., cleavage of a tetra-decanoic acid
residue between the ester oxygen and GlcN or of a 3-hydroxytetradecanoic acid
group between the oxygen and the carbonyl group of the hydrocarbon chain.
The reason for this may be found in the limited mass resolution of the
applied TOF analyzer and in the uncertainty of the mass scale calibration of
approximately 1 amu. It must be pointed out that the spectrum lacks peaks
deriving from sugar ring fragmentations of the kind described[21] or from the
oxonium ion of the nonreducing lipid A subunit (Glcn II). Nevertheless,
essential features of the chemical structure of *E. coli* lipid A can be
deduced from this spectrum.

In addition, information about the intrinsic biological heterogeneity
is obtained. An example for this is given in Figure 2 showing the positive
ion LD-mass spectra of dephosphorylated free lipid A of *Chromobacterium
violaceum* (A) and of *Xanthomonas sinensis* (B).[26] The strong quasimolecular
ions $(M + K)^+$ at m/z 1479 and 1495 (A), and at m/z 1479, 1495, 1511, 1525,

1539, 1553, and 1567 (B) represent glucosamine disaccharide acylated with 6 moles (per 2 moles GlcN) of fatty acids. The other peaks correspond to disaccharides with four or five fatty acids (e.g., m/z 1297 (A) and m/z 1357, 1173 (B)) or represent acylated glucosamine monosaccharides (m/z 768 and 784 (A) and peak assemblies around m/z 798 and 614 (B)). The great number of quasimolecular ions of *X. sinensis* hexaacyl lipid A is indicative for the expresses high heterogeneity concerning the type of acyl residues. With the mass spectrometric analysis of *C. violaceum* (Figure 2A) we could show for the first time the existence of a symmetrically acylated lipid A carrying three fatty acid residues at each glucosamine residue. The relative low heterogeneity of this lipid A could be explained by the presence of only molecular species differing in one hydroxyl group (a dodecanoic acid is replaced by a 2-hydroxydodecanoic acid).

Similarly, the influence of the chemical pretreatment of an LPS to prepare free lipid A or of ambient conditions like growth temperature on the fatty acid composition can be studied using LD-MS.[24,27] As an example Figure

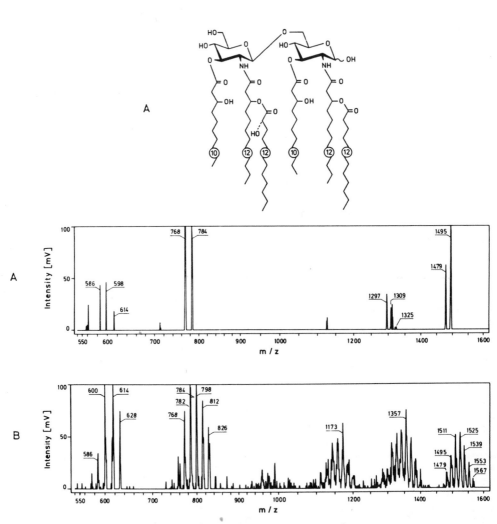

Figure 2. Positive ion LD-mass spectrum of dephosphorylated free lipid A of *C. violaceum* (A) and of *X. sinensis* (B) mixed with KCl. Reproduced with permission from reference 26.

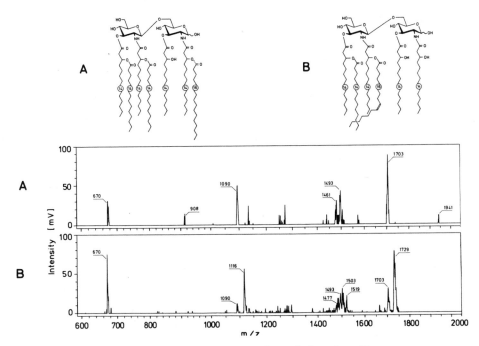

Figure 3. Chemical structure and positive ion LD-mass
spectra of dephosphorylated free lipid A of
P. mirabilis strain R 145 grown at 37°C (A) and
at 12°C (B) mixed with KCl. Reproduced with
permission from reference 27.

3 shows the positive ion LD-mass spectrum of dephosphorylated free lipid A
obtained from LPS of *Proteus mirabilis* grown at 37°C (A) and at 12°C (B),
respectively. Comparison of the two spectra shows that in the 12°C
preparation additional abundant mass peaks are present at m/z 1116
(1090 + 26) and at m/z 1729 (1703 + 26). These peaks can be interpreted as
exchange of a tetradecanoic (present in a 37°C preparation) by hexadecenoic
acid (Δ 9-cis-16:1). From the fact that the peak at m/z 670 (GlcN I)
remains unaffected it can be concluded that this alteration in fatty acid
composition is restricted to GlcN II. The spectra furthermore show that the
heptaacyl lipid A species (expressed in the 37°C preparation as ion signal at
m/z 1941) and the respective acylated monosaccharide at m/z 908 are missing
in the 12°C preparations.

Detailed studies of the parameters influencing the fragmentation under the
particular conditions of laser-sample interaction in the LAMMA instrument[28,29]
have shown that from purified bacterial lipid A or from synthetic lipid A
preparations and lipid A analogues preferentially the quasimolecular ions are
produced when small amounts of Cs- or K-salts (e.g., CsI, KI) are added. The
admixture of larger amounts of Li- or Na-salts (LiI, NaI) leads to the
generation of fragment ions including those from sugar ring fragmentation and
to the cleavage of ester-bound fatty acid residues allowing the assignment of
the reducing and non-reducing sugar, respectively, and of the distribution of
the fatty acid residues on these.[30,31] This is demonstrated for the synthetic
phosphate-free *E. coli* type lipid A (compound 503, T. Shiba and S. Kusumoto,
Univ. of Osaka, Japan) in Figure 4. The different influence of the nature
and amount of the added cation species can be explained by the different
binding energies and affinities of the cations to the molecule.[32]

LD-MS as utilized by the LAMMA instrument has also been applied to the structural elucidation of lipid A precursors[33], of core oligosaccharide partial structures and of complete LPS.[34] Figure 5 gives the chemical structure and the positive ion LD-mass spectrum of a highly purified de-O-acylated 1-dephospho LPS of a *E. coli* Re-mutant (F 515) after methylation and peracetylation. The mass spectrum shows the abundant quasimolecular peak at m/z 2049 and supports the chemical structure derived by analytical work.[35]

In Figure 6 an example is given for the structural analysis of the core region of *Acinetobacter calcoaceticus* (NCTC 10305) LPS, a member of the *Neisseriaceae*. The positive ion LD-mass spectrum of the acid-degraded,

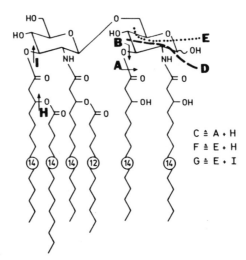

$$C \triangleq A + H$$
$$F \triangleq E + H$$
$$G \triangleq E + I$$

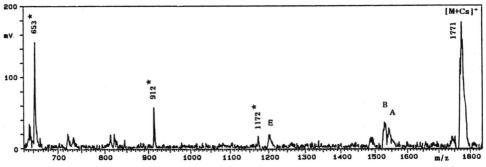

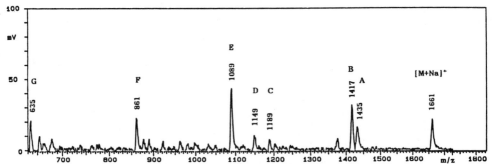

Figure 4. Chemical structure with fragmentation pathways and the positive ion LD-mass spectra of synthetic phosphate-free *E. coli* lipid A (compound 503) mixed with CsI (top) or NaI (bottom).

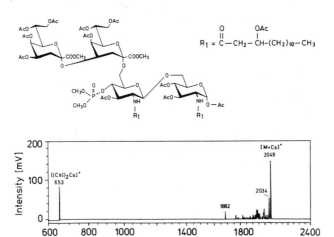

Figure 5. Chemical structure and positive ion LD-mass spectrum of de-O-acylated 1-dephospho LPS of the *E. coli* Re-mutant F 515 after methylation and peracetylation mixed with CsI.

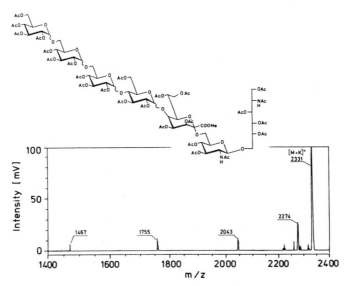

Figure 6. Chemical structure and positive ion LD-mass spectrum of a heptasaccharide-derivative of LPS of *A. calcoaceticus* mixed with CsI.

reduced and deacylated LPS (in the carbomethylated and peracetylated form) yielded a quasimolecular $(M + K)^+$ ion at m/z 2331. This value was by 58 amu higher than that which would have been expected if a KDO unit would provide the linkage between core oligosaccharide and lipid A as is the case in other LPS.[3] Chemical analyses then showed that in A. *calcoaceticus* a 2-octulosonic acid positionally replaces the KDO adjacent to the lipid A.[36] The LD-mass spectrum furthermore revealed that the main component of the analyzed compound is a heptasaccharide. The low intensity peaks at m/z 2045, 1757, and 1469 must originate from minor subfractions differing in the number of repeating acetylated glucose units (m/z 288), further supporting a linear structure of the purified derivative. Fragmentation during laser desorption can be excluded because the residues in terminal positions have higher molecular weights due to additional acetyl groups.

To complete the survey over the application of liquid SIMS and LD-MS to the structural elucidation of LPS and lipid A the analysis of fatty acids in the negative ion mode and in the low mass range must be mentioned.[27,37] This mode yields detectable ions only of ester-linked fatty acids, but from a combination of this direct information on the ester-linked acyl chains with the structural data from the high mass region in either the negative or positive ion mode, the complete information on the identity of all fatty acids can be derived. In this way we could, for example, explain the shifts in quasimolecular peaks by 26 amu between the 37°C and the 12°C preparations of P. *mirabilis* free lipid A (see above). One must, however, be aware, of the limitations of this method for a quantification and an identification of isomers. The latter restriction has been overcome[38] with the use of a soft ionization source in combination with an MS-MS-detection system.

CONCLUDING REMARKS

This chapter has concentrated on the description of soft ionization mass spectrometric techniques and their application to the elucidation of the complex structure of bacterial lipopolysaccharides. It has been shown that these techniques can provide information on parameters which are not accessible by chemical analysis, i.e., the molecular weight and the fatty acid distribution. With nonpurified LPS preparations also information on the biological heterogeneity can be obtained. The analysis of LPS, however, represents only a small section out of the wide range of applications of soft ionization techniques in microbiology which include peptide, oligosaccharide, and phospholipid analyses. Increased knowledge on the molecular basis of the desorption/ionization mechanisms and experimental experience with different compounds will, in the future, certainly provide more profound and extensive information. The introduction of new matrix compounds, especially to LD-MS, will help to overcome problems inherent to highly polar substances. It is to be expected that soft ionization mass spectrometry will become an essential tool in the characterization of complex biomolecules supplying complementary information to results obtained by GLC-MS and NMR. As the amount of sample material for the mass spectrometric analysis with either liquid SIMS or LD-MS is extremely low (20 μg or less) these methods provide valuable first information for the aimed application of other techniques and will thus possibly reduce tedious and time-consuming chemical separation and derivatization procedures. Quantitative analyses, however, will remain a domain of the established chemical methodologies.

REFERENCES

1. O. Luderitz, M. A. Freudenberg, Ch. Galanos, V. Lehmann, E. Th. Rietschel, and D. H. Shaw, Lipopolysaccharides of Gram-negative bacteria, in: "Current Topics in Membranes and Transport 17," S. Razien and S. Rottem, eds., Academic Press, New York (1982).

2. E. Th Rietschel, Z. Sidorczyk, U. Zahringer, H.-W, Wollenweber, and O. Luderitz, Analysis of the primary structure of lipid A, in: "Bacterial Lipopolysaccharides: Structure, Synthesis, and Biological Activities, ACS Symposium Series 231," L. Anderson and F.M. Unger, eds. American Chemical Society, Washington, D. C. (1983).

3. E. Th. Rietschel, L. Brade, K. Brandenburg, H. D. Flad, J. Jong-Leuvenink, K. Kawahara, B. Lindner, H. Loppnow, Th. Luderitz, U. Schade, U. Seydel, Z. Sidorczyk, A. Tacken, U. Zahringer, and H. Brade, Chemical structure and biologic activity of bacterial and synthetic lipid A, Rev. Infect. Dis. 9, Suppl. 5:S527 (1987).

4. Th. Luderitz, K. Brandenburg, U. Seydel, A. Roth, C. Galanos, and E. Th. Rietschel, Structural and physicochemical requirements of endotoxins for the activation of arachidonic acid metabolism in mouse peritoneal macrophages in vitro, Eur. J. Biochem. 179:11 (1988).

5. M. Barber, R. S. Bordoli, G. J. Elliott, R. D. Sedgwick, A. N. and Tyler, Fast atom bombardment mass spectrometry, Anal. Chem. 54:645A (1982).

6. R. J. Cotter, Laser mass spectrometry: An overview of techniques, instruments and applications, Anal. Clin. Acta. 196:45 (1987).

7. N. Qureshi, K. Takayama, P. Mascagni, J. Honovich, R. Wong, and R. J. Cotter, Complete structural determination of lipopolysaccharide obtained from deep rough mutant of Escherichia coli, J. Biol. Chem. 263:11971 (1988).

8. A. L. Burlingame, T. A. Baillie, and Derrick, Mass Spectrometry, Anal. Chem. 58:165R (1986).

9. S. S. Wong, K. P. Wirth, and F. W. Rollgen, Sputtering from liquid and solid organic Matrices, in: "Ion formation from organic solids IFOS III, Springer Proceedings in Physics 9," A. Benninghoven, ed., Springer-Verlag, Berlin-New York (1986).

10. C. Fenselau and R. J. Cotter, Chemical aspects of fast atom bombardment, Chem. Rev. 87:501 (1987).

11. F. Hillenkamp, Laser induced ion formation from organic solids, in: "Ion Formation From Organic Solids, Springer Series in Chemical Physics 25," A. Benninghoven, ed., Springer-Verlag, Berlin New-York (1983).

12. R. J. Cotter, and J.-C Tabet, Laser desorption MS for nonvolatile organic molecules, Am. Lab. 16:86 (1984).

13. U. Seydel and B. Lindner, Mass spectrometry of organic compounds (< 2000 amu) and tracing of organic molecules in plant tissue with LAMMA, in: "Ion Formation From Organic Solids, Springer Series in Chemical Physics 25," A. Benninghoven, ed., Springer-Verlag, Berlin-New York (1983).

14. N. Qureshi, K. Takayama, D. Heller, and C. Fenselau, Position of ester groups in the lipid A backbone of LPS obtained from Salmonella typhimurium, J. Biol. Chem. 258:12947 (1983).

15. N. Qureshi, N., P. Mascagni, E. Ritoi, and K. Takayama, Monophosphoryl lipid A obtained from LPS of Salmonella minesota R 595, J. Biol. Chem. 260:5271 (1985).

16. N. Qureshi, R. J. Cotter, and K. Takayama, Application of fast atom bombardment mass spectrometry and nuclear magnetic resonance on the structural analysis of purified lipid A, J. Microbiol. Methods 5:65 (1986).

17. Ch. R. H. Raetz, S. Purcell, M. V. Meyer, N. Qureshi, and Takayama, Isolation and characterization of eight lipid A precursors from a KDO-deficient mutant of Salmonella typhimurium, J. Biol. Chem. 260:16080 (1985).

18. K. Takayama, N. Qureshi, K. Hyver, J. Honovich, R. J. Cotter, D. Mascagni, and H. Schneider, Characterization of structural series of lipid A obtained from the LPS of Neisseria gonorrhoeae, J. Biol. Chem. 261: 10624 (1986)

19. J. Peter-Katalinic, U. Zahringer, and H. Egge, Identification of phosphate substitution in lipopolysaccharide from Escherichia coli

and *Salmonella minnesota* by fast atom bombardment mass spectrometry, XIVth Int. Carbohydrate Symposium (Stockholm, Sweden) A3:23 (1988).

20. K. Takayama, and N. Qureshi, Structures of lipid A, its precursors, and derivatives, Microbiology 1986:5 (1986).

21. R. J. Cotter, J. Honovich, N. Qureshi, and K. Takayama, Structural determination of lipid A from gram negative bacteria using laser desorption mass spectrometry, Biomed. Environ. Mass Spectrom. 14:591 (1987).

22. N. Qureshi, J. Honovich, H. Hara, and R. J. Cotter, Location of fatty acids in lipid A from lipopolysaccharide of *Rhodopseudomonas sphaeroides* ATCC 17023, J. Biol. Chem. 263:5502 (1988).

23. M. L. Coates, and C. L. Wilkins, Laser desorption Fourier transform mass spectra of malto-oligosaccharides, Biomed. Mass Spectrom. 12:424 (1985).

24. U. Seydel, B. Lindner, H.-W Wollenweber, and E.-Th Rietschel, Structural studies on the lipid A component of enterobacterial lipopolysaccharides by laser desorption mass spectrometry. Location of acyl groups at the lipid A backbone, Eur. J. Biochem. 145:505 (1984).

25. U. Seydel, B. Lindner, U. Zahringer, E. Th. Rietschel, S. Kusumoto, and T. Shiba, Laser desorption mass spectrometry of synthetic lipid A-like compounds, Biomed. Mass Spectrom. 11:132 (1984).

26. H. -W., Wollenweber, U. Seydel, B. Lindner, O. Luderitz and E. Th. Rietschel, Nature and location of amide-bound (R)-3-acyloxyacyl groups in lipid A of LPS from various gram-negative bacteria, Eur. J. Biochem. 145:265 (1984).

27. B. Lindner and U. Seydel, Laser desorption mass spectrometry of complex biomolecules at high laser power density, in: "Secondary Ion Mass Spectrometry, Springer Series in Chemical Physics 36, A. Benninghoven," J. Okano, R. Shimizu, and H.W. Werner, eds., Springer-Verlag, Berlin New-York (1984).

28. B. Lindner and U. Seydel, Laser desorption mass spectrometry of nonvolatiles under shock wave conditions, Anal. Chem. 57: 895 (1985).

29. B. Lindner and U. Seydel, On different desorption modes in LDMS's, in: "Ion formation from organic solids IFOS III, Springer Proceedings in Physics 9," A. Benninghoven, ed., Springer-Verlag, Berlin-New York (1986).

30. I. M. Helander, B. Lindner, H. Brade, K. Altmann, A. A. Lindberg, E. Th. Rietschel, Chemical structure of the lipopolysaccharide of *Haemophilus influenzae* strain I-69 Rd⁻/b⁺, Eur. J. Biochem. 177:483 (1988).

31. J. H. Kraus, U. Seydel, J. Weckesser, and H. Mayer, Structural analysis of the nontoxic lipid A of *Rhodobacter capsulatus* 37b4, Eur. J. Biochem. 180:519 (1989).

32. L. M. Mallis and D. H. Russell, Fast atom bombardment-tandem MS of organo-alkali metal ions of small peptides, Anal. Chem. 58:1076 (1986).

33. Th. Hansen-Hagge, V. Lehmann, U. Seydel, B. Lindner, and U. Zahringer, Isolation and structural analysis of two lipid A precursors from a KDO deficient mutant of *Salmonella typhimurium* differing in their hexadecanoic acid content, Arch. Microbiol. 141:353 (1985)

34. U. Seydel, B. Lindner, and U. Zahringer, LDMS of natural membrane lipids: on the origin of "fragments", in: "Advances in mass spectrometry 1985, Part B," J. F. J. Todd, ed., John Wiley, Chichester-New York (1985).

35. U. Zahringer, B. Lindner, U. Seydel, E. Th. Rietschel, H. Naoki, F. M. Unger, M. Imoto, S. Kusumoto, and T. Shiba, Structure of de-O-acylated LPS from *Escherichia coli* Re mutant-stain F 515, Tetrahedron Letters 26:6321 (1985).

36. K. Kawahara, H. Brade, E. Th. Rietschel, and U. Zahringer, Studies of the chemical structure of the core lipid A region of the LPS of *Acinetobacter calcoaceticus* NCTC 10305, Eur. J. Biochem. 163: 489 (1987).

37. R. C. Seid, W. M. Bone, and L. R. Phillips, Identification of ester-linked fatty acids of bacterial endotoxin by negative ion fast atom bombardment mass spectrometry, Anal. Biochem. 155:168 (1986).

38. K. B. Tomer, N. J. Jensen, and M. L. Gross, Fast atom bombardment and tandem mass spectrometry for determining structural modification of fatty acids, Anal. Chem. 58: 2429 (1986).

CHAPTER 11

STRUCTURAL DETERMINATION OF UNSATURATED LONG CHAIN FATTY ACIDS FROM
MYCOBACTERIA BY CAPILLARY GAS CHROMATOGRAPHY AND COLLISION ACTIVATION
DISSOCIATION MASS SPECTROMETRY

Jean-Claude Prome, Helene Aurelle, Francois Couderc,
Daniele Prome, Arlette Savagnac, and Michel Treilhou

Centre de Recherche de Biochimie et Genetique Cellulaires
C.N.R.S.
118 route de Narbonne
31062 Toulouse Cedex, France

INTRODUCTION

Mycobacteria synthesize considerable quantities of a wide variety of
fatty acids. In the mycolic acid series, for example, fatty acids containing
over 80 carbon atoms have been detected. They represent the major lipid
components of the cell walls of these organisms, and many studies have been
devoted to their characterization. The shorter chain fatty acids (below C_{40})
have also been studied extensively, especially the branched chain series.[1]
In fact mycobacteria produce highly complex mixtures of fatty acids. In
Mycobacterium phlei, for example, more than 40 different fatty acids from C_{12}
to C_{36} have been detected by gas chromatography-mass spectrometry (GC-MS).[2]

Although it is not too difficult to characterize the saturated fatty
acids by GC-MS using electron impact (EI) ionization, the localization of
double bonds, cyclopropane rings or branch points is more complex. EI
isomerizes double bonds and cyclopropane rings, making their spectra hard to
resolve.

A widely used strategy is to introduce heteroatoms onto these
functional groups by chemical means before subjecting them to GC-MS analysis.
The fragmentation pattern is markedly influenced by the presence of these
heteroatoms, and the positions of the ring structures and double bonds can be
deduced from the characteristic EI spectra. However, the various methods of
derivatization have a number of drawbacks. First, chemical reactions with
rather unreactive functional groups are involved, which in some cases require
several stages. Moreover, these reactions must be carried out quantitatively
on a micro-scale, which is not always easy to achieve. The reactions may
also favor particular components of a mixture. In addition, secondary
reactions can arise if double bonds and carboxyl groups are close to each
other. Second, the orientation of the fragmentation by the new group can
mask other details, such as branching of the aliphatic chain. Some authors
have attempted to transform the double bonds into functional groups in the
gas phase by ion-molecule reactions. Cycloaddition using methyl vinyl
ether,[3,4] addition of NO^+,[5] or complexation by Fe^+,[6,7] have been described.

Analytical Microbiology Methods
Edited by A. Fox *et al.*
Plenum Press, New York, 1990

The spectra obtained are generally complex, and the reagents employed may contaminate the ion source or attack the filament. A recent innovation is a combination of gas phase capillary chromatography with Fe^+ ionization using laser production of reagent ions, followed by Fourier transform mass spectrometry on the resulting ions. However, as yet, this technique cannot be used for routine work due to its rather low sensitivity.[7]

EI ionization of nitrogen-containing derivatives of the carboxylic function has also been employed. Andersson and Holman[8] have demonstrated that the EI spectra (70 eV) of the pyrolidides of monounsaturated fatty acids contain ions derived from the rupture of all the carbon-carbon bonds in the chain. A straight polymethylene sequence gives rise to 14 mass unit intervals between fragments, whereas a double bond is located by a 12 mass unit interval between two consecutive ion fragments. Thus, in contrast to the methyl esters, where EI leads to isomerization of the double bonds, nitrogenous derivatives of the carboxylic function avoid such isomerization by localizing charge and radicals over the carboxylic group. The major disadvantage of the pyrolidide derivatives is the low intensity of the characteristic ion fragments. Other more efficient derivatives have been described. For example, the 3-picolinyl esters produce much larger quantities of ion fragments,[9,10] and they possess good chromatographic properties on non-polar capillary columns. The main drawback is the difficulty of interpretation of the spectra if the double bond is close to the carboxyl group.

It has been shown recently that mass analyzed kinetic energy spectrometry (MIKE) after remote site fragmentation of certain types of ion by collision activation dissociation (CAD) can be used to determine the structure of aliphatic chains.[11] Although this technique has not yet been combined with chromatographic methods, we have developed a similar approach for the analysis of bacterial fatty acids.

REMOTE SITE FRAGMENTATION OF CLOSED-SHELL IONS FROM FATTY ACIDS

Principle of the method

Jensen et al.[11] have shown that the fragmentation of aliphatic chains possessing a stable ionized group does not involve charge participation. Fragmentation occurs after increasing the internal energy by collision (CAD) and appears to involve a unique thermal mechanism. The great advantage of this process is that the corresponding spectra are highly characteristic of the position of double bonds, cyclopropane moieties and chain branching. For fatty acid analysis, the most widely used "closed-shell ions" are either carboxylate anions[12,13] or molecular ions cationized by alkali metals.[14] The CAD spectra of the [M − H]⁻ ions of the methyl esters can also be used to locate double bonds or branchings, although the interpretation of the signals is less straightforward.[15] The [M − H]⁻ ions of the acetates of the unsaturated aliphatic alcohols also give rise to characteristic CAD spectra. Fragmentation of these closed-shell ions results from cleavages of the aliphatic chain with simultaneous elimination of alkenes and hydrogen (equivalent to a loss of alkanes) (Figure 1). For example, the CAD spectrum of the anion derived from deprotonation of stearic acid contains a series of regularly spaced signals 14 mass units apart, corresponding to the formal elimination of alkanes of increasing size. The overall charge remains on the carboxylate group (Figure 2). For unsaturated chains, another pattern is observed. Cleavage of allyl bonds is favored, and relatively few ions derive from vinyl cleavage or double bond breakage. The CAD-MIKE spectrum of the oleate anion (Figure 3) displays a characteristic pattern of two strong signals (allyl cleavages) surrounding three weak signals (vinyl cleavages and breakage of the double bond). The position of the double bond is deduced

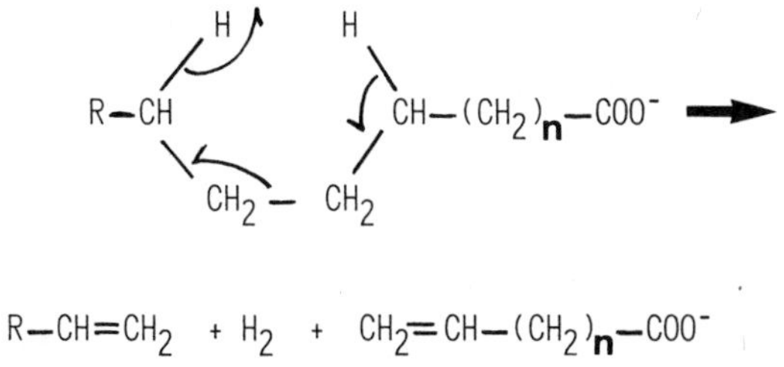

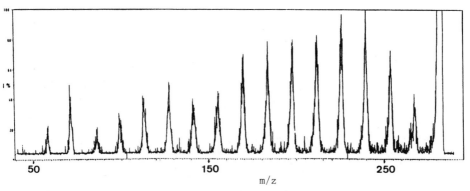

$$R-CH=CH_2 \ + \ H_2 \ + \ CH_2=CH-(CH_2)_n-COO^-$$

Figure 1. Mechanism of fragmentation of aliphatic chains of
carboxylates by collision activation dissociation.
Reprinted with permission from reference 11.

from the number of signals between the main beam and the dip representing the
three weak signals. This gives the number of methylene groups between the
methyl terminal and the double bond.

A methyl branch is indicated by the absence of a unique signal in the
pattern. The positions of the *iso-* and *anteiso-* groups, a difficult task
with conventional mass spectrometry, can be clearly identified by this
method.[13]

Chromatography combined with tandem mass spectrometry (GC-MS-MS)

In experiments on carboxylates[12] or cationized complexes of fatty
acids,[14] ions submitted to CAD fragmentation were produced by fast atom
bombardment. Typically, 10-20 ng of fatty acid are employed, and the spectra
result from an accumulation of 20 to 40 twenty second scans. In another
series of experiments the [M - H]$^-$ ions of the methyl esters[15] were produced
by chemical ionization using OH$^-$. However, to obtain a reliable CAD spectrum
under these conditions, the scan time in the second sector must not be less
than 15 s. [M - H]$^-$ ions from the fatty acid acetates have also been

Figure 2. CAD-MIKE spectrum of the [M-H]- ion (m/z 283) of
stearic acid. This ion was produced by fast atom
bombardment of stearic acid in a triethanolamine
matrix.

165

generated by OH⁻ chemical ionization. Their analysis requires long recording times (20 to 40, 2 s scans) due to the low intensity of the CAD signals.[16]

The low intensity of the CAD signals from anions is due to the competing process of electron detachment. The long recording times due to the low ionic yield of the collision process makes such methods incompatible with capillary GC. We reasoned that it might be possible to compensate for the low yield from collision-induced decompositions of fatty acid carboxylates by increasing the quantity of carboxylates produced from the original fatty acids.[17] It is known that electron capture reactions of halogenated compounds occur in high yield, enabling detection of these compounds in the femtomole range. We therefore devised a method for the production of carboxylates by dissociative electron capture of the fatty acid pentafluorobenzyl esters (PFB) according to the scheme:

$$R\text{-}COOCH_2C_6F_5 \quad + \quad e^- \quad \rightarrow \quad R\text{-}COO^- \quad + \quad CH_2C_6F_5 \cdot$$

The PFB derivatives have excellent chromatographic properties. Their electron capture spectra contain a unique ion, corresponding to the carboxylate. Only a few ng of the PFB derivatives give rise to quite characteristic CAD spectra (Figure 3b). Moreover, scan times (1 s) perfectly compatible with capillary GC can be employed. The following method was used.

The spectrometer was a reverse geometry (B-E) ZAB 2F instrument (VG-Analytical, Manchester, UK) with a collision chamber situated between the two sectors (second field-free zone). It was connected to a DANI 3800

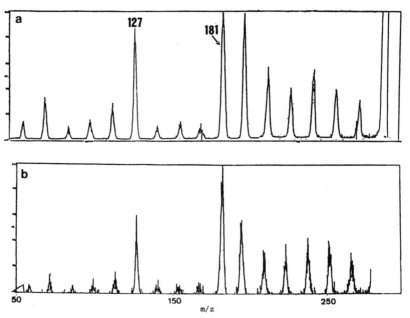

Figure 3. CAD-MIKE spectrum of the oleate anion (m/z 281).
(a) generation of the precursor ion by fast atom bombardment from 10 μg of oleic acid in 2 μl of triethanolamine; 10 accumulations of 10 s scans.
(b) generation of the precursor ion by dissociative electron capture of the pentafluorobenzyl ester of oleic acid. The ester (10 ng) was introduced in the capillary GC coupling. A single 1.5 s scan.
Reprinted with permission from reference 27.

chromatograph (Monza, Italy). The 25 m x 0.32 mm i.d. OV-1 coated fused silica column (Spiral, Dijon, France) ended directly in the source. A solid injector was used. The signals were analyzed by a Data System 2035 based on a PDP-8 computer (VG-Analytical) using dedicated MIKE software for acquisition of the CAD spectra. The energy spectra were directly converted into mass units by the program.

Ionization by electron capture dissociation was carried out by bombardment of methane. The source was raised to -8 kV with respect to ground. The collision-induced dissociation was carried out by introducing sufficient helium into the collision chamber to attenuate the main beam by 50%. The MIKE spectra were recorded by scanning the electric field from 0.98 E_0 to 0.05 E_0 over 1.5 s (E_0 is the value of the electric field transmitting the main beam).

A preliminary analysis of the PFB esters of fatty acids was carried out by conventional GC-MS. These experiments established the retention times of the various constituents and the masses of the corresponding carboxylate ions.

During the recording of the GC-MS-MS CAD spectra, just before the arrival of the chromatographic peak, the magnetic field was adjusted manually to the value required for transmission of the carboxylate ion. The electric field was scanned continuously. Consecutive portions of the energy spectra were then converted by the computer into mass spectra. These portions corresponded to samples of precursor ions of identical mass. Injection of less than 10 ng of the PFB derivative of the fatty acid onto the column gave rise to a characteristic CAD spectrum. Under the chromatographic conditions employed (temperature programmed from 160°C to 290°C at 1°C/min), 3 to 5 spectra could be recorded for each peak. Alterations in consecutive spectra could also be observed in cases where the constituents were only partially separated, enabling their individual characterization.

APPLICATION OF REMOTE SITE FRAGMENTATION OF CARBOXYLATES FOR ANALYSIS OF UNSATURATED FATTY ACIDS FROM MYCOBACTERIUM PHLEI

Due to their potential interest as intermediates in the biosynthesis of very long chain fatty acids (mycolic acids), the unsaturated fractions of mycobacterial fatty acids have been investigated by several research groups.[18-21] However, only partial analysis has been achieved due to the complexity of the mixtures. Depending on the exact conditions used, it is probable that some constituents are missed. For example, in *Mycobacteria phlei*, Asselineau et al.[18] have detected several fatty acids from C_{22} to C_{27}, either branched or linear, with a double bond in the 5 position. Takayama et al.[20] have also found Δ5 unsaturated acids and their elongation products in *Mycobacterium tuberculosis* H37 Ra. In these studies, the structures were established by identification of the products from oxidative cleavage after chromatographic enrichment of fractions containing compounds with the same number of carbon atoms. On the other hand, in *Mycobacterium phlei*, Cervilla and Puzo[21], using direct MS-MS analysis of amino alcohol derivatives ionized chemically, detected a predominance of elongation products of oleic acid rather than the Δ5 unsaturated acids. This discrepancy is probably due to secondary reactions during preparation of the amino alcohol derivatives. These tend to occur in compounds in which the double bond and the carboxyl group are within reactive range of each other. Another drawback is that chromatographic pre-purification before oxidative cleavage is somewhat tedious and requires a large amount of material. In our method, only small quantities of material are needed, and a structural analysis of both the saturated and unsaturated fatty acids can be achieved in a few days.

Production of unsaturated fatty acid fractions

After saponification of diethyl ether-ethanol (1:1) extract of bacteria from a 500 mL culture, followed by acidification, the ether soluble mixture is methylated using diazomethane. The non-hydroxylated esters are separated from the other components by silicic acid column chromatography (1 g silicic acid, elution with hexane-diethyl ether 98:2). This fraction is separated on a silver nitrate-impregnated silicic acid column (4 cm x 0.5 cm). A hexane-diethyl ether (98:2) mixture is used to elute the saturated esters, while the monounsaturated esters are eluted with pure diethyl ether. A third fraction eluted with diethyl ether-methanol (95:5) contains the polyunsaturated acids. Each fraction is saponified, then transformed into the pentafluorobenzyl esters by the following procedure. An aliquot of each fraction (0.1 to 0.5 mg) is dissolved in a mixture of dry methanol (10 μL) and acetonitrile (50 μL). Pentafluorobenzyl bromide (2 μL) and diisopropylethylamine (2 μL) are added. After 1 h at ambient temperature, the mixture is evaporated and the residue is dissolved in hexane (1 mL).

Analysis of the polyunsaturated fraction

M. phlei produce polyunsaturated fatty acids called phleic acids[22] found in the form of trehalose polyesters.[18,23] The structure of these long chain acids (C_{32} to C_{50}) is unusual in that each double bond is separated from the next by two methylene groups (Figure 4). Although the classical methylene interrupted double bond sequence of polyunsaturated fatty acids of plant and animal origin is not easily determined by remote site fragmentation methods,[25] the phleic acids should give rise to characteristic spectra due to the presence of allylic bonds in the consecutive unsaturated bonding. However, the relatively high molecular mass of these compounds, especially as PFB derivatives, makes analysis by capillary GC rather more difficult. Since isomeric structures have not been reported, the phleic acids can in principle be analyzed by direct MS-MS. The mixture of the PFB esters of the polyunsaturated fraction is introduced in a crucible and progressively vaporized in the source. The magnetic field is arranged to transmit the carboxylate ions of the various homologues in sequence, while the electric field is scanned. Typical spectra are illustrated in Figure 4. The spectra have a similar shape. From high to low mass, one observes a series of regularly spaced peaks separated by 14 mass units derived from cleavage of the polymethylene chain. This series ends at the first allyl linkage. After three small peaks, a large peak due to allyl cleavage between the double bonds is observed. This is followed by three small peaks and another large peak due to a further allyl cleavage. This sequence repeats for the number of allyl linkages present in the compound, terminating with the allyl linkage beta to the carboxylate group. This latter peak is much smaller than the previous allyl cleavage peaks. This is partially due to the low kinetic energy of the corresponding ion. The collection and conversion yields by the collector dynode are lower than those for the high energy ions.

These typical sequences enable an unambiguous characterization of the ethylene interrupted sequence of the double bonds in the phleic acids. The major constituent ($C_{36:5}$), and the various homologues ($C_{32:4}$, $C_{34:5}$, $C_{38:6}$, and $C_{40:6}$) can be determined in a single experiment, which only requires about 1 μg of a mixture of the phleic acids.[26]

Analysis of monounsaturated acids

The monounsaturated acid fraction is highly complex. Figure 5 shows a typical capillary gas chromatogram of the PFB esters of such a mixture. The major components are $C_{16:1}$ and $C_{18:1}$, although there is at least one

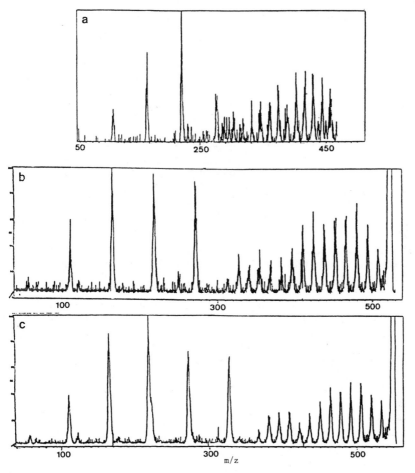

Figure 4. Characterization of various phleic acid homologues from the CAD-MIKE spectra of their carboxylates.

$$CH_3 - (CH_2)_x - (CH_2 - CH=CH-CH_2)_y - CH_2COOH$$

(a) x = 13, y = 4
(b) x = 13, y = 5   Reprinted with permission
(c) x = 11, y = 6   from reference 26.

representative of all the intermediates between C_{16} and C_{27}. Multiple isomers are indicated by the presence of peak complexes at a given number of C atoms. Direct analysis of such mixtures by MS-MS is therefore not feasible. Capillary chromatographic separation combined with CAD-MIKE of the carboxylate derivatives is therefore required. Fortunately, most of the spectra obtained are readily interpretable. For example, the predominant $C_{16:1}$ acid is $\Delta 10$ $C_{16:1}$ (Figure 6d), but small quantities of $\Delta 5$ and $\Delta 9$ $C_{16:1}$ are also detectable (Figures 6a and 6c). The $\Delta 6$ and $\Delta 7$ $C_{16:1}$ acids, while poorly separated by chromatography (peak b) can be identified in this way.

A methyl branch at a distance from the double bond can be detected and localized. Figure 7 shows a spectrum of the main component of $C_{25:1}$ in which the double bond is at $\Delta 5$, and the methyl branch is on the ninth carbon from the methyl terminal. Typical results are shown in Tables 1 and 2. The hexadecenoic acids mainly contain the $\Delta 10$ isomer. The predominant isomer of the octadecenoic acids is oleic acid ($\Delta 9$). The eicosenoic acid fraction is more complex. The main component is $\Delta 11$, an elongation product of oleic acid, but numerous other isomers ($\Delta 5$, $\Delta 9$-13) are also found. In addition, there are two isomers with branches close to the double bond which have not yet been completely characterized. The $C_{22:1}$ fraction contains mainly elongation products of eleic acid ($\Delta 13$), but a $\Delta 5$ unsaturated acid is also present. In the C_{24} fraction, the $\Delta 5$ and $\Delta 15$ unsaturated acids predominate, although an unusual $\Delta 4$ acid with a methyl on the 15 position has been detected. The C_{26} fraction contains mostly the $\Delta 5$ isomer (Figure 8c), although two branched acids are also present: a $\Delta 4$ acid homologue of the C_{24} series with a methyl on position 17 (Figure 8b), an elongation product of the C_{24} $\Delta 4$ acid, and the $\Delta 6$ acid likewise branched at position 17 (Figure 8a). In this series, the elongation product of oleic acid is only found in small quantities (Figure 8d).

Table 2 shows the results for the C_{21} to C_{27} acids with an odd number of carbon atoms. However, for some of these acids, interpretation of the spectra is complicated. The observed pattern does not correspond to the classical picture from an isolated double bond (three small signals surrounded by two strong ones). This is especially noticeable for the $C_{19:1}$ acids whose spectra all indicate a double bond between C_8 and C_{10} together with a branch in this region.

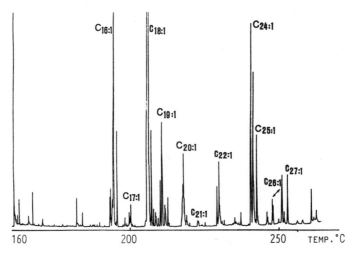

Figure 5.   GC trace of GC/MS analysis of the pentafluorobenzyl ester derivatives of the monounsaturated fatty acids from *M. phlei*. Reprinted with permission from reference 27.

170

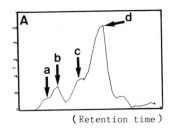

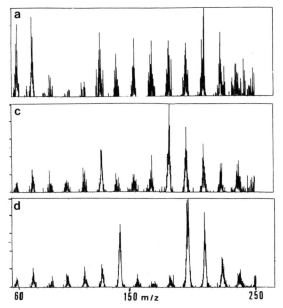

Figure 6. GC/MS-MS analysis of monounsaturated acids from
M. phlei: hexadecenoic acids.

A (at top): Reconstituted mass chromatogram from MS-MS
analysis of m/z 253 ions

a, c, d (at bootom): CAD-MIKE spectra of the various
monounsaturated hexadecenoates

a: $CH_3-(CH_2)_9-CH=CH-(CH_2)_3-COOH$

c: $CH_3-(CH_2)_5-CH=CH-(CH_2)_7-COOH$

d: $CH_3-(CH_2)_4-CH=CH-(CH_2)_8-COOH$

To obtain more information on the relative abundance of the ion fragments in this zone, a further analysis of the $C_{19:1}$ acids was carried out by scanning over a more limited range of energy, followed by an accumulation of 4 consecutive spectra. The ions of weak intensity were thus sampled for a longer period, leading to a more reliable statistic for each signal. Figure 9 shows the pattern obtained. We interpreted these figures on the assumption that a weaker signal in a dip due to 4 weak signals indicated the presence of a methyl group on a double bond. The peaks b, c, d, and e in Figure 9 correspond to the various compounds with a branch on the double bond. The methyl group is always at position 9 from the methyl terminal. The 4 peaks stem from the cis/trans isomers which cannot be resolved using this technique. In peak f, a dip of three signals with a particularly weak one in the center is observed. This corresponds to a methylenic double bond, an expected intermediate in the biosynthesis of tuberculostearic acid. The peak g corresponds to the unbranched Δ9 acid. The interpretation of peak a is less obvious. The dip of four identical signals might have been due to a methyl in an allylic position (an acid with a disubstituted cyclopropane would produce a similar pattern, but it should not be present in this fraction of the unsaturated acids). Two possible structures could account

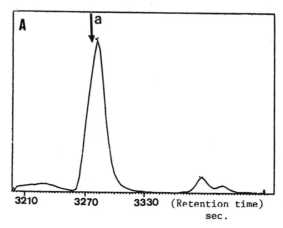

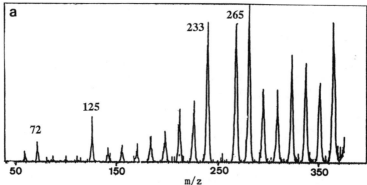

Figure 7. GC/MS-MS analysis of monounsaturated acids from *M. phlei*: pentacosenoic acids.

A: Reconstituted mass chromatogram from MS-MS analysis of m/z 379 ions.

a) CAD-MIKE spectra of the main component.
a: $CH_3-(CH_2)_7-CH(CH_3)-(CH_2)_9-CH=CH-(CH_2)_3-COOH$
Reprinted with permission from reference 27.

Table 1. Principal monounsaturated fatty acids with an even number of carbon atoms from *M. phlei*.

L = unbranched aliphatic chain

B = methyl branched chain

* = unidentified compound with a branch close to the double bond

Even Carbon Number	4	5	6	7	8	9	10	11	12	13	14	15	16	17
16				L		L	L							
18				L		L								
20**				L		L	L	L	L	L				
22				L						L	L	L		
24		B15		L								L	L	
26		B17		L	B17									L

The major components are double underlined; less abundant components are underlined once; and the minor components are dashed underlined. The relative abundances are with respect to the total of a series with an identical number of carbon atoms.

Table 2. Principal unsaturated fatty acids with an odd number of carbon atoms from *M. phlei*, from C_{21} to C_{27} (for the C_{19} acids see Figure 9). See Table 1 for comments.

Odd Carbon Number

21	$\Delta 11, \Delta 12, *$
23	$\Delta 4 , \Delta 14, *, *$
25	16-methyl-$\Delta 5$
27	18-methyl-$\Delta 5$

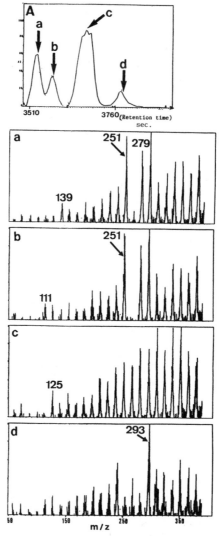

Figure 8.   GC/MS-MS analysis of monounsaturated acids from
M. phlei: hexacosenoic acids.   Reprinted with
permission from reference 27.
A: Reconstituted mass chromatogram from MS-MS analysis
of m/z 393 ions

a to d: CAD-MIKE spectra of the various components

a: $CH_3-(CH_2)_7-CH(CH_3)-(CH_2)_9-CH=CH-(CH_2)_4-COOH$

b: $CH_3-(CH_2)_7-CH(CH_3)-(CH_2)_{11}-CH=CH-(CH_2)_2-COOH$

c: $CH_3-(CH_2)_{19}-CH=CH-(CH_2)_3-COOH$

d: $CH_3-(CH_2)_7-CH=CH-(CH_2)_{15}-COOH$

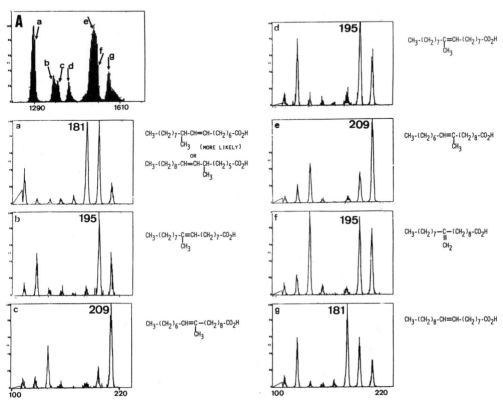

Figure 9. GC/MS-MS analysis of monounsaturated acids from
M. phlei: nonadecenoic acids

A: Reconstituted mass chromatogram from MS-MS analysis
of m/z 295 ions

a to g: partial CAD-MIKE spectra of the various
components. The scanned energy range corresponds
to m/z 100-220.

for these results. To distinguish them, we hydrogenated (H_2/Pt) the monounsaturated fraction and then analyzed the structure of the $C_{19:0}$ acids using the same technique. Only two compounds were detected, one corresponding to the unbranched acid, the other to 10-methyl stearic acid (tuberculostearic acid, Figure 10). Thus out of the two possible structures giving rise to peak a in Figure 9, only the 10 methyl $\Delta 8$ acid can account for the results obtained.

CONCLUSION

Accurate analysis of bacterial fatty acids can be achieved by tandem mass spectrometry of their carboxylate derivatives, either with or without prior capillary chromatography. The carboxylates are produced by electron capture ionization of the pentafluorobenzyl esters. Sensitivity is adequate, since a complete analysis can be carried out on 1 μg of a complex mixture of the fatty acids. For example, in the mixture illustrated in Figure 5, the minor components were detected at levels of less than 10 ng. A complete

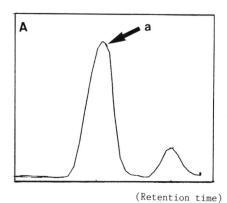

(Retention time)

Figure 10.   GC/MS-MS analysis of monounsaturated acids from *M. phlei* after hydrogenation: nonadecanoic acids.

A: Reconstituted mass chromatogram from MS-MS analysis of m/z 297 ions

a: CAD-MIKE spectrum of the methyl branched component (largest GC peak)

$CH_3-(CH_2)_7-CH(CH_3)-(CH_2)_8-COOH$

analysis can thus be obtained on the contents of a single culture flask (100 mL of culture medium). Investigation of numerous strains can be attempted since the isolation and characterization of almost all the fatty acids can be completed within a few days for any given mixture. The instrumentation is relatively unsophisticated since double sector reverse geometry mass spectrometers are now commonplace. The only practical difficulty is the quick manual adjustment of the magnetic field required when the chromatographic peaks of compounds of different masses are very close together. In fact, a simple instrument with no magnet, and only equipped with an electrostatic analyzer (like that used for IKE measurements), would probably be adequate. There is no need for mass selection of the primary beam since each spectrum only contains one ion and the number of signals in the CAD spectrum gives the total number of carbon atoms. This technique would therefore be a valuable adjunct in studies on the chemical taxonomy of bacterial strains.

REFERENCES

1. J. Asselineau, Branched-fatty acids of mycobacteria, Ind. J. Chest Dis. 24:143 (1982).
2. I. M. Campbell and J. Naworal, Composition of the saturated and monounsaturated fatty acids of *Mycobacterium phlei*, J. Lipid Res. 10:593 (1969).
3. A. J. V. Ferrer-Correira, K. R. Jennings, and D. K. Sen-Sharma, The use of ion-molecule reactions in the mass spectrometric location of double bonds, Org. Mass Spectrom. 11:867 (1976).
4. R. Chai and A. G. Harrison, Location of double bonds by chemical ionization mass spectrometry, Anal. Chem. 53:34 (1981).
5. A. Brauner, H. Budzikiewicz, and W. Francke, Chemical ionization (NO) spectra of n-alkenoic acids and their esters, Org. Mass Spectrom. 20:578 (1985).
6. D. A. Peake and M. L. Gross, Iron (I) chemical ionization and tandem mass spectrometry for locating double bonds, Anal. Chem. 57:115 (1985).
7. D. A. Peake, S. K. Huang, and M. L. Gross, Iron (I) chemical ionization for analysis of alkene and alkyne mixtures by tandem sector mass spectrometry or gas chromatography/Fourier transform mass spectrometry, Anal. Chem. 59:1557 (1987).
8. B. A. Andersson and R. T. Holman, Pyrolidides for mass spectrometric determination of the position of the double bond in monounsaturated fatty acids, Lipids 9:185 (1974).
9. D. J. Harvey, Picolinyl esters as derivatives for the structural determination of long chain branched and unsaturated fatty acids, Biomed. Mass Spectrom. 9:33 (1982).
10. W. W. Christie, E. Y. Brechany, and R. T. Holman, Mass spectra of the picolinyl esters of isomeric mono and dienoic fatty acids, Lipids 22:224 (1987).
11. N. J. Jensen, K. B. Tomer, and M. L. Gross, Gas-Phase ion decompositions occurring remote to a charge site, J. Am. Chem. Soc. 107:1863 (1985).
12. K. B. Tomer, F. W. Crow, and M. L. Gross, Location of double bond position in unsaturated fatty acids by negative ion MS/MS, J. Am. Chem. Soc. 105:5487 (1983).
13. N. J. Jensen and M. L. Gross, Fast atom bombardment and tandem mass spectrometry for determining *iso-* and *anteiso-* fatty acids, Lipids 21:362 (1986).
14. J. Adams and M. L. Gross, Tandem mass spectrometry for collisional activation of alkali-metal-cationized fatty acids: a method for determining double bond location, Anal. Chem. 59:1576 (1987).

15. M. Bambagiotti, S. A. Coran, V. Gianellini, F. F. Vincieri, S. Daolio, and P. Traldi, Structural identification of fatty acid methyl esters by collision spectra of their [M-H]- species, Org. Mass Spectrom. 19:577 (1984).

16. G. Takeuchi, M. Weiss, and A. G. Harrison, Location of double bonds in alkenyl acetates by negative ion tandem mass spectrometry, Anal. Chem. 59:918 (1987).

17. J. C. Prome, H. Aurelle, F. Couderc, and A. Savagnac, Structural determination of unsaturated fatty acids in complex mixtures by capillary GC/MS-MS, Rapid Comm., Mass Spectrom. 1:50 (1987).

18. C. P. Asselineau, C. S. Lacave, H. L. Montrozier, and J. C. Prome, Relations structurales entre les acides mycoliques insatures et les acides inferieurs insatures synthetises par Mycobacterium phlei. Implications metaboliques, Eur. J. Biochem. 14:406 (1970).

19. J. G. C. Hung and R. W. Walker, Unsaturated fatty acids of Mycobacteria, Lipids 5:720 (1970).

20. K. Takayama, N. Qureshi, and H. K. Schnoes, Isolation and characterization of the monounsaturated long chain fatty acids of Mycobacterium tuberculosis, Lipids 13:575 (1978).

21. M. Cervilla and G. Puzo, Determination of double bond position in monounsaturated fatty acids by mass analyzed ion kinetic energy spectrometry-collision-induced dissociation after chemical ionization of their amino alcohol derivatives, Anal. Chem. 55:2100 (1983).

22. C. P. Asselineau, H. L. Montrozier, and J. C. Prome, Presence d'acides polyinsatures dans une bacterie: isolement, a partir des lipides de Mycobacterium phlei, d'acide hexatriacontapentaene-4,8,16,20-oique et d'acides analogues, Eur. J. Biochem. 10:580 (1969).

23. C. P. Asselineau, H. L. Montrozier, J. C. Prome, A. M. Savagnac, and M. Welby, Etude d'un glycolipide polyunsature synthetise par Mycobacterium phlei, Eur. J. Biochem. 28:102 (1972).

24. C. P. Asselineau and H. L. Montrozier, Etude du processus de biosynthese des acides phleiques, acides polyunsatures synthetises par Mycobacterium phlei, Eur. J. Biochem. 63:509 (1976).

25. N. J. Jensen, K. B. Tomer, and M. L. Gross, Collisional activation decomposition mass spectra for locating double bonds in polyunsaturated fatty acids, Anal. Chem. 57:2018 (1985).

26. H. Aurelle, M. Treilhou, D. Prome, A. Savagnac, and J. C. Prome, Analysis of mycobacterial polyunsaturated fatty acids ("Phleic acids") by remote site fragmentation, Rapid Comm., Mass Spectrom. 1:65 (1987).

27. F. Coudrec, H. Aurelly, D. Prome, A. Savagnac, and J. C. Prome, Analysis of fatty acids by negative ion gas chromatography/tandem mass spectrometry: structural correlations between α-mycolic acid chains and Δ5-monounsaturated fatty acids from Mycobacterium phlei, Biomed. Environ. Mass Spectrom. 16:317 (1988).

CHAPTER 12

PYROLYSIS GC/MS PROFILING OF CHEMICAL MARKERS FOR MICROORGANISMS

Stephen L. Morgan, Bruce E. Watt, and Kimio Ueda

Department of Chemistry
University of South Carolina
Columbia, SC 29208

Alvin Fox

Department of Microbiology & Immunology
School of Medicine
University of South Carolina
Columbia, SC 29208

INTRODUCTION

Analytical pyrolysis is an attractive tool for the rapid classification, identification, or structural characterization of microorganisms. Pyrolysis-based analytical methods thermally fragment samples in the absence of oxygen to produce volatile components (pyrolyzates) that can then be separated on-line by capillary GC with flame ionization detection (Py-GC-FID), separated by GC and detected by MS (Py-GC/MS), or detected directly by MS (Py-MS). The success of analytical pyrolysis for microbial characterization is ultimately based on detecting chemical markers -- compounds that are unique or prominent in a group of organisms and that can be used to identify those organisms (see Chapter 1). The mere presence of a particular substance in the pyrolysis product mixture from a microorganism does not qualify it as a chemical marker; discriminating information that is relevant to taxonomic differences must be provided. This chapter describes an approach for the validation of pyrolysis products as chemical markers and their chemical identification. Instrumental aspects and background of analytical pyrolysis have been discussed in Chapter 2 as well as by other sources.[1,2] Several reviews on analytical pyrolysis in microbial analysis have also been published.[3-5] The use of short capillary columns combined with ion trap mass spectrometry for the rapid characterization of microbes is described in Chapter 12.

Capillary GC and GC/MS using derivatization can analyze the structural composition of bacterial cells (see Chapters 4-6), but these techniques require the sample to be volatile. Derivatization generally requires one depolymerization step and one or more volatilization steps to produce components amenable to GC analysis. Thermal fragmentation simultaneously depolymerizes and volatilizes bacterial components in a single instrumental step. Derivatization schemes can be designed to manipulate components of the microorganism into a form that retain or reflect the original functionality and stereochemistry. Sample clean-up procedures capable of concentrating particular classes of components and eliminating background are often incorporated into a derivatization scheme. Pyrolysis has traditionally been considered to be an on-line process in which, for simplicity and rapidity, clean-up steps are omitted. Sample clean-up or other fractionation steps prior to pyrolysis can add selectivity.[6-9] Indeed, lack of sample preparation in analytical pyrolysis may be a substantial source of variability.[10] On-line derivatization with pyrolysis has been used for profiling fatty acid components of bacteria.[11]

Pyrolysis of a microorganism generates an exceedingly complex product mixture, potentially consisting of contributions from all components present (including polysaccharides, nucleic acids, and proteins). The complexity of the pyrolysis product mixture presents both advantages and disadvantages.

It is difficult to predict the pyrolysis products that will be generated from a particular microbial component. Pyrolysis does not just involve breaking bonds to form smaller volatile fragments of the parent structure. Complete structural integrity of monomeric components is often not preserved. Rather than breaking bonds at the connections between subunits (e.g., glycosidic linkages or peptide bonds), thermal fragmentation may produce structures derived from a combination of several monomeric units. Rearrangements, dehydration, and the formation of new bonds both within and between intermediate pyrolysis products are common, producing new chemical components and complicating interpretation of the pyrogram.[1]

Despite its complexity, pyrolysis of a microbial sample is reproducible and completely determined by the structure and accessibility of components to depolymerization and volatilization. Validating a pyrolysis product as a chemical marker requires recognizing its origin from an identifiable structure within the microorganism. Certain pyrolysis products, particularly small organic molecules, represent fragments of the original monomeric structures (such as acetamide) or do not retain the original stereochemistry (such as furfuryl alcohol).[12, 8-9, 13] Such pyrolysis products may originate from many different sources. Other pyrolysis products (such as dianhydroglucitol) are less common and more closely reflect the original parent structure (including functionality or stereochemistry) from which they were derived.[14,15] Only a limited number of possible origins in the microorganism may exist for these more specific chemical markers.

Although many microbial pyrolyzates have been identified, few have been validated in the above sense of a chemical marker. A large number of studies employ non-specific GC detectors (e.g., flame ionization detector (FID)) which are incapable of providing chemical identification. Interpretation of mass spectra can provide not only the class of chemical compound, but often the structural identity of a pyrolyzate. Confident identification can be difficult if standard samples of pure components are not available, as is the case for many pyrolysis products of bacteria. If a non-selective detector is used, often even the general chemical class to which a particular pyrolyzate belongs can not be determined. Incomplete pyrolysate identification may produce confusion between laboratories when there is uncertainty as to whether the same pyrolysis product is under investigation.

Pyrolysis GC can neither resolve differentiating chemical markers (without high resolution capillary GC) nor identify them (without MS). In Py-MS, the pyrolyzate mixture is swept directly into the ion source of the MS without prior separation. Although Py-MS is rapid, the results are difficult to interpret since the mass pyrogram consists of a mixture of contributions from many different parent structures. Pyrolysis GC/MS and pyrolysis MS/MS offer the potential for chemical identification of pyrolysis products not achievable by GC or MS alone. Modular analytical pyrolysis units for interfacing to GC or MS are available (Chapter 2). An automated Py-MS instrument has also been described[16].

Chemical markers generated by analytical pyrolysis can be validated by high resolution capillary Py-GC/MS of selected groups of model compounds, cell fractions, and microorganisms. The ultimate objective is decisive information on the probable origin of relevant chemical markers. Recognizing that a particular marker might exist requires, in addition to a knowledge of microbial chemistry, knowing the taxonomic significance of specific chemical components and how these structures might be altered during pyrolysis. We propose the systematic search for and validation of microbial chemical markers to take the following steps:

(1) Appropriate groups of organisms should be selected that differ in particular chemotaxonomic features. Pyrolysis of a random group of bacteria possessing unrelated differences may not permit the significance of a particular chemical marker to be evaluated. It is essential that multiple strains from each group be available to confirm reproducibility of results within that group of organisms. Replicate analyses of each sample should also be run to assess instrumental variability.

(2) It should be demonstrated that a pyrolysis product is generated that differentiates the groups from one another based on taxonomic features. This step can often be aaccomplished by simple visual comparison of pyrograms, but with large data sets computer-assisted pattern comparison and multivariate statistics can aid the process.[17-21]

(3) Having recognized discriminating GC peaks, mass spectra of these components should be examined in the various bacterial pyrograms and structural interpretations made.

(4) Knowing the identity of the relevant pyrolysis products, the next step is to relate their origin to known chemotaxonomic features. GC/MS data from pyrolysis of whole bacterial cells should be correlated with data from the pyrolysis of model compounds for known biomarkers and/or appropriate cell fractions.

Ideally, Py-GC/MS data on markers present in a wide variety of organisms could be correlated to develop libraries analogous to those available for fatty acid profiling (Chapter 4). Automated comparison of profiles and library searching for pattern matching is a long-term goal of importance. However, further research in the identification of chemical markers for microorganisms is needed.

SOURCES OF VARIABILITY IN ANALYTICAL PYROLYSIS OF MICROORGANISMS

Analytical pyrolysis should be subject to the same rigorous controls and quality checks as any other technique. Results should be reproducible both within and between laboratories. The development of methods based on pyrolysis should also take ruggedness, simplicity, and ease of interpretation into account. Sources of variability in pyrolysis involve three areas: the preparation of microbial samples, thermal fragmentation, and separation, and detection of volatile pyrolysis products.

Even organisms of the same species exhibit natural variability in chemical composition. If a chemical marker is to be useful it must be shown to be an invariant feature, which can only be demonstrated by the analysis of multiple strains or isolates of the same species. Invariant chemical features of the organism will provide effective discrimination if distinctive pyrolysis products can be generated under a wide range of experimental conditions. Furthermore, the discrimination thus achieved should be reproducible between different instruments and different laboratories.

Strains should be cultured in a well defined manner and systematically characterized by traditional microbiological tests. Only in this manner can quality control be assured. In our recent study of streptococci, a sample labelled as a Group F strain was found to contain the chemical marker previously detected only in Group B organisms.[17] Routine tests in our laboratory (hippurate hydrolysis and serological grouping) identified the organism as a group B streptococcus. It was later confirmed that this sample had been incorrectly identified. Bacteria are not simple pure chemical substances; misassignments or sample contamination are not uncommon. Routine microbiological controls are necessary, as in any study. This is not a criticism of interlaboratory studies as much as a comment on the complexity of performing work at the interface between analytical chemistry and microbiology.

Although not commonly performed, it is often not sufficient to only have samples characterized biologically. It may be necessary to know the chemical composition with respect to relevant carbohydrate, amino acid, or fatty acid chemical markers.

The form of the microbial sample is also of importance. Microbes are often transported between laboratories in a killed form. Naturally, traditional microbiological characterization should be performed before killing as most of these tests require viable cells. Organisms should be prepared and killed in a similar fashion whenever possible. In our laboratory, samples are usually physically killed with heat (at 60 to 120 °C in air, depending on the susceptibility of the organism). Heat killing is not unreasonable prior to pyrolysis which exposes the sample to even greater heating effects. Heating might be performed under an inert gas to avoid the possibility of oxidation. Preparing samples in widely different fashions may not permit hypotheses concerning observed chemical markers to be tested.

The conditions under which microorganisms are stored, the growth media, incubation temperatures, and times may not always be critical if an invariant chemical marker is under study. If the chemical marker is directly related to a taxonomic feature, the presence of that feature should be constant within a certain taxon and be relatively unaffected by growth conditions. The literature has often been equivocal on the effect of growth conditions on pyrolysis results.[22-24] What is important is that these microbiological culture variables are known, held constant for all samples if possible, and standardized for repeatability. The media itself may contribute to the pyrogram if not removed by washing the cells, however, additional sample handling and possible sources of variability are involved. If contamination from the media is not a problem, sampling directly off agar growth plates onto the pyrolysis probe allows rapid analysis.[25]

Commercial instrumentation can adequately control the experimental variables affecting the thermal dissociation of an analytical sample. In our work we have used a resistively heated filament pyrolyzer made by Autoclave Engineers-- Oxford, CDS Instruments (Oxford, PA). A close-up photo of a ribbon and a coil probe is shown in Figure 1A. The sample may be coated directly on the pyrolysis filament or wire. Alternatively, the sample may be loaded into a quartz sampling tube that is inserted into the coil probe. Following loading of sample, the probe is inserted into an heated interface connected to the GC injection port (Figure 1B). The ribbon or coil active element is rapidly heated to the final pyrolysis temperature by discharge of a capacitor. Knowledge of the actual temperature of the pyrolyzer element for a resistively heated system may require calibration.[26] Additionally, the temperature experienced by the sample itself may differ somewhat from the temperature setting on the instrument, depending on the amount of sample used, positioning of the sample on the ribbon or coil, and whether or not a quartz sampling tube is employed. Results from different pyrolysis systems

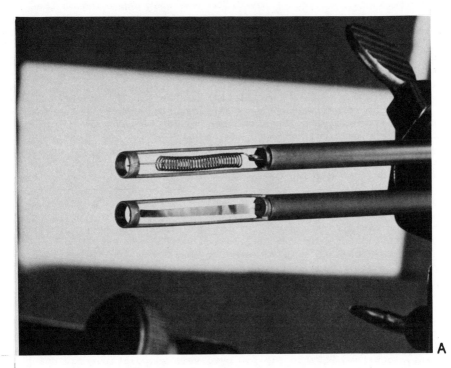

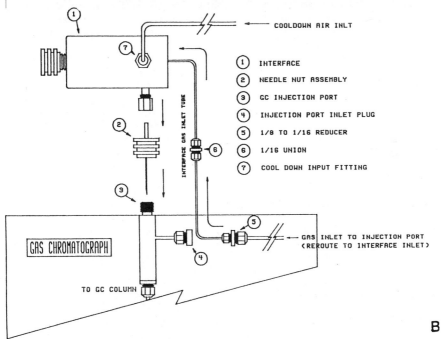

Figure 1.   (A) A platinum ribbon and a platinum coil sampling probe.
These probes are resistively heated by controlled discharge
of a capacitor.   (B) Typical pneumatic connections for
Pyroprobe interface.   Photo and schematic courtesy of
Autoclave Engineers-- Oxford, CDS Instruments (Oxford, PA).

(e.g., Curie-point pyrolyzers) may differ somewhat depending on the pyrolysates examined; while overall amounts of pyrolysates generated may vary, the absence or presence of a chemotaxonic marker should not depend dramatically on the choice of pyrolysis system. Also influencing reproducibility is the eventual deterioration of resistive filaments or sample wires. Components exposed to sample during pyrolysis often require acid cleaning and solvent washing.

Preliminary experiments are usually required to choose the pyrolysis conditions. The pyrolysis literature may guide the researcher, but optimal conditions for a particular sample must be determined empirically.[1,2] To achieve reproducibility, sample sizes for analytical pyrolysis should be in the lower μg range (80 μg). Larger sample sizes make controlled heat transfer and uniform heating difficult to achieve; secondary reactions are promoted, resulting in more complicated and less reproducible pyrograms. Small sample sizes allow formation of thin films and permit effective and rapid heat transfer from the heat source. The question of appropriate heating rates and final temperatures has been addressed by other authors.[27-30] Pyrolysis heating conditions interact with sample size and effects are usually system-dependent (i.e., resistive heating vs. Curie-point, wires vs. ribbons vs. quartz tubes).

With microorganisms, temperatures that produce pyrolyzates characteristic of the parent sample may range from 400-800 °C or even 1000 °C. Generally, lower pyrolysis temperatures induce less fragmentation and decrease the total amount of pyrolyzates produced. Higher temperatures ensure more complete fragmentation and thus might reduce the structural information obtained. Extremely low pyrolysis temperatures may not produce significant amounts of volatile fragments from intractable biomaterials and may lead to the accumulation of nonvolatile residues contaminating the analytical system. Pyrolysis at a variety of different temperatures may provide selective information about degradation patterns and thereby reveal different aspects of structure. A time profile of volatile products generated at different temperatures by linear programmed thermal degradation may also provide characteristic structural information.[2,31-33]

Once the sample has been pyrolyzed, volatile fragments are swept by a flow of carrier gas into the analysis system (usually a GC, a GC/MS, or a MS). Chromatographic or MS instrumentation introduce their own sources of variability and poor choices or usage may comprise analytical information.

Popular choices for Py-GC include columns packed with porous polymers, conventional packed columns, and open tubular capillary columns. Porous polymer columns are suitable for analysis of low molecular weight volatile pyrolyzates but limit information on higher molecular weight fragments. For example, 1.5 m x 4 mm columns packed with Chromosorb 104 were used in an isothermal repetitive mode to profile volatile pyrolysates of oral streptoccocci.[34,35] Because microbial pyrolysates are often labile polar species, the column material and chromatographic should be as inert as possible. Fused silica capillary columns coated with "bonded" phases represent the best chromatographic approach offering improved resolution, increased inertness, and better analytical precision than packed columns. Flexible fused silica columns can be attached directly to the pyrolysis interface and threaded straight through to the FID flame tip or the MS ion source, thus minimizing contact with active surfaces. Superior resolution per unit time available with capillary columns means that adequate resolution can often be achieved using short (5-10 m) columns. With optimized capillary chromatographic conditions, analysis times should be no greater than 10-30 min.

Life expectancy of chromatographic columns is always of concern when pyrolyzing microbial samples. Sample sizes that are too large (approaching 1 mg) may contaminate the column. Column lifetime can be extended by reducing the sample amount to 100 μg or less. Crosslinked or bonded phase columns can sometimes be rejuvenated by rinsing with organic solvent such as methylene chloride and then reconditioning. If contamination is a problem, resolution can often be restored by breaking off a few inches at the head of the column.

MS analysis system design also influences results. Active metal or other sorption sites may effectively filter polar or reactive components from the chromatogram. Condensation losses due to cold spots in the combined chromatographic-mass spectrometric can also severely limit the ability to observe higher molecular weight and less volatile components. In Py-MS, whether or not an expansion chamber is included can affect sensitivity and transmission of pyrolyzates and the ion source iteself may play a role in long term reproducibility.[2,28,29]

APPLICATIONS OF THE CHEMICAL MARKER CONCEPT TO ANALYTICAL PYROLYSIS

Differentiation of Gram types

Whether analytical pyrolysis is capable of providing information similar to that provided by the Gram stain was the question addressed in several articles.[8,9,13] Gram positive and Gram negative are two major categories of bacterial cell envelopes (Chapter 1). Peptidoglycan comprises between 30-70 % of the dry weight of Gram positive cell walls but less than 10 % of the dry weight of Gram negative cell walls. Peptidoglycan consists of a glycan backbone that is a repeating polymer of N-acetylglucosamine and N-acetylmuramic acid; attached covalently to the lactyl group of the muramic acid in peptidoglycan are tetra- and pentapeptides consisting of repeating L- and D-amino acids crosslinked by peptide bridges. Variations in the nature and conformation of the peptide structures occur in microorganisms and, in many cases, additional polymers may be bound to peptidoglycan.[36] On the basis of these known structural aspects of bacteria, it was speculated that chemical markers for peptidoglycan if they could be found would be present at higher levels in Gram positive than Gram negative bacteria. Further these markers would be expected to be absent in fungi which do not contain peptidoglycan.

Among the first studies reporting systematic identification of bacterial pyrolysis products by GC/MS was that of Simmonds and coworkers, who were interested in the use of pyrolysis GC to detect extraterrestrial life.[12,37] They attempted to correlate the formation of specific pyrolysates with their site of origin in the structure of microorganisms by pyrolyzing three Gram positive bacterial species (*Micrococcus luteus*, *Bacillus subtilis*, and *Streptomyces longisporoflavus*). Because of the lack of model compounds to simulate actual structures present in microorganisms, many of their assignments are rather tentative. Acetamide and propionamide were suggested to originate from N-acetyl groups (Figure 2) and the lactyl-peptide portion of peptidoglycan respectively. Acetamide has also been reported in Py-MS of mycobacteria.[38]

A trend toward increasing amounts of acetamide and propionamide produced upon pyrolysis of Gram positive as compared to Gram negative organisms or fungi was also reported by Hudson, et al.[8,9,13] who pyrolyzed single strains from six species of fungi, four species of Gram negative bacteria, four species of Gram positive bacteria, and some purified cell wall fractions. The approach adopted in these and other studies in our laboratory was to examine pyrolysis products from model compounds representing components or analogs of substructures within peptidoglycan. Relatively low levels of acetamide were generated by amino acids and peptides (isoglutamine, glutamine, glutamylglutamic acid, asparagine, triglycine, and trialanine) and sugars (glucose, glucosamine, muramic acid). N-acetylated sugars (N-acetyl glucosamine, N-acetylmuramic acid) produced much larger amounts of acetamide on pyrolysis. The minimally active subunit or monomer for peptidoglycan, N-acetyl muramyl-L-alanine-D-isoglutamine (muramyl dipeptide), also produced high amounts of acetamide. These results demonstrated that N-acetyl groups on the glycan backbone are the major source of acetamide within peptidoglycan (Figure 2). Other sources for acetamide within microorganisms were not ruled out; however, N-acetyl groups on the peptidoglycan backbone appear to be one major source of acetamide.

Muramyl dipeptide was shown to produce a high amount of propionamide when pyrolyzed. Only small amounts of propionamide were produced when muramic acid, N-acetylmuramic acid, or other model compounds not containing the lactyl-peptide moiety were pyrolyzed. Pyrolysis GC/MS of additional compounds modelling aspects of the structure of the lactyl-peptide bridge of peptidoglycan were performed.[9] The butanoic acid analog of muramyl dipeptide (with one less carbon in the lactyl-peptide bridge) and the ethanoic acid analog of muramyl dipeptide (with one extra carbon) produced 5-10 times less propionamide than did muramyl dipeptide, which has a structure most closely corresponding to peptidoglycan.

Another prominent pyrolyzate from microorganisms is furfuryl alcohol, which had been assigned to a carbohydrate origin.[12,37,39] The presence of furfuryl alcohol was also reported in the Py-MS spectra of DNA and attributed to deoxyribose.[40] Results from our laboratory also found furfuryl alcohol to be a major pyrolyzate of DNA and RNA.[9] The deoxypentose, 2-deoxyribose, produced almost 20 times as much furfuryl alcohol as ribose, and 100 times more than glucose or rhamnose. These results are not surprising when viewed in terms of the structural similarity of 2-deoxyribose (a partially dehydrated furanose ring structure) to furfuryl alcohol (a dehydrated furanose ring compound) (see Figure 3). The pyrogram of DNA is dominated by furfuryl alcohol. DNA is probably a major source of furfuryl alcohol in pyrograms of bacteria. DNA constitutes 2-3 % of the dry weight of bacteria.[41]

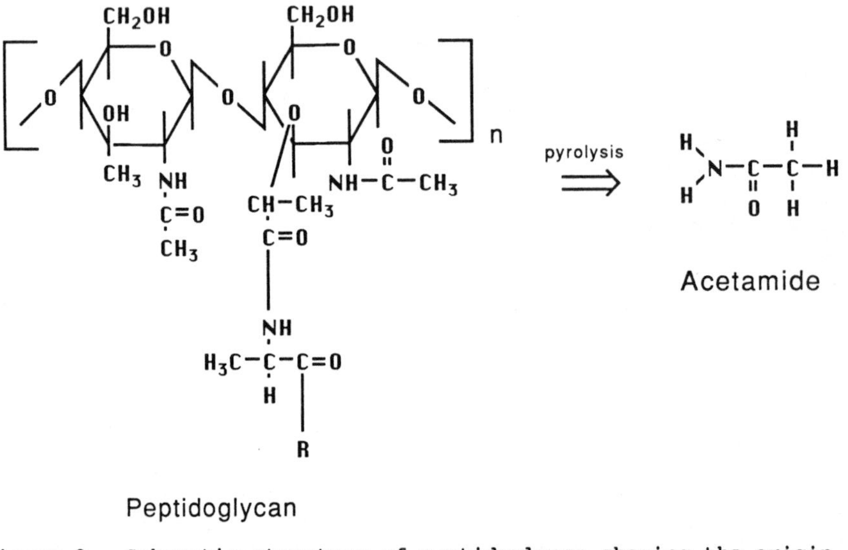

Peptidoglycan

Figure 2. Schematic structure of peptidoglycan showing the origin of acetamide from N-acetyl groups on the glycan backbone.

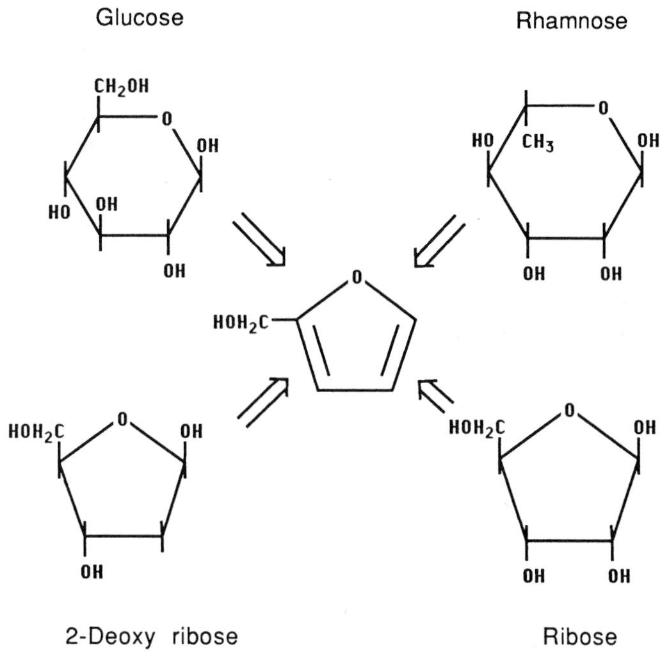

Figure 3.   Structures of common sugars that produce furfuryl
             alcohol on pyrolysis.

Although these results are promising, the critical question is whether
the levels of these pyrolysis products correlate with the actual amount of
peptidoglycan and DNA present in cells.

Quantitative Py-GC/MS on the relative amounts of acetamide,
propionamide, furfuryl alcohol, and other pyrolysis products have
demonstrated that groups of Gram positive bacteria, Gram negative bacteria,
and fungi can be separated with limited overlap of clusters.[9,13] Both
acetamide and furfuryl alcohol provide discriminating information for Gram
type. Although propionamide was demonstrated to have a plausible origin
within bacteria, it was not useful as a chemical marker. Propionamide is
produced at similar levels by Gram positive and Gram negative bacteria and at
an order of magnitude lower than acetamide.

As fungi do not contain peptidoglycan, the levels of a pyrolysis
product generated from fungi provide a measure of non-peptidoglycan derived
sources. The amount of acetamide is typically much higher for Gram positive
bacteria than fungi, with Gram negative bacteria having only slightly higher
levels than fungi. The peptidoglycan content of Gram negative bacteria is
low, making it difficult to discriminate between Gram negative organisms and
fungi on the basis of the small amount of acetamide generated alone. The
higher amounts of peptidoglycan present in some Gram positive organisms
elevate the levels of acetamide produced upon pyrolysis above the background
levels produced from other sources in Gram negative organisms and fungi. The
inclusion of fungi in such studies is valuable as a measure of background
derived from sources other than peptidoglycan.

Furfuryl alcohol is typically higher in the pyrograms of Gram negative
bacteria; however, Gram positive bacteria can not be discriminated well from
fungi on the basis of the small amounts of furfuryl alcohol generated. The
prominent furfuryl alcohol peak in the pyrogram of S. pyogenes whole cells is

187

considerably diminished in pyrograms of isolated cell walls. Pyrograms of two LPS isolates from different organisms (*E. coli* and *S. typhimurium*) also did not contain appreciable furfuryl alcohol, indicating that LPS is probably not a major source of furfuryl alcohol in Gram negative bacteria.[9] As the only major carbohydrate that distinguishes Gram negative from Gram positive bacteria is LPS, it is surprising that LPS is not a major source for furfuryl alcohol.

How can one rationalize that furfuryl alcohol can discriminate Gram negative bacteria from Gram positive bacteria or fungi? All contain appreciable amounts of DNA and should generate similar amounts of furfuryl alcohol. In analogy to recent work in fast atom bombardment of microbial cells, the accessibility of cellular components may determine efficiency of volatilization.[42] Gram negative bacteria have relatively thin cell walls in comparison to Gram positive bacteria and fungi. For example, it is difficult to disrupt Gram positive bacterial and fungal cell walls by sonication, whereas Gram negative bacteria are readily disrupted. We speculate that levels of furfuryl alcohol may be more representative of DNA accessibility rather than the amount present.

Figure 4 presents the results of a previously unpublished study of the pyrolysis of a group of microorganisms. Relative amounts of acetamide and furfuryl alcohol are plotted for each sample analyzed. The amount of each pyrolysis product was measured as the integrated reconstructed ion intensity (m/z 59 for acetamide, m/z 98 for furfuryl alcohol) at the appropriate retention time as a percentage of the total ion intensity. This procedure differs from previous studies in that samples were not weighed and the assumption was made that sample weight was proportional to total ion count. In this study absolute amounts of the pyrolysis products were also not determined by external calibration with a standard.

In Figure 4, replicate pyrolysis measurements using the same strain are connected by lines to give an indication of variability. Fungal samples produced low levels of acetamide and furfuryl alcohol and clustered close to the origin. Gram positive bacterial samples produced higher levels of acetamide and moderate levels of furfuryl alcohol in comparison to the Gram negative samples. Gram negative bacteria produced the highest amounts of furfuryl alcohol, but only moderate amounts of acetamide. Diagonal dividing lines can be drawn separating the three groups, although some samples are close to these discriminators.

Figure 5 shows representative Py-GC/MS results from three different bacterial species: (A) *Bacillus anthracis* (a Gram positive bacterium), (B) *Escherichia coli* (a Gram negative bacterium) and (C) *Pseudomonas aeruginosa* (also Gram negative). The *B. anthracis* pyrogram contains a high amount of acetamide and only a small amount of furfuryl alcohol; the *P. aeruginosa* pyrogram shows the reverse with a higher furfuryl alcohol peak. These two pyrograms may be located in Figure 4 in the lower right and upper left quadrants respectively. The *E. coli* pyrogram shows intermediate levels of acetamide and furfuryl alcohol and in Figure 4 points representing *E. Coli* lie close to the diagonal separating line between Gram positive and Gram negative bacteria.

Another pyrolysis product that recently attracted our attention is a peak that we tentatively identified as 2-acetylamino-2-butenoic acid (Figure 5). This potential marker was of interest because it could conceivably represent a larger fragment that is more representative of peptidoglycan. A degradation scheme rationalizing the production of this potential marker from the peptidoglycan backbone is shown in Figure 6. Figure 7 presents a three-

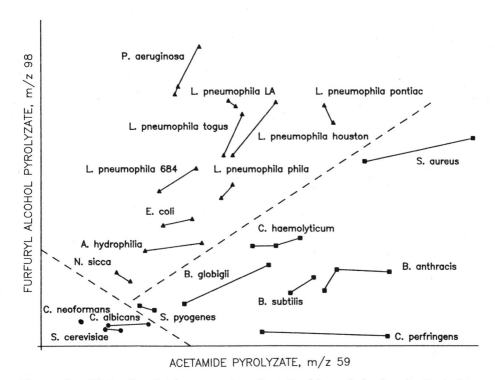

Figure 4. Plot of relative amounts of acetamide and furfuryl alcohol
in a group of microorganisms. Amounts are measured as the
integrated reconstructed ion intensity (m/z 59 for acetamide,
m/z 98 for furfuryl alcohol) at the appropriate retention
time as a percentage of the total ion intensity. Gram
positive bacteria are represented by squares, Gram negative
bacteria by triangles, and fungi by circles.

dimensional perspective plot of the levels of acetamide (ion m/z 59),
furfuryl alcohol (ion m/z 98), and 2-acetylamino-2-butenoic acid (ion m/z
83). From this display, it can be seen that the levels of 2-acetylamino-2-
butenoic acid do not add to the discrimination previously achieved with
acetamide and furfuryl alcohol alone (Figure 4). Figure 8 shows mass spectra
of this compound found in the pyrograms of *Clostridium perfringens* (a Gram
positive organism) and *Candida albicans* (a fungus). Obviously, 2-
acetylamino-2-butenoic acid can be generated from N-acetylglucosamine
(present as chitin in *Candida*) in addition to N-acetylmuramic acid and thus
does not represent a good choice for a marker that is characteristic of the
peptidoglycan structure.

The production of a specific pyrolysis product from a microorganism
does not automatically qualify it as a chemical marker. Useful
discrimination must be clearly demonstrated. At this time, no larger
molecular weight pyrolysis products from peptidoglycan have been conclusively
shown to be highly discriminating. Although Gram positive bacteria, Gram
negative bacteria, and fungi can be distinguished as groups on the basis of
their acetamide and furfuryl alcohol pyrolysis products, it remains to be
demonstrated whether individual bacteria can be reliably Gram typed by this
procedure.

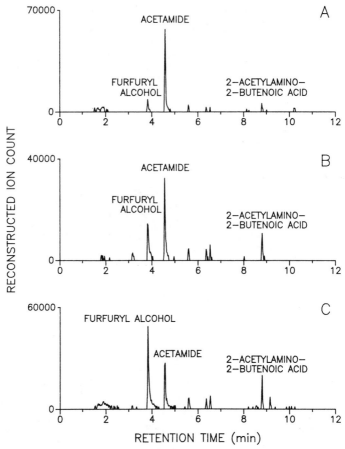

Figure 5. Total ion abundance GC/MS pyrograms for (A) *Bacillus anthracis* (Gram positive), (B) *Escherichia coli* (Gram negative) and (C) *Pseudomonas aeruginosa* (Gram negative).

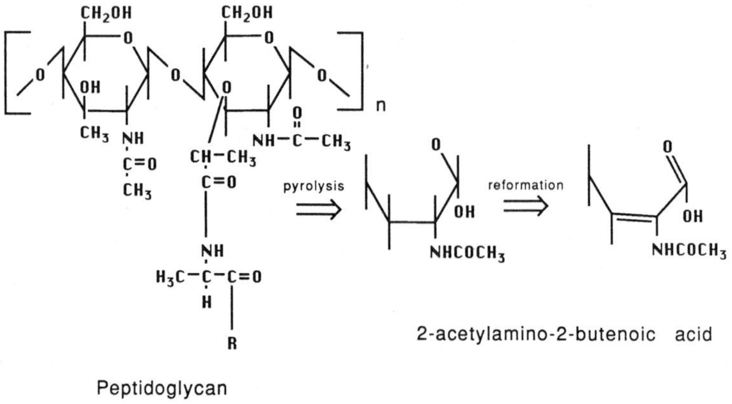

Peptidoglycan

2-acetylamino-2-butenoic acid

Figure 6. Possible pyrolysis scheme rationalizing the production of 2-acetylamino-2-butanoic acid from the peptidoglycan backbone.

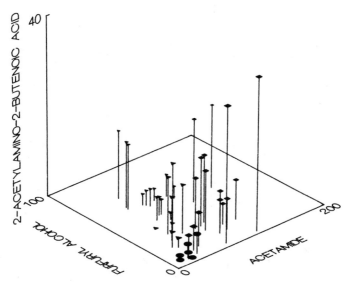

Figure 7.   Three-dimensional perspective plot of the levels of
acetamide (ion m/z 59), furfuryl alcohol (ion m/z 98),
and 2-acetylamino-2-butenoic acid (ion m/z 83) for the
same group of bacteria shown in Figure 4.

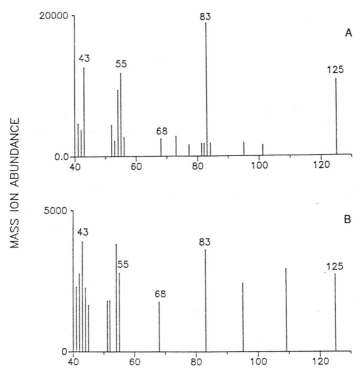

Figure 8.   EI Mass spectra of the 2-acetylamino-2-butanoic acid
pyrolysis product at a retention time of 8.9 min in
pyrograms of (A) *Clostridium perfringens* (a Gram
positive organism) (B) *Candida albicans* (a fungus).

Differentiation of Group B Streptococci by the Generation of a Specific Pyrolytic Chemical Marker

In contrast to pyrolysis products such as acetamide and furfuryl alcohol which could conceivably be generated from multiple sources within microorganisms, we have found that group B streptococci can be differentiated from groups A, C, F, and G by the presence of a unique chemical marker appearing only in pyrograms of group B organisms.[14,21,25] The differentiation of streptococci has been of interest to a number of investigators. Curie point Py-GC was used to differentiate a streptococcal strain from its mutant lacking the type III antigen by the presence or absence of peaks attributed to a polysaccharide antigen.[43] Oral streptococci from deposits on teeth and from blood cultures have also been discriminated by Curie point Py-GC.[44-46] Reproducible differentiation of *Streptococcus mutans* isolates and other bacteria has also been achieved by Py-GC.[34,35] Nonspecific detection (e.g., FID) was usually employed and the chemical identities of peaks not established.

Figure 9 shows reconstructed ion (mass m/z 86) pyrograms of a group A and a group B streptococcus. The chemical marker peak eluting at 15.6 minutes is unique to group B streptococci. The repeatability of this differentiation is shown by the fact that the instrumentation used to produce these results (CDS Pyroprobe 120 interfaced to a Hewlett-Packard 5988 GC/MS system) was entirely different from that used three years previously.[14] A different pyrolysis unit (although still a CDS Pyroprobe) was used on a different GC/MS instrument (previous work used a Hewlett-Packard 5970 mass slective detector interfaced to a Hewlett-Packard 5880 GC); the chromatographic columns in both instances were 25 m fused silica capillaries coated with free fatty acid phase (Quadrex, New Haven, CT).

The group A specific polysaccharide consists of a polyrhamnose backbone with side chains of single N-acetylglucosamine residues, the group B specific polysaccharide consists of a backbone of rhamnose and glucitol phosphate residues and trisaccharide side chains composed of rhamnose, galactose, and N-acetylglucosamine. Our interpretation of the Py-GC/MS results was based in part on derivatization GC/MS studies of streptococci.[47-49] Electron impact

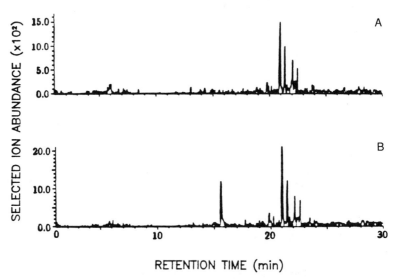

Figure 9. Reconstructed ion (m/z 86) pyrograms produced from (A) group A and (B) B streptococci. The dianhydroglucitol chemical marker elutes at 15.6 min.

(EI) MS and comparison of reconstructed ion pyrograms of glucitol, purified group B antigen, and whole streptococcal cells suggested that the chemical marker was derived from glucitol phosphate residues in the group-specific polysaccharide. The same chemical marker peak appeared at the appropriate retention time and with the same mass spectrum in all pyrograms.

Further work has employed both EI and methane chemical ionization (CI) MS to characterize this chemical marker.[15] The pyrogram of glucitol in Figure 10A indeed contains a matching peak (peak 2 at 15.6 min) at the correct retention time, but also contains several other peaks eluting immediately before and after the chemical marker peak. We have proposed that the group B chemical marker peak is dianhydroglucitol, derived by loss of two water molecules from glucitol.

The exact structure of the chemical marker is difficult to resolve using MS because of the possibility of a variety of isomeric forms and different pyrolytic degradation pathways. For example, in the first dehydration step a cyclic ether can be formed between non-adjacent carbons to produce a structure of the form:

$$
\begin{array}{c}
\text{CH}_2 \\
| \\
\text{H-C-OH} \\
| \\
\text{HO-C-H} \\
| \\
\text{H-C} \\
| \\
\text{H-C-OH} \\
| \\
\text{CH}_2\text{OH}
\end{array}
\qquad \text{[1]}
$$

Another pathway involves the loss of water from two adjacent hydroxyls to form an enol which can tautomerize to the keto form, leaving a ketone on one carbon and hydrogens on the other carbon to produce a straight chain structure of the form:

$$
\begin{array}{c}
\text{CH}_3 \\
| \\
\text{C=O} \\
| \\
\text{HO-C-H} \\
| \\
\text{H-C-OH} \\
| \\
\text{H-C-OH} \\
| \\
\text{CH}_2\text{OH}
\end{array}
\qquad \text{[2]}
$$

Derivatization GC/MS studies showed that glucitol was present in group B polysaccharide as covalently bound glucitol phosphate in an open chain structure which upon elimination of the phosphate group by acid hydrolysis formed a cyclic 1,4-anhydroglucitol.[48] The formation of cyclic 1,4-anhydroglucitol (structure [1] above) from glucitol phosphate under acid hydrolysis conditions has been employed with methanolysis to produce trimethylsilyl derivatives for GC analysis.[48,50,51] The production of anhydrosugars as an important step in the production of volatiles from carbohydrates is also well known.[1,52-54] The pyrogram of glucose typically contains 2-furaldehyde (or furfural), 5-hydroxymethyl-2-furaldehyde, levoglucosenone, levulinic acid, and 1,4:3,6-dianhydroglucopyranose. Pyrolysis of cellulose (β-1,4 glucose polymer) is also known to involve the production of anhydrosugars through dehydration. Alternate pathways may be promoted depending on the conditions.[53,54] A major degradation pathway for cellulose involves the formation of levoglucosan (1,6-anhydro-β-D-glucopyranose, a monoanhydrosugar).[1,52] Levoglucosan can lose two additional molecules of water to form levoglucosenone (a trianhydrosugar). This step involves a dehydration process similar to structure [2]. Levoglucosenone contains a ketone on carbon 2 and a double bond between carbon 3 and 4. Another path produces 1,4:3,6-dianhydroglucopyranose as an intermediate between cellulose and levoglucosenone. Anhydrosugars have been identified as pyrolysis products of chitin and N-acetylglucosamine.[55] The production of 3-acetamido-5-acetylfuran from chitin was suggested to occur following rearrangement and loss of two molecules of water. Py-GC/MS of beech wood,[56] agarose,[57] and other polyhexoses[58] have also produced anhydrohexoses and anhydropentoses.

The structure of the dehydration products of glucitol (cyclic or straight chain forms, and the position of aldehyde or ketone groups) is not

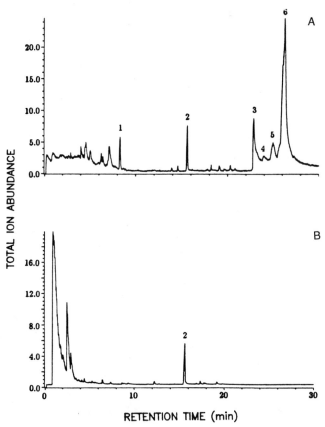

Figure 10. Total ion abundance pyrograms of (A) glucitol, and (B) glucitol-6-phosphate. The dianhydroglucitol chemical marker is peak 2, eluting at 15.6 min. Peak 1 is a trianhydroglucitol. Peaks 3-6 are monoanhydroglucitols.

readily apparent from EI and CI mass spectra. Other analytical methods (perhaps NMR or IR) are required for more confident identification. For simplicity, the straight chain structure is adopted for discussion here.

The EI mass spectrum of the dianhydroglucitol chemical marker from the total ion pyrogram of glucitol-6-phosphate is shown in Figure 11A. The small peak at m/z 146 represents the molecular ion of dianhydroglucitol (M^+). A molecular ion, $(M + H)^+$, at m/z 147 in the CI mass spectrum of Figure 11B also indicates a molecular weight of 146 daltons. EI fragment ions of m/z 103, m/z 73, m/z 43, and m/z 29 can result from a straight chain dianhydroglucitol structure as follows:

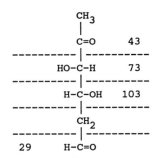

EI fragment ions of m/z 86 and m/z 69 may be formed in various ways, one of which involves loss of hydroxyl groups from the primary fragment ion at m/z 103. The molecular ion may lose another water molecule to form trianhydroglucitol, which has a molecular weight of 128 daltons, and give rise to ions of m/z 57, m/z 58, and m/z 85. A fragment ion at m/z 129 in the CI spectrum of Figure 11B can be formed by the loss of water from the protonated molecular ion at m/z 147.

EI and CI mass spectra have enabled us to identify the other pyrolysis products in the pyrogram of glucitol, Figure 10A.[15] These other peaks are dehydration products from glucitol having lost one molecule of water (monoanhydroglucitol, peak 1) and three molecules of water (trianhydroglucitol, peaks 3-6). When phosphate is bound to glucitol, the formation of monoanhydroglucitol and trianhydroglucitol dehydration products is suppressed. The dianhydroglucitol chemical marker is the only dehydration product of glucitol appearing in the pyrogram of glucitol-6-phosphate shown in Figure 10B. In the pyrogram of purified group B polysaccharide, the only glucitol dehydration product present is the dianhydroglucitol.[10] Of the six pyrolysis products from the model compound glucitol, only one peak is useful for microbial differentiation.

CONCLUSIONS

In this chapter the application of analytical pyrolysis GC/MS to the differentiation of microorganisms has been discussed. Control of the many sources of variability associated with growth, sampling, and instrumental measurement is critical for reproducibility. Steps in validating chemical markers for microbial differentiation have been illustrated by two major examples: the differentiation of Gram types, and the identification of a specific microbial species, *Streptococcus agalactiae*.

Microorganisms contain many unique structural components that are useful in chemotaxonomic characterization, both by classical biochemical and modern GC/MS methods. Some of these chemical components are unique enough to produce characteristic pyrolysis products. We have demonstrated here the

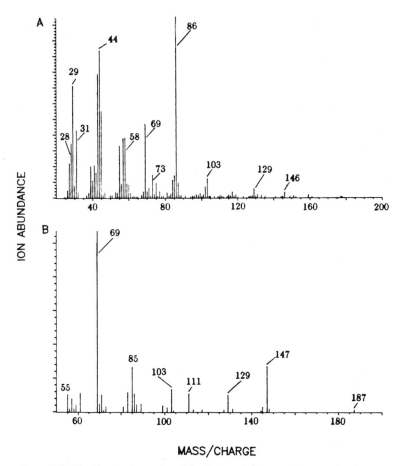

Figure 11. (A) Electron impact (70 eV) and (B) methane chemical ionization mass spectra from the dianhydroglucitol chemical marker peak in a standard sample of glucitol-6-phosphate.

potential of Py-GC/MS for rapid detection of some of these chemical markers. That a microbial structural constituent produces a particular pyrolyzate is by itself not proof of the value of that compound for chemotaxonomy. Validation of chemical markers, including the use of appropriate model compounds, isolated microbial structures, and sufficiently large numbers of strains, is required. The ideal discriminator between two groups of organisms at any taxonomic level is the absolute presence of a unique chemical marker in one group and and its absence in the other. There are, however, numerous examples of chemical components that provide useful discrimination among bacterial groups (e.g., species) despite not being completely unique to one group (e.g., acetamide and furfuryl alcohol). Computer-assisted data handling is invaluable for dealing with the large and complex data sets resulting from analytical pyrolysis of microorganisms and for recognizing subtle patterns that can differentiate groups. Multivariate statistical methods, however, are complementary to chemical approaches for pyrolysate identification and should not subtitute for determining the structure of pyrolysates and their biochemical origin within the microbial cell.

ACKNOWLEDGEMENTS

This work was supported by the U. S. Army Office of Research and by a Department of Defense University Research Instrumentation grant. The authors acknowledge the assistance of Dr. David Pritchard (University of Alabama-Birmingham) for helpful discussions and a gift of group B streptococcal polysaccharide. The helpful assistance of James C. Rogers with data presentation is appreciated.

REFERENCES

1. W. J. Irwin, "Analytical Pyrolysis," Marcel Dekker, Inc., New York (1982).
2. H. L. C. Meuzelaar, J. Haverkamp, and F. D. Hileman, "Pyrolysis Mass Spectrometry of Recent and Fossil Biomaterials," Elsevier, Amsterdam (1982).
3. C. S. Gutteridge and J. R. Norris, The application of pyrolysis techniques to the identification of microorganisms, J. Appl. Bacteriol. 47:5 (1979).
4. W. J. Irwin and J. A. Slack, Analytical pyrolysis in biomedical studies, Analyst 103:673 (1978).
5. F. L. Bayer, and S. L. Morgan, The analysis of biopolymers by analytical pyrolysis gas chromatography, in: "Pyrolysis and GC in polymer analysis," E. Levy and S. A. Liebman, eds., Marcel Dekker, New York (1985).
6. Huis In't Veld, H. L. C. Meuzelaar, A. Tom, Analysis of streptococcal cell wall fractions by Curie-point pyrolysis gas-liquid chromatography, Appl. Microbiol. 26:92 (1973).
7. G. Dahlen and I. Ericsson, Differentiation between Gram-negatiave anaerobic bacteria by pyrolysis gas chromatography of lipopolysaccharides, J. Gen. Microbiol. 129:557 (1983).
8. J. R. Hudson, S. L. Morgan, and A. Fox, Quantitative pyrolysis gas chromatography-mass spectrometry of bacterial cell walls, Anal. Biochem. 120:59 (1982).
9. L. W. Eudy, M. D. Walla, J. R. Hudson, S. L. Morgan, and A. Fox, Gas chromatography-mass spectrometry studies on the occurence of acetamide, propionamide, and furfuryl alcohol in pyrolyzates of bacteria, bacterial fractions, and model compounds, J. Anal. Appl. Pyrol. 7:231 (1985).
10. G. Montaudo, Current problems in pyrolysis, J. Anal. Appl. Pyrol. 13:1 (1988).
11. G. Holzer, T. F. Bourne, and W. Bertsch, Analysis of in situ methylated fatty acid constituents by curie-point pyrolysis gas chromatography-mass spectrometry, J. Chromatogr. 468:181 (1989).
12. P. G. Simmonds, Whole microorganisms studied by pyrolysis-gas chromatography-mass spectrometry: Significance for extraterrestrial life detection experiments, Appl. Microbiol. 20:567 (1970).
13. L. W. Eudy, M. D. Walla, S. L. Morgan, and A. Fox, Gas chromatographic-mass spectrometric determination fo muramic acid content and pyrolysis profiles for a group of Gram-positive and Gram-negative bacteria, Analyst 110:381 (1985).
14. C. S. Smith, S. L. Morgan, C. D. Parks, A. Fox, and D. G. Pritchard, Chemical marker for the differentiation of group A and group B streptococci by pyrolysis gas chromatography-mass spectrometry, Anal. Chem. 59:1410 (1987).
15. K. Ueda, S. L. Morgan, and A. Fox, The origin of dianhydroglucitol, a carbohydrate chemical marker generated by pyrolysis from group B streptococci, Anal. Chem., submitted (1989).

16. R. E. Aries, C. S. Gutteridge, and T. W. Ottley, Evaluation of a low-cost, automated pyrolysis-mass spectrometer, J. Anal. Appl. Pyrolysis. 9:81 (1986).

17. A. M. Harper, H. L. C. Meuzelaar, G. S. Metcalf, and D. L. Pope, Numerical techniques for processing pyrolysis mass spectra data, in: Analytical Pyrolysis, techniques and applications," K. J. Voorhees, ed., Butterworths, London (1984).

18. W. Windig and H. L. C. Meuzelaar, Numerical extraction of components from mixture spectra by multivariate analysis, in: "Computer-enhanced analytical spectroscopy", H. L. C. Meuzelaar and T. L. Isenhour, eds., Plenum, New York (1987).

19. H. J. H. MacFie and C. S. Gutteridge, Comparative studies on some methods for handliong quantitative data generated by analytical pyrolysis, J. Anal. Appl. Pyrolysis 4:175 (1982).

20. K. J. Voorhees, S. L. Durfee, and D. M. Updegraff, Identification of diverse bacteria grown diverse conditions using pyrolysis-mass spectrometry, J. Microbiol. Methods 8:315 (1988).

21. C. S. Smith, S. L. Morgan, and A. Fox, Discrimination and clustering of streptococci by pyrolysis gas chromatography-mass spectrometry, J. Anal. Appl. Pyrolysis, in press (1990).

22. E. Reiner and W. H. Ewing, Chemotaxonomic studies of some Gram negative bacteria by means of pyrolysis-gas-liquid chromatography, Nature 217:191 (1968).

23. C. S. Gutteridge and J. R. Norris, Effect of different growth conditions on the discrimination of three bacteria by pyrolysis gas-liquid chromatography, Appl. Environ. Microbiol. 40:462 (1980).

24. H. Engman, H. T. Mayfield, T. Mar, and W. Bertsch, Classification of bacteria by pyrolysis-capillary column gas chromatography-mass spectrometry and pattern recognition, J. Anal. Appl. Pyrolysis 6:137 (1984).

25. J. Gilbart, A. Fox and S. L. Morgan, Carbohydrate profiling of bacteria by gas chromatography-mass spectrometry: chemical derivatization and analytical pyrolysis, Eur. J. Clin. Micro. 6:715 (1987).

26. G. Wells, K. J. Voorhees, and J. H. Futrell, Heating profile curves for resistively heated filament pyrolyzers, Anal. Chem. 52:1782 (1980).

27. R. L. Levy, Trends and advances in design of pyrolysis units for gas chromatography, J. Gas Chromatogr. 5:107 (1967).

28. W. Windig, P. G. Kistemaker, J. Haverkamp, and H. L. C. Meuzelaar, The effects of sample preparation, pyrolysis and pyrolyzate transfer conditions on pyrolysis mass spectra, J. Anal. Appl. Pyrolysis 1:39 (1979).

29. W. Windig, P. G. Kistemaker, J. Haverkamp, and H. L. C. Meuzelaar, Factor analysis of the influence of changes in experimental conditions in pyrolysis mass spectrometry, J. Anal. Appl. Pyrolysis 2:7 (1980).

30. A. van der Kaaden, R. Hoogerbrugge, and P. G. Kistemaker, Effect of sample layer thickness and temperature rise time on the pyrolysis temperature of cellulose, J. Anal. Appl. Pyrolysis 9:267 (1986).

31. J. A. Adkins, T. H. Risby, J. J. Scocca, R. E. Yasbin, and J. W. Ezzell, Linear-programmed thermal degradation methane chemical-ionization mass spectrometry. I. Peptidoglycan, cell walls, and related compounds from *Bacillus*, J. Anal. Appl. Pyrol. 7:15 (1984).

32. W. Windig, E. Jakab, J. M. Richards, and H. L. C. Meuzelaar, Self-modelling curve resolution by factor analysis of a continuous series of pyrolysis mass spectra, Anal. Chem. 59:317 (1987).

33. W. Windig, S. A. Liebman, M. B. Wasserman, and A. P. Snyder, Fast self-modelling curve resolution for time resolved mass spectral data, Anal. Chem. 60:1503 (1988).

34. G. L. French, I. Phillips, S. Chin, Reproducible pyrolysis-gas chromatography of micro-organisms with solid stationary phases and isothermal oven conditions, J. Gen. Microbiol. 125:347 (1981).

35. G. L. French, H. Talsania,and I. Philips, Identification of viridans stretococci by pyrolysis-gas chromatography, Med. Microbiol. 29:19 (1989).

36. K. H. Schleifer and O. Kandler, Peptidoglycan types of bacterial cell walls and their taxonomic implications. Bacteriol. Rev. 36:407 (1972).

37. E. E. Medley, P. G. Simmonds, and S. L. Manatt, Pyrolysis-gas chromatography mass-spectrometry study of *Actinomycete streptomyces-longisporoflavis*, Biomed. Mass Spectrom. 2:261 (1975).

38. J. Haverkamp, G. Wieten, A. J. H. Boerboom, J. W. Dallinga, and N. M. M. Nibbering, Pyrolysis-collisionally activated dissociation mass spectrometry of organic model compounds and bacterial samples, in: "Analytical Pyrolysis-- Techniques and Applications", K. J. Voorhees, ed., Butterworths, London, p. 305 (1984).

39. K. Kato, Pyrolysis of cellulose. Part III. Comparative studies of the volatile compounds from pyrolysates of cellulose and its related compounds, Agr. Biol. Chem., 31:657 (1967).

40. M. A. Posthumus, N. M. M. Nibbering, A. J. Boerboom, and H.-R. Schulten, Pyrolysis mass-spectrometric studies on nucleic-acids, Biomed. Mass Spectrom. 1:352 (1974).

41. A. H. Rose, "Chemical Microbiology," Plenum, New York (1976).

42. C. Fenselau and R. Cotter, Chemical aspects of fast atom bombardment, Chem. Rev. 87:501 (1987).

43. Huis In't Veld, H. L. C. Meuzelaar, A. Tom, Analysis of streptococcal cell wall fractions by Curie-point pyrolysis gas-liquid chromatography, Appl. Microbiol. 26:92 (1973).

44. Stack, M. V.; Donoghue, H. D.; Tyler, J. E., Discrimination between oral streptococci by pyrolysis gas-liquid chromatography, Appl. Environ. Microbiol. 35:45 (1980).

45 M. V. Stack, H. D. Donoghue, J. E. Tyler, M. Marshall, Comparaison of oral streptococci by pyrolysis gas-liquid chromatography, in: "Analytical Pyrolysis" C. E. R. Jones, C. A. Cramers, eds., Elsevier, Amsterdam, p. 57 (1977).

46. M. V. Stack, H. D. Donoghue, and J. E. Tyler, Differentiation of *Streptococcus mutans* serotypes by discriminant analysis of pyrolysis-gas-liquid chromatographic data, J. Anal. Appl. Pyrolysis 3:221 (1981/1982).

47. D. Pritchard, J. E. Colligan, S. E. Speed, and B. M. Gray, Carbohydrate fingerprints of streptococcal cells, J. Clin. Microbiol. 13:89 (1981).

48. D. G. Pritchard, G. B. Brown, B. M. Gray, and J. E. Coligan, Glucitol is present in the group-specific polysaccharide of group B streptococcus, Current Microbiol. 5:283 (1981).

49. D. G. Pritchard, B. M. Gray, and H. C. Dillon, Characterization of the group-specific polysaccharide of group B streptococcus, Arch. Biochem. Biophys. 235:385 (1984).

50. J. Szafranek and A. Wisniewski, Gas chromatographic and mass spectrometric analyses of the acid-catalyzed dehydration reactions of D-mannitol, J. Chromatogr. 161:213 (1978).

51. G. J. Gerwig, J. P. Kamerling, and J. F. G. Vliegenthart, Anhydroalditols in the sugar analysis of methanolysates of alditols and oligosaccharide-alditols, Carbohydr. Res. 129:149 (1984).

52. A. Ohnishi, K. Kato, and E. Takagi, Curie-point pyrolysis of cellulose, Polymer J. 7:431 (1975).

53. F. Shafizadeh, Introduction to pyrolysis of biomass, J. Anal. Appl. Pyrol. 3:283 (1982).

54. A. D. Pouwels, G. B. Eijkel, and J. J. Boon, Curie-point pyrolysis-capillary gas chromatography-high-resolution mass spectrometry of microcystalline cellulose, J. Anal. Appl. Pyrol. 14:237 (1989).

55. R. A. Franich, S. J. Goodin, and A. L. Wilkins, Acetamidofurans, acetamidopyrones, and acetamidoacetaldehyde from pyrolysis of chitin and N-acetylglucosamine, J. Anal. Appl. Pyrol. 7:91 (1984).

56. A. D. Pouwels, A. Tom, G. B. Eijkel, and J. J. Boon, Characterisation of beech wood and its holocellulose and xylan fractions by pyrolysis-gas chromatography-mass spectrometry, J. Anal. Appl. Pyrol. 11:417 (1987).

57. R. J. Helleur, E. R. Hayes, W. D. Jamieson, and J. S. Craigie, Analysis of polysaccharide pyrolysate of red algae by capillary gas chromatography-mass spectrometry, J. Anal. Appl. Pyrol. 8:333 (1985).

58. A. van der Kaaden, J. J. Boon, and J. Haverkamp, The analytical pyrolysis of carbohydrates. 2-- Differentiation of homopolyhexoses according to their linkage type, by pyrolysis-mass spectrometry and pyrolysis-gas chromatography/mass spectrometry, Biomed. Mass Spectrom. 11:486 (1984).

CHAPTER 13

RAPID CHARACTERIZATION OF MICROORGANISMS BY CURIE-POINT PYROLYSIS IN COMBINATION
WITH SHORT-COLUMN CAPILLARY GAS CHROMATOGRAPHY AND ION TRAP MASS SPECTROMETRY

A. Peter Snyder

U. S. Army Chemical Research,
Development and Engineering Center
Aberdeen Proving Ground, MD 21010-5423

William H. McClennen and Henk L. C. Meuzelaar

Center for Micro-Analysis and Reaction Chemistry
University of Utah
Salt Lake City, UT 84112

INTRODUCTION

Most conventional methods for identification of microorganisms involve
characterizing intact, viable organisms under physiological conditions.
Physicochemical characterization methods such as gas chromatography (GC),
mass spectrometry (MS), or combined GC/MS necessarily involve a more
artificial set of experimental parameters. Because of the chemical
complexity of most physiological media, microorganisms usually need to be
separated from the growth matrix. The separated organisms are then generally
submitted to extraction and derivatization methods which may take several
hours, followed by GC or GC/MS analyses requiring between 30 and 60 minutes.

Profiling of monomeric carbohydrate and fatty acid content in
microorganisms provides chemotaxonomic information.[1-7] A (recently
discontinued) commercial system[8] takes advantage of a derivatization method
with which a GC pattern of fatty acid methyl esters is produced from a 10-15
mg sample of organisms.[9] Profiling of bacterial constituents after
derivatization is discussed in further detail in Chapters 4-6.

Mass spectrometry techniques used in the analysis of complex, high
molecular weight substances, including whole microorganisms, have been
predominantly based on pyrolysis methods such as Curie-point pyrolysis,[10-14]
resistively-heated filament pyrolysis,[15-16] quartz tube pyrolysis,[17-23] oven
pyrolysis,[24-26] and direct insertion probe pyrolysis.[27-35] Mass spectra of
microorganisms and high molecular weight biopolymers obtained with pyrolysis
inlets with relatively cold reactor walls or involving so-called "expansion
chambers" are predominantly characterized by low mass range features (e.g.,
below m/z 200). However, direct probe pyrolysis MS techniques which minimize
cold spots tend to produce high mass features (e.g., in the 200-1000 amu mass
range) as well.[30-35] However, as has been pointed out before[36], the price to
be paid for observing large, nonvolatile pyrolysis products is often much
more rapid contamination of the ion source, thereby endangering prospects of
long-term reproducibility.

Analytical Microbiology Methods
Edited by A. Fox *et al.*
Plenum Press, New York, 1990

Highly sophisticated sample introduction techniques for MS that have shown great promise in producing key information from high molecular weight biological substances are pyrolysis field desorption (Py-FD),[37,38] fast atom bombardment (FAB),[27,28,39-42] secondary ion MS (SIMS),[27,43] laser desorption (LD),[44,45] and plasma desorption (PD).[27,41,46] However, the sample preparation and introduction procedures into the MS ion source tend to be less straightforward. Moreover, the MS hardware and operating procedures involved are generally more complex and costly than those associated with the above mentioned thermal procedures. Various lipids, including triglycerides and diglycerides in the m/z 700-1000 molecular weight range, have been observed in FAB, LD, and PD spectra obtained from intact microorganisms.[40,41] Treatment of whole cells with toluene provides greater mass spectral response for lipids. Cell membrane lipids are covered by the cell wall and are not readily accessible. Toluene presumably releases the lipids, thereby making them availbale for desorption or pyrolysis. Features in the thousands of daltons mass range are routinely found in desorption mass spectra. Such desorption techniques can characterize not only lipids but also polypeptides, hormones, enzymes, deoxyribonucleotides, and polysaccharides.[27,44,46]

Le Clerq et al.[47] and Yost et al.[48] have presented data on the advantages of using short (1.0-5.0 m) capillary GC columns interfaced to the low pressure source of a mass spectrometer. Particularly attractive characteristics of such short columns (as compared to columns 10 m or longer) are (1) dramatically reduced analysis times due to increased carrier gas velocities; (2) surprisingly little loss of resolution; and (3) elution of labile compounds at substantially lower temperatures.

McClennen et al.[49] and Richards et al.[50] have reported the use of special Curie-point pyrolysis techniques in combination with 4-5 m fused silica capillary columns coupled to so-called "Ion Trap Detector" and "Ion Trap Mass Spectrometry" (ITMS) devices. The data obtained provide structural information for a wide range of complex biological compounds including proteins, nucleic acids, carbohydrates, and phospholipid-derived diglycerides. In most cases, a useful degree of GC separation can be obtained in less than 5 min. Furthermore, sample requirements are minimal, typically in the 10^{-6}-10^{-7} g range.

This chapter explores the application of short capillary column Py-GC/MS techniques using an "ion trap" mass spectrometer for rapid characterization of microorganisms.

EXPERIMENTAL

A total of twelve strains representing eight bacterial species were studied. *Bacillus anthracis* (strains BO463 and BO464), *B. cereus* (BO037), *B. thuringiensis* (BO158 and BO150), *B. licheniformis* (BO017 and BO089), *B. subtilis* (BO014 and BO095), *Staphylococcus aureus*, and *Esherichia coli* (type 0127) were supplied by Drs. Tony P. Phillips (CDE, Porton Down, UK) and Leslie A. Shute (University of Bristol, Bristol, UK). *Legionella pneumophila* serogroup 1 was from Luc Berwald (FOM Institute, Amsterdam, The Netherlands). Most bacteria were grown in Lab M nutrient broth for 3 days at 37°C. *Bacillus anthracis* required 3 days for adequate growth. Some bacteria were cultured for shorter time periods. The cells were harvested by centrifugation, washed with sterile water and then resuspended in water. Cells were killed by adding an equal volume of 6% H_2O_2 and incubating overnight. Samples were then centrifuged, washed, and freeze-dried. *Legionella pneumophila* (1 mg/mL suspension) was heat-killed at 120°C and then lyophilized.

Suspensions of the organisms (1-1.7 mg/mL) were prepared by adding 0.15 mL of methanol to 0.5 mg of the lyophilized bacteria, sonicating to effect a uniform dispersion, and then adding 0.15 mL of deionized distilled water. Three microliters, or approximately 5 μg of bacteria, were applied to the tip of a Curie-point wire, and the suspension dried in a stream of warm (40°C) air.

Py-GC/MS experiments were performed on a system including a Hewlett Packard 5890 GC (Palo Alto, CA) and a Finnigan MAT ion trap mass spectrometer (San Jose, CA). The pyrolyzer consisted of our own (BPC-1000) low dead volume Curie-point pyrolysis GC inlet and a 1 MHz, 1.5 kW Curie-point power supply (Fischer 0310). The sample on the ferromagnetic wire was placed in the 280°C inlet and then heated to the 610°C Curie-point temperature in a 1 s pyrolysis. A 5 m X 0.32 mm i.d. capillary column coated with a 0.25 μm film of dimethylsilicone (J & W Scientific, SE-30) was operated with a He carrier gas at 170 cm/s linear velocity. The column was temperature programmed from 100 to 300°C at 40°C/min and held isothermal for 5 min. The column was connected directly into the ITMS vacuum through a 1 m transfer line maintained at 300°C with the vacuum manifold at 180°C. The ion trap was run without the normal teflon spacer rings to allow better conductance of the high (8 mL/min) carrier gas flow. Spectra were accumulated over the mass range 100 to 620 m/z at a rate of 1 scan/s for the 8 min analysis.

RESULTS AND DISCUSSION

Figure 1 presents the total ion current chromatogram (TIC) of a 6 min Py-GC/ITMS analysis of an avirulent strain of *Bacillus anthracis* (BO464). Despite the visual simplicity of the TIC, far more information is present than can be readily extracted and presented in a comprehensive manner. However, an informative, initial impression can be achieved by examining reconstructed ion chromatograms (RIC's) of selected m/z values from the complex total ion patterns.

Lipids constitute an important constituent of all bacteria, and because of their documented diversity and abundance, the GC/MS pattern of *B. anthracis* (BO464) was investigated for the presence of lipid components. High molecular weight profiles (m/z 494-564) were observed (Figure 1) at regular 14 amu intervals at progressively longer elution times. Since CH_2 groups have an m/z of 14 these were hypothesized to represent fragments of the fatty acid moieties of the diglyceride-H_2O lipid components. Direct probe and GC/MS data on model compounds were consistent with this hypothesis. Other lower molecular weight compounds (see Figure 1) appear in the same elution envelope and appeared to include monoglyceride-H_2O, presumably representing more extensive pyrolysis products of diglycerides.

The cluster of diglyceride-H_2O peaks is conveniently removed from the majority of smaller bacterial pyrolysis products and appears to be among the last of the eluting pyrolyzate materials. Several extracted mass spectra of individual diglyceride-H_2O features are shown in Figure 2. Characteristic mass spectra are obtained for all major diglyceride-H_2O peaks within their respective elution windows. This is especially true for the m/z 564 species (compare the top and bottom spectra in Figure 2). The mass spectra in Figure 2 are similar to those of diglyceride standards.[31-35] Some of the dehydrated monoglyceride and diglyceride features seen in Figures 1 and 2 as well as different lipid species have been observed in direct probe[33] and Curie-point[31-35] pyrolysis mass spectra. Solid probe work by Risby and Yergey has provided ion profiles in the m/z 259-543 range, but no attempt was made to identify the underlying compounds.[51] The addition of a GC dimension, however, allows a more precise identification of individual diglyceride-H_2O

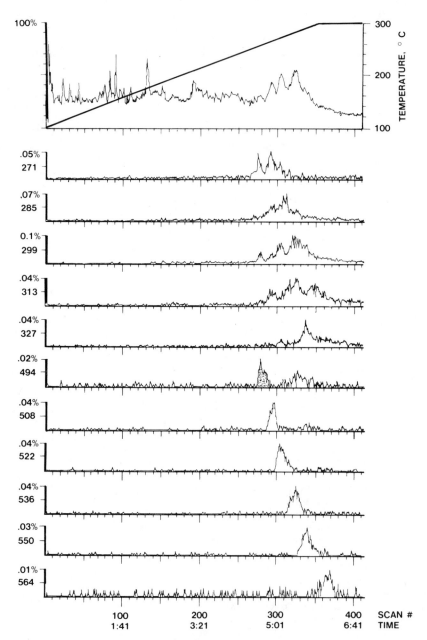

Figure 1. The top panel displays the Py-GC/ITMS total ion current of a 3 day culture of *B. anthracis* (BO464). Subsequent panels show selected reconstructed ion current profiles as percentages of the TIC.

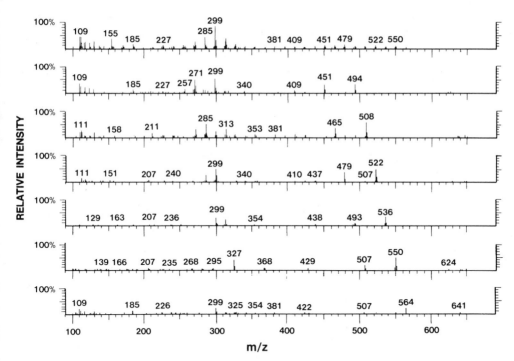

Figure 2. Background-subtracted, integrated mass spectra
that correspond to the individual features observed in
the Figure 1 m/z RICs in the following scan regions
from top to bottom: scans 250-390, 274-286, 287-298,
299-315, 316-332, 333-352, and 356-374.

species. We believe our study represents the first Py-GC/MS characterization
of underivatized mono- and diglycerides obtained by pyrolysis of intact
bacteria.

Reproducibility. Figure 3 shows replicate TIC's for 3 different
strains of bacilli, and each respective set displays very similar
chromatograms. Sharper, narrower peaks within each set, however, show some
variability, e.g., in the 70-100 mass scan region of the B. thuringiensis
(Figure 4b). and B. subtilis (Figure 4c) chromatograms. One of the replicate
chromatograms for B. thuringiensis (Figure 3b) displays a broader feature
(shaded) in the 240-260 mass scan region. The TICs in Figure 3 however
appear remarkably similar in each set of replicates. The data in Figure 3
demonstrate that short-term reproducibility of Py-GC-MS of bacteria is good
and appears to be comparable to that in other studies of microorganisms using
Py-GC[17,21,52,53] and Py-GC-MS.[20]

Culture age. The effect of culture age on appearance of chromatograms
was investigated. TIC's of 24 h and 3 day cultures of the B. anthracis, B.
subtilis, and B. thuringiensis are compared in Figure 4. At less than 250
mass scans, B. anthracis (Figure 4a) and B. subtilis (Figure 4b) display
minor differences with culture age while B. thuringiensis (Figure 4c) shows
considerably more age-related variation. The lipid region of the TIC
displays a noticeable and reproducible difference in relative abundance
(shaded portions) for each species in Figure 4. In each case there appears
to be a relatively greater amount of lipid material in the younger (24 h)
cultures. Other studies of the influence of culture age (over one to seven
days) on Py-GC and Py-GC/MS of microorganisms have shown significant
variation in only one or two peaks in the chromatograms. Organisms studied

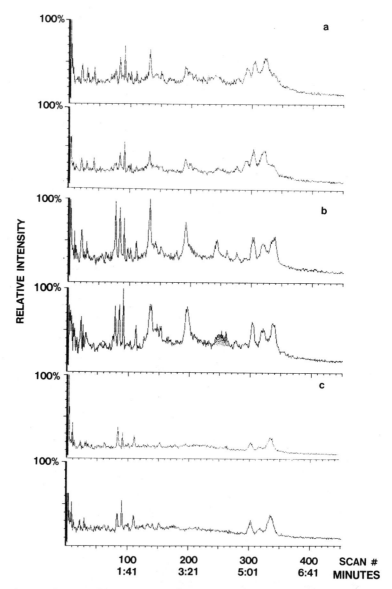

Figure 3. Replicate Py-GC/ITMS TIC plots of 3 day cultures
of (a) *B. anthracis* (BO464) (b) *B. thuringiensis*
(BO150) (c) *B. subtilis* (BO095).

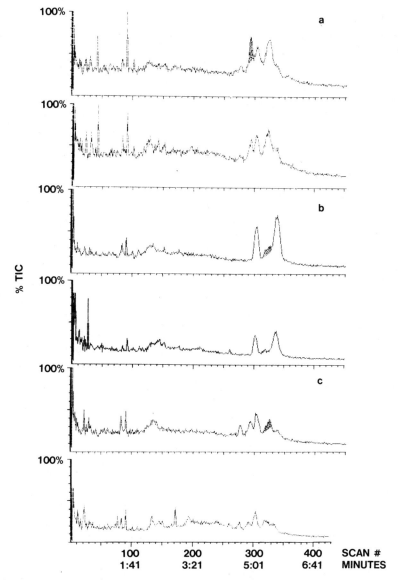

Figure 4. Py-GC/ITMS TIC plots as a function of culture age. The top and bottom panels represent 24 h and 3 day cultures respectively for the following bacterial strains: (a) *B. anthracis* (BO463) (b) *B. subtilis* (BO014) (c) *B. thuringiensis* (BO158).

included *F. meningosepticum*[20] and *B. cereus*, *B. firmus*, and *B. subtilis* var. *niger*.[54] The different chromatographic conditions compared to the present studies resulted in different pyrolysis profiles. However, all such studies demonstrate a high degree of similarity among profiles of organisms cultured for different time periods.

Diglyceride analysis. Extraction and derivatization procedures have been used successfully for the discrimination of microorganisms by probing their lipid content both qualitatively and quantitatively. We performed a detailed study of 12 bacterial strains (eight species, cultured 3 days) using selected features of their TIC's after pyrolysis and GC separation. The pyrolysis TIC's of these strains displayed a high degree of reproducibility.

Figures 5-6 show diglyceride-H_2O information from 2 strains of each of 4 different species of bacilli and from one strain from a fifth (*B. cereus*). Displayed in Figures 5-6 are the portions of the TIC relevant to the lipid profile and the m/z 494, 508, and 564 RIC's for each. Figure 7 presents similar data from a Gram positive organism (*Staphylococcus aureus*) and two Gram negative organisms (*E. coli* and *L. pneumophila*). RIC's for m/z 522, 536, and 550 for all strains were very similar to that of *B. anthracis* (BO464) (compare with Figure 1). The maximum intensity in each chromatogram, regardless of its retention time, is plotted as a fraction of the total ion intensity for 6 m/z features for each *Bacillus* strain. A similar procedure was performed for *S. aureus* and the two Gram negative bacteria (Figure 7). The TIC profile, the distribution plot of the maximum intensity of each RIC, as well as the relative abundance or intensity distribution and retention time (t_R) of selected m/z (RIC) profiles were then used as criteria for visually determining the relative similarity/difference of the bacterial strains and species with each other (Figures 5-7).

B. anthracis. The TIC profiles of both strains are very similar; however, they have a unique intensity distribution in comparison to that of other organisms. Of all the Bacillus strains tested, only *B. anthracis* has the m/z 564 feature, and both *B. anthracis* strains can be distinguished by the abundance pattern of the 6 selected diglyceride-H_2O compounds.

B. cereus. The TIC profile of this organism is easily distinguished visually from that of *B. anthracis*. The reconstructed ion chromatogram of *B. cereus* (B0037) showed two features in both the m/z 494 and the m/z 508 RIC's. Although the distribution of maximum reconstructed ion intensities were similar to that of the *B. anthracis* strains, the m/z 564 feature was absent and different retention time values for m/z 494 and 508 features were seen.

B. subtilis. Both strains showed similar total and reconstructed ion profiles. More information is required to distinguish these two strains.

B. thuringiensis. The total ion chromatograms, retention times in the m/z 494 RIC, and the m/z 508 RIC for the two strains are quite different. The BO150 strain is similar in m/z intensity distribution compared to the *B. subtilis* strains, except for the presence of an extra peak in the m/z 508 RIC along with a significantly different TIC.

B. licheniformis. The TIC of both strains are very different, even though they display similar retention times for the five m/z features and corresponding peak intensities in the RIC profiles. The intensity distribution of the five high molecular weight masses of each *B. licheniformis* strain is similar to that of both *B. thuringiensis* strains. However, *B. thuringiensis* (BO158) has different m/z 494/508 t_R values; secondly, *B. thuringiensis* (BO150) has an extra m/z 508 RIC peak; *B. licheniformis* and *B. thuringiensis* strains have sufficiently different TIC's

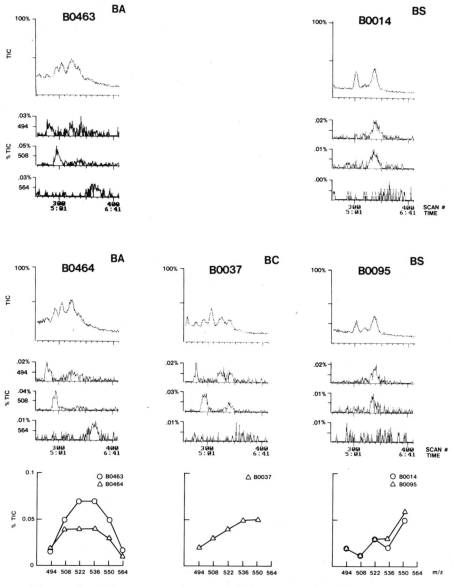

Figure 5.   TIC, selected RIC profiles, and the RIC maximum
intensity distribution plots for strains of
B. anthracis (BA), *B. cereus* (BC), and *B. subtilis*
(BS).

209

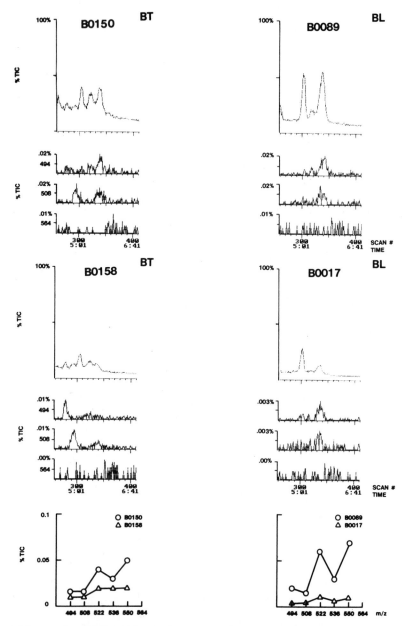

Figure 6.   TIC, selected RIC profiles, and the RIC maximum
intensity distribution plots for strains of *B.*
thuringiensis (BT) and *B. licheniformis* (BL).

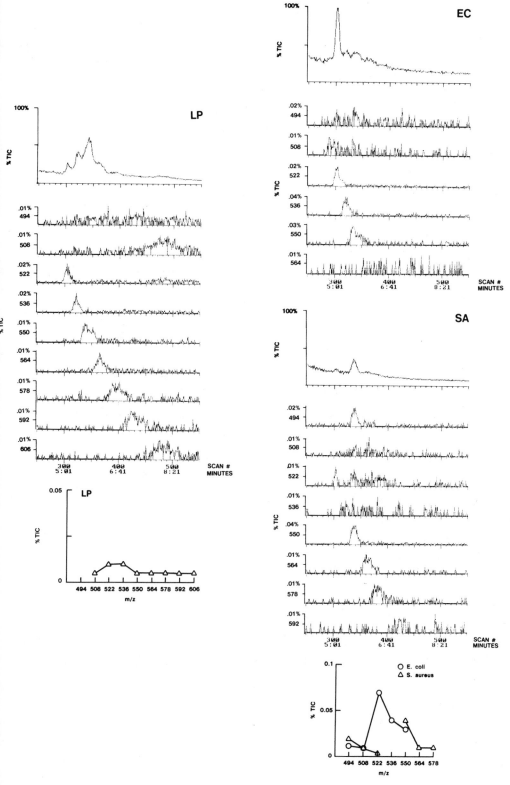

Figure 7. TIC, selected RIC profiles, and the RIC maximum
intensity distribution plots for *Legionella pneumophila*
(LP), *E. coli* (EC), and *S. aureus* (SA).

to enable visual discrimination. All four *B. subtilis* and *B. licheniformis* strains show similar retention times at each respective m/z value. The TIC of *B. licheniformis* (BO089) and both *B. subtilis* strains are similar, but the *B. subtilis* strains collectively differ from both *B. licheniformis* strains by the intensity distribution profiles of the five high molecular weight lipid species.

In general, it appears the m/z intensity distribution of the *Bacillus* strains increases with increasing m/z with the exception that *B. anthracis* strains show a parabolic distribution about the 494-564 amu region.

E. coli and S. aureus. The TIC's of these organisms are different compared to the bacilli. *E. coli* appears to contain a greater majority of relatively low molecular weight diglyceride-H_2O species while *S. aureus* contains a greater amount of relatively high molecular weight diglyceride-H_2O compounds.

L. pneumophila. The RIC profile maximum intensity distribution for this organism is similar to that of *B. anthracis* except that m/z 494 is absent and distinct differences in lipid distribution are observed in the m/z 578, 592, and 606 RIC profiles. Furthermore, the lipid region of the TIC for *L. pneumophila* and *B. anthracis* are markedly different in appearance.

Classification of bacteria using lipid components

Microbiological classification of bacilli by lipid composition is based on the Kaneda groups.[55,56] The proportion of unsaturated fatty acids along with the carbon chain length serve to categorize bacilli into A-F groups. *B. licheniformis* and *B. subtilis* belong to Group A and *B. anthracis*, *B. licheniformis*, and *B. thuringiensis* comprise Group E. The degree of unsaturation and the presence of polar lipids distinguish *B. subtilis* from *B. licheniformis* in Group A and *B. cereus* from *B. anthracis* and *B. thuringiensis* in Group E.

Profiling of fatty acids as methyl ester derivatives by GC has been used in the classification and discrimination of *Bacillus* species. Although long analysis times are sometimes required,[57,58] chromatograms of *B. thuringiensis*, *B. anthracis*, and *B. subtilis* can be visually discriminated. Capillary GC followed by principal components analysis of the fatty acid profiles of 25 different bacilli readily distinguished between *B. subtilis*, *B. licheniformis*, *B. cereus*, and *B. thuringiensis*.[59]

Pyrolysis-MS followed by canonical variate analysis and principal component analysis has also been used to differentiate sets of bacilli with various degrees of success. Canonical variate analysis of Py-MS data from sporulating and non-sporulating strains showed that *B. cereus*, *B. thuringiensis*, and *B. subtilis* could be separated from each other when data from *B. pumilus* and *B. megaterium* were included in the data set.[60] In this Py-MS study of non-sporulating, vegetative organisms, data points representing *B. pumilus* and *B. megaterium* overlapped with other *Bacillus* data points (Figure 16 in reference 60). In a study of sporulating organisms, however, greater discrimination was achieved: of the five organisms studied, only *B. subtilis* and *B. megaterium* data points overlapped in a plot of the first two canonical variates (Figure 17 in reference 60). Canonical variate analysis distinguished between a *B. anthracis* strain and pathogenic and nonpathogenic *B. cereus* strains.[60] The Py-MS patterns of *B. licheniformis* (BO017 and BO089) and *B. subtilis* (BO014 and BO095) (same as the species used in the present study) were also distinguished from *B. pumulis* and *B. amyloliquefaciens*.[13,21,61]

In the present study, lipid TIC's alone have been demonstrated to have potential in the differentiation of species of bacilli. Intraspecies (strain) differences were in some instances as variable as that in interspecies comparisons. If the interspecies and intraspecies differences can be better defined (with larger numbers of strains for each species), both sets of information could be useful. The intensity distribution and t_R values of certain m/z RIC profiles both contribute substantially to the discrimination of the different species by providing salient features on which to base relative degrees of similarity and difference.

CONCLUSIONS

Curie-point pyrolysis combined with a 5 m capillary GC column and ITMS detection has been shown to be a promising new tool for the characterization of microorganisms. The ion trap MS used in this study is commercially available and relatively inexpensive. Using only microgram amounts of bacteria, lipid components were profiled and the analyses were completed in less than 6 min. Information fundamental to all microorganisms was produced by analyzing a test set of 12 bacterial strains (eight species) in Py-GC/MS experiments. Dehydrated diglycerides, which may represent pyrolysis products of bacterial phospholipids, were observed in all bacterial chromatograms. The diglyceride-H_2O peaks constituted some of the last eluting components in the bacterial chromatograms, and their abundance and individual m/z retention times proved to be diverse enough to allow ready visual differentiation. Only a fraction of the total lipid information available was used for the characterization of these bacteria; additional reconstructed ion profiles and individual mass spectra available in the data were not used. The Py-GC/ITMS approach presented here has potential as a rapid prescreening technique for producing information which can serve as a guide for other, more definitive (but perhaps, more time consuming), analytical or microbiological studies of bacteria.

ACKNOWLEDGEMENT

The authors thank Ms. Linda Jarvis and Mrs. Melinda Van for the preparation and editing of the manuscript.

REFERENCES

1. M. D. Walla, P. Y. Lau, S. L. Morgan, A. Fox, and A. Brown, Capillary gas chromatography-mass spectrometry of carbohydrate components of legionellae and other bacteria, J. Chromatogr. 288:399 (1984).
2. L. Larsson and P. -A. Mardh, Gas chromatographic of mycobacteria: Analyis of fatty acids and trifluoroacetylated whole cell methanolysates, J. Clin. Microbiol. 3:81 (1976).
3. C. W. Moss and S. B. Dees, Cellular fatty acids of *Flavobacterium meningosepticum* and *Flavobacterium* species, J. Clin. Microbiol. 8:772 (1978).
4. C. W. Moss, S. B. Dees, and G. O. Guerrant, Gas-liquid chromatography of bacterial fatty acids with a fused silica capillary column, J. Clin. Microbiol. 12:127 (1980).
5. B. D. Kerger, P. D. Nichols, C. P. Antworth, W. Sand, E. Bock, J. C. Cox, T. A. Langworthy, and D. C. White, Signature fatty acids in the polar lipids of acid-producing *Thiobacillus* spp.: methoxy, cyclopropyl, α-hydroxy-cyclopropyl and branched and normal monoenoic fatty acids, FEMS Microbiol. Ecology 38:67 (1986).

6. R. H. Findlay and D. C. White, A simplified method for bacterial nutritional status based on the simultaneous determination of phospholipid and endogenous storage lipid poly-β-hydroxyalkanoate, J. Microbiol. Methods 6:113 (1987).

7. A. Fox, P.-Y. Lau, A. Brown, S. L. Morgan, Z.-T. Zhu, M. Lema, amd M. D. Walla, Capillary gas chromatographic analysis of carbohydrates of *Legionellaceae*, J. Clin. Microbiol. 19:326 (1984).

8. Microbial Identification System, 1985, Hewlett-Packard Co., Palo Alto, CA.

9. L. Miller and T. Berger, "Bacteria identification by gas chromatography of whole cell fatty acids," Hewlett-Packard Application Note 228-41 (1985).

10. J. J. Boon and J. W. de Leeuw, Amino acid sequence information in proteins and complex proteinaceous material revealed by pyrolysis-capillary gas chromatography-low and high resolution mass spectrometry, J. Anal Appl, Pyrolysis 11:313 (1987).

11. J. M. Hindmarch and J. T. Magee, The Staphylococci: A classification and identification study using pyrolysis-gas-liquid chromatography, J. Anal. Appl, Pyrolysis 11:527 (1987).

12. A. P. Snyder, J. H. Kremer, H. L. C. Meuzelaar, W. Windig, and K. Taghizadeh, Curie-point pyrolysis atmospheric pressure chemical ionization mass spectrometry: Preliminary performance data for three biopolymers, Anal. Chem. 59:1945 (1987).

13. L. A. Shute, C. S. Gutteridge, J. R. Norris, and R. C. W. Berkeley, Curie-point mass spectrometry applied to characterization and identification of selected *Bacillus* species, J. Gen. Microbiol. 130:343 (1984).

14. C. S. Gutteridge and J. R. Norris, The application of pyrolysis techniques to the identification of micro-organisms, J. Appl. Bacteriol. 47:5 (1979).

15. L. W. Eudy, M. D. Walla, S. L. Morgan, and A. Fox, Gas chromatographic-mass spectrometric determination of muramic acid content and pyrolysis profiles for a group of Gram-positive and Gram-negative bacteria, Analyst 110:381 (1985).

16. A. Fox and S. L. Morgan, The chemotaxonomic characterization of microorganisms by capillary gas chromatography and gas chromatography-mass spectrometry, in: "Instrumental Methods for Rapid Microbiological Analysis," W. H. Nelson, ed., VCH, Deerfield Beach, pp. 135-164 (1985).

17. G. Papa, P. Balbi, and G. Audisio, Preliminary study of fungal spores by pyrolysis-gas chromatography, J. Anal. Appl. Pyrolysis 11:539 (1987).

18. C. S. Smith, S. L. Morgan, C. D. Parks, A. Fox, and D. G. Pritchard, Chemical marker for the differentiation of Group A and Group B streptococci by pyrolysis-gas chromatography-mass spectrometry, Anal. Chem. 59:1410 (1987).

19. R. J. Helleur, E. R. Hayes, J. S. Craigie, and J. L. McLachlan, Characterization of polysaccharides of red algae by pyrolysis capillary gas chromatography, J. Anal. Appl. Pyrolysis 8:349 (1985).

20. H. Engman, H. T. Mayfield, T. Mar, and W. Bertsch, Classification of bacteria by pyrolysis-capillary column gas chromatography-mass spectrometry and pattern recognition, J. Anal. Appl. Pyrolysis 6:137 (1984).

21. A. G. O'Donnell, J. R. Norris, R. C. W. Berkeley, D. Claus, T. Kaneko, N. A. Logan, and R. Nozaki, Characterization of *Bacillus subtilis*, *Bacillus pumilus*, *Bacillus* licheniformis, and *Bacillus amyloliquefaciens* by pyrolysis-gas liquid chromatography, deoxyribonucleic acid-deoxyribonucleic acid hybridization, bichemical tests, and API systems, Int. J. Systematic Bacteriol. 30:448 (1980).

22. E. E. Medley, P. G. Simmonds, and S. L. Manatt, A pyrolysis gas chromatography mass spectrometry study of the actinomycete *Streptomyces longisporoflavus*, Biomed. Mass Spectrom. 2:261 (1975).

23. R. J. Helleur, Characterization of the saccharide composition of heteropolysaccharides by pyrolysis-capillary gas chromatography-mass spectrometry, J. Anal. Appl. Pyrolysis 11:297 (1987).

24. S. Tsuge and H. Matsubara, High-resolution pyrolysis-gas chromatography of proteins and related materials, J. Anal. Appl. Pyrolysis 8:49 (1985).

25. J. M. Haddadin, R. M. Stirland, N. W. Preston, and P. Collard, Identification of *Vibrio cholerae* by pyrolysis-gas liquid chromatography, Appl. Microbiol. 25:40 (1973).

26. E. Reiner, J. J. Hicks, R. E. Beam, and H. L. David, Recent studies on mycobacterial differentiation by means of pyrolysis-gas-liquid chromatography, Am. Rev. Resp. Disease 104:656 (1971).

27. R. M. Caprioli, Enzymes and mass spectrometry: A dynamic combination, Mass Spectrom. Rev. 6:237 (1987).

28. R. L. Cerny, K. B. Tomer, and M. L. Gross, Desorption ionization combined with tandem mass spectrometry: Advantages for investigating complex lipids, disaccharides and organometallic complexes, Org. Mass Spectrom. 21:655 (1986).

29. H. -R. Schulten, Advances in field desorption mass spectometry, in: "Soft Ionization Biological Mass Spectrometry," H. R. Horris, ed., Heyden, London, pp. 6-38 (1981).

30. H. U. Winkler, R. J. Beuhler, and L. Friedman, Field desorption mass spectrum of glucagon, Biomed. Mass Spectrom. 3:201 (1976).

31. A. C. Tas, J. van der Greef, J. de Waart, J. Bouwman, and M. C. ten Noever de Brauw, Comparison of direct chemical ionization and direct-probe electron impact/chemical ionization pyrolysis for characterization of *Pseudomonas* and *Serratia* bacteria, J. Anal. Appl. Pyrolysis 7:249 (1985).

32. J. A. Adkins, T. R. Risby, J. J. Scocca, R. E. Yasbin, and J. W. Ezzell, Linear-programmed thermal degradation methane chemical ionization mass spectrometry. II. Defined compounds and lipid-containing envelope constituents from *Salmonella*, J. Anal. Appl. Pyrolysis 7:35 (1984).

33. J. P. Anhalt and C. Fenselau, Identification of bacteria using mass spectrometry, Anal. Chem. 47:219 (1975).

34. R. J. Buehler, E. Flanigan, L. J. Greene, and L. Friedman, Proton transfer mass spectrometry of underivatized peptides, Biochemistry 24:5060 (1974).

35. A. C. Tas, J. de Waart, J. Bouwman, M. C. ten Noever de Brauw, and J. van der Greef, Rapiud characterization of *Salmonella* strains with direct chemical ionization pyrolysis, J. Anal. Appl., Pyrolysis 11:329 (1987).

36. H. L. C. Meuzelaar, J. H. Haverkamp, and F. D. Hileman, "Curie-point Pyrolysis Mass Spectrometry of Recent and Fossil Biomaterials; Compendium and Atlas," Elsevier, Amsterdam (1984).

37. M. Langhammer, I. Luderwald, and A. Simons, Analytical pyrolysis of proteins, Fresenius Z. Anal. Chem. 324:5 1986).

38. H. -R. Schulten and W. Gortz, Curie-point pyrolysis and field ionization mass spectrometry of polysaccharides, Anal. Chem. 50:428 (1978).

39. J. W. Kelly, New tools for probing protein structures, Bio/Technology 6:125 (1988).

40. D. N. Heller, R. J. Cotter, C. Fenselau, and O. M. Uy, Profiling of bacteria by fast atom bombardment mass spectrometry, Anal. Chem. 59:2806 (1987).

41. D. N. Heller, C. Fenselau, R. J. Cotter, P. Demirov, J. K. Olthoff, J. Honovich, M. Uy, T. Tanaka, and Y. Kishimoto, Mass spectral analysis of complex lipids desorbed directly from lyophilized membranes and cells, Biochem. Biophys. Res. Commun. 142:194 (1987).

42. K. B. Tomer, N. J. Jensen, and M. L. Gross, Fast atom bombardment and tandem mass spectrometry for determining structural modification of fatty acids, Anal. Chem. 58:2429 (1986).

43. D. F. Hunt, J. R. Yates III, J. Shabanowitz, S. Winston, and C. R. Hauer, Protein sequencing by tandem mass spectrometry, Proc. Natl. Acad. Sci. USA 83:6233 (1986).

44. P. G. Kistemaker, G. J. Q. van der Peyl, and J. Haverkamp, Laser desorption mass spectrometry, in: "Soft Ionization Biological Mass Spectrometry," H. R. Morris, ed., Heyden, London, pp. 120-136 (1981).

45. R. J. Cotter, J. Honovich, N. Qureshi, and K. Takayama, Structural determination of lipid-A from Gram negative bacteria using laser desorption mass spectrometry, Biomed. Environ. Mass Spectrom. 14:591 (1987).

46. A. Viari, J. -P. Ballini, P. Vigny, D. Shire, and P. Dousset, ^{252}Cf-plasma desorption mass spectrometry of unprotected tri- and tetra-deoxyribonucleotides, in: "Mass Spectrometry in the Analysis of Large Molecules," C. J. McNeal, ed., John Wiley and Sons, New York pp. 199-205 (1986).

47. C. A. Cramers, G. J. Scherpenzeel, and P. A. Leclercq, Increased speed of analysis in directly coupled GC/MS systems; capillary columns at sub-atmospheric outlet pressures, J. Chromatogr. 203:207 (1981).

48. M. L. Trehy, R. A. Yost, and J. G. Dorsey, Short open tubular columns in gas chromatography/mass spectrometry, Anal. Chem. 58:14 (1986).

49. W. H. McClennen, J. M. Richards, and H. L. C. Meuzelaar, "Curie-point Desorption, Short Column GC/MS of Large Polar Molecules," presented at the 36th American Society for Mass Spectrometry Conf. on Mass Spectrom. and Allied Topics, June 5-10, 1988, San Francisco, CA, pp. 403-404.

50. J. M. Richards, W. H. McClennen, J. A. Bunger, and H. L. C. Meuzelaar, "Analysis of Synthetic and Natural Polymers Using Pyrolysis Short Column GC/MS," presented at the 36th American Society for Mass Spectrometry Conf. on Mass Spectrom. and Allied Topics, June 5-10, 1988, San Francisco, CA, pp. 547-548.

51. T. H. Risby and A. L. Yergey, Identification of bacteria uisng linear programmed thermal degradation mass spectrometry. The preliminary investigation, J. Phys. Chem. 80:2839 (1976).

52. G. S. Oxborrow, N. D. Fields, and J. R. Puleo, Preparation of pure microbiological samples for pyrolysis-gas liquid chromatography studies, Appl. Environ. Microbiol. 32:306 (1976).

53. G. L. French, C. S. Gutteridge, and I. Phillips, Pyrolysis gas chromatography of Pseudomonas and Acinetobacter species, J. Appl. Bacteriol. 49:505 (1980).

54. G. S. Oxborrow, N. D. Fields, and J. R. Puleo, Pyrolysis gas-liquid chromatography studies of the genus Bacillus. Effect of growth time on pyrochromatogram reproducibility, in: "Analytical Pyrolysis," C. E. R. Jones and C. A. Cramers, eds., Elsevier Scientific Publishing Co., Amsterdam, pp. 69-76 (1977).

55. D. E. Minnikin and M. Goodfellow, Lipids in the classification of Bacillus and related taxa, in: "The Aerobic Endospore-forming Bacteria," R. C. W. Berkeley and M. Goodfellow, eds., Academic Press, London, pp. 59-90 (1981).

56. T. Kaneda, Fatty acids of the genus Bacillus: an example of branched-chain preference, Bacteriol. Rev. 41:391 (1977).

57. T. Kaneda, Biosynthesis of branched chain fatty acids. I. Isolation and identification of fatty acids from Bacillus subtilis (ATCC 7059), J. Biol. Chem. 238:1222 (1963).

58. T. Kaneda, Fatty acids in the genus Bacillus. II. Similarity in the fatty acid compositions of Bacillus thuringiensis, Bacillus anthracis, and Bacillus cereus, J. Bacteriol. 95:2210 (1968).

59. A. G. O'Donnell, Numerical analysis of chemotaxonomic data, in: "Computer-Assisted Bacterial Systematics," M. Goodfellow, D. Jones, and F. G. Priest, eds., Academic Press, London, pp. 403-414 (1985).

60. A. G. O'Donnell and J. R. Norris, Pyrolysis gas-liquid chromatographic studies, in: "The Aerobic Endospore-forming Bacteria," R. C. W. Berkeley and M. Goodfellow, eds., Academic Press, London, pp. 141-179 (1981).

61. C. S. Gutteridge, A. J. Sweatman, and J. R. Norris, Potential applications of Curie-point pyrolysis mass spectrometry with emphasis on food science, in: "Analytical Pyrolysis," K. J. Voorhees, ed., Butterworths, London, pp. 324-348 (1984).

CHAPTER 14

ULTRASENSITIVE DETERMINATION OF LIPID BIOMARKERS BY GAS CHROMATOGRAPHY AND MASS SPECTROMETRY IN CLINICAL AND ECOLOGICAL MICROBIOLOGY

Lennart Larsson

Department of Medical Microbiology
Lund University Hospital
Lund
Sweden

Goran Odham

Department of Ecology
Division of Chemical Ecology
Lund University
Lund
Sweden

INTRODUCTION

Gas chromatography/mass spectrometry (GC/MS) has consolidated its position as a highly sensitive and selective technique for determining lipid biomarkers in microbial environments. Since the publication of the volume Gas Chromatography/Mass Spectrometry Applications in Microbiology in 1984[1], the literature has seen a significant increase in papers on the subject. The intent of this chapter is to update the present situation with special reference to ultrasensitive detection in clinical and ecological microbiology.

As will be shown, the continuous systematic studies on the preparation and detection of halogenated derivatives of simple lipid constituents (fatty acids) have proved particularly fruitful. These derivatives provide means for measurements of electron capture processes using the sensitive electron capture (EC) detector and the mass spectrometer (MS) operating in the negative ion (NI)-selected ion monitoring (SIM) mode. The halogenated derivative and the detector of choice allow an increase in method sensitivity by a factor of 100 to 1000 in routine operations compared with use of the more conventionally applied flame ionization (FI) detector.

THE LIPID BIOMARKERS

Lipid biomarker survey

Lipids, polar as well as non-polar, are present in the cell envelopes of bacteria. Both free (e.g., glycolipids, phospholipids, and lipopolysaccharides) and bonded (e.g., mycolic acids) lipids are encountered. Over 50 different bacterial fatty acids, largely bonded to other structures

Analytical Microbiology Methods
Edited by A. Fox *et al.*
Plenum Press, New York, 1990

via ester or amide linkages, have been identified. A substantial accumulated volume of data on the cellular fatty acid compositions of various bacterial species is available in the literature; such data can be used to assist in the identification of studied organisms.[2] Indeed, cellular fatty acid profiling is now a firmly established routine for differential diagnosis of bacteria (see chapter 4). Unusual fatty acids are occasionally found. Such constituents may serve as chemical marker substances for determining microbes in complex matrices. A wide variety of fatty acid markers with different degrees of specificity have been identified and applied for use in clinical[3] and ecological[4] microbiology. Methods for preparation and detection of particularly halogen-containing derivatives of bacterial fatty acids by GC with EC and MS will be described in the following.

Hydrolysis and micromanipulations

Bacterial fatty acid profiling is usually conducted after conversion of the acids to methyl esters. Simple and reproducible methods for preparing these derivatives have been elaborated. One frequently applied technique involves heating the bacteria in alkaline methanol:water solution for saponifying the lipids; then, the liberated fatty acids are methylated by heating in methanol under acidic conditions. This method cleaves ester-linked fatty acids quantitatively but leaves an appreciable proportion of the amide bonds intact.[5] Acid methanolysis is efficient in releasing amide-bound (hydroxy) acids but may lead to degradation of cyclopropane-substituted fatty acids. The reader should consult Chapter 4 in this volume for further information on selection of hydrolysis conditions.

Capillary vessels are very useful when handling materials in micro-scale. The samples, comprising volumes of some hundreds of microliters, are transferred to glass capillary tubings having inner diameters of 2-3 mm and subsequently evaporated to dryness either under vacuum or by using a stream of nitrogen. The capillary tubing allows make-up with as little as 5 μL of solvent prior to sampling for injections (1-2 μL) into the gas chromatograph.[6]

Derivatives for positive ion MS detection

Methyl ester derivatives have been extensively employed in GC analysis utilizing MS detection. A thorough discussion on electron impact (EI) mass spectra of methyl esters of the various bacterial fatty acid types (saturated, unsaturated, cyclopropane-substituted, hydroxylated) is provided in Chapter 3 of the first volume.[7] Chemical ionization (CI) spectra of positive ion (PI) type typically show an abundant $(M + 1)^+$ or $(M + 18)^+$ peak depending upon the reagent gas used.[8] Whether an improved sensitivity is achieved when using CI compared with EI (monitoring at the molecular adduct ion and the molecular ion, respectively) must be determined in each individual case, however.

Alternative fatty acid derivatives include *tert*-butyl-dimethylsilyl (t-BDMS) esters. Preparation of these derivatives was first reported by Phillipou et al.[9] Prominent ions are formed at m/z = $(M - C_4H_9)^+$ upon EI ionization. Monitoring this abundant ion in SIM analyses was found to increase the sensitivity considerably compared with molecular ion monitoring of the methyl ester derivative.

Maitra et al.[10,11] used the trimethylsilyl (TMS)-methyl ester derivative of 3-hydroxymyristic acid in a rather elaborate method for determining bacterial lipopolysaccharides (LPS, endotoxins). In brief, the samples were hydrolyzed by heating in 8 M hydrogen chloride for 4.5 h at 110°C. The free fatty acids were extracted and esterified; the sample was then applied to a silica gel column for removing methyl esters of various saturated and unsaturated fatty acids from the more polar hydroxy acid esters. Silylated ether derivatives were prepared by adding TMS reagent and allowing the mixture to react for 2 min. The derivatives were extracted with pentane and re-dissolved in acetonitrile. The corresponding pentadeuterated derivative was used as an internal standard, and the analyses were carried out using SIM at EI conditions.

Halogenated derivatives for EC and NI MS detection

A variety of halogenated derivatives of fatty acids suitable for EC and NI-SIM detection have been described. We found that pentafluorobenzyl (PFB) esters were easily prepared.[12] The lipid material is hydrolyzed under alkaline conditions. After acidification the liberated fatty acids are extracted, evaporated, and dissolved in acetonitrile. PFB bromide (35% v/v, in acetonitrile) and triethylamine are added, and after 10 min at room temperature, the mixture is dried in an evacuated desiccator. The derivatives are dissolved in hexane prior to analysis.

The CI-NI mass spectra show the carboxylate ion (from loss of the PFB radical) as the base peak with little additional fragmentation. The 2-OH PFB esters behave as unsubstituted or alkyl branched saturated and unsaturated esters whereas the 3-OH PFB esters may lose a water molecule in addition to the PFB radical producing m/z (M − ($H_2CC_6F_5$ + 18))⁻. Compared with trichloroethyl ester derivatives which have been used extensively in conjunction with EC detection,[13,14] the PFB esters showed considerably higher sensitivity in NI-SIM.[15]

Rosenfeld et al.[16,17] described an elegant method for preparing the PFB derivatives by using a PFB bromide-impregnated resin as a solid-support reagent. This technique lends itself to automation as all sample preparation is based on filtration, and a satisfactory recovery can be achieved.

Several methods for preparing halogenated derivatives of hydroxy-substituted bacterial fatty acids as well as their detection by NI-SIM were compared.[15] In the recommended procedure, the methyl ester is first prepared by heating the sample in methanolic hydrogen chloride followed by hexane extraction. After evaporation, the methyl esters are dissolved in methylene chloride, and pyridine and pentafluorobenzoyl chloride are added. The sample is then heated at 80°C for 5 min. After cooling, methylene chloride and a phosphate buffer solution are added, and after extraction, the organic phase is evaporated to dryness. Prior to GC/EC or GC/MS analyses, the samples are made up in hexane. The recovery is practically quantitative. The pentafluorobenzoyl(PFBO)-methyl ester derivative was found to be chemically stable, easy to prepare, provide high sensitivity and produce the molecular ion as the base peak in the mass spectrum. When the technique was applied to determine hydroxy fatty acids in LPS, disposable silica gel columns were used for sample purification.

Sud and Feingold[18] studied the heptafluorobutyryl-butyl ester derivative of 3-hydroxylauric acid of *Neisseria gonorrhoea* LPS. Dried samples were heated in butanol-HCl at 80°C during 2 h. The butyl esters were extracted using petroleum ether, evaporated, and purified by thin-layer chromatography. Heptafluorobutyric anhydride and hexane were added and the reaction was allowed to proceed at room temperature for 1 h. The analyses were made using EC detection after sample clean-up with preparative GC.

A method for combined derivatization of both hydroxylated and non-hydroxylated bacterial fatty acids using halogenated reagents was reported[19] and recently improved.[20] In the latter procedure, the fatty acids are liberated from lipid material by hydrolysis, extracted, and evaporated to dryness under dry nitrogen. The acids are then re-dissolved in triethylamine-containing (1% v/v) acetonitrile before the addition of PFB bromide (35% v/v, in acetonitrile). After 15 min at room temperature, pentafluoropropionic anhydride is added; after another 15 min, hexane and 1 M phosphate buffer solution (pH 7) are added. After extraction and centrifugation, the hexane phase is applied to a disposable silica gel column; the PFB and pentafluoropropionyl-PFB esters are eluted using methylene chloride. The product is evaporated and re-dissolved in heptane prior to analysis.

It was found that the PFB and pentafluoropropionyl/PFB esters of typical bacterial non-hydroxy and hydroxy acids, respectively, separated well on a capillary column containing a stationary phase of medium polarity. Use of other halogenated derivatives, e.g., trichloroethyl instead of PFB esters and/or heptafluorobutyryl instead of pentafluoropropionyl esters, resulted either in lowered sensitivity or poor separation between some of the acids. Apart from the dramatic increase of sensitivity encountered when using EC detection of these halogenated derivatives compared with FI detection of methyl esters, the chromatographic profiles did not differ which made it possible to "match" bacterial fatty acid profiles obtained by these different techniques.[21]

SEPARATION

Capillary columns. Use of column switching.

Use of capillary columns, particularly of fused-silica type, is today routine for the separation of derivatives of bacterial fatty acids. Both split, splitless and on-column injection techniques are practiced. Split injection is easy to apply although quantitative data may be difficult to achieve due to possible discrimination of compounds of low or high boiling points. When ultrahigh sensitivity is a prerequisite, either on-column or splitless injection should be applied. The reader should consult specialized literature for further information.[22]

An efficient separation also improves detection sensitivity. The theoretically achievable sensitivity of EC detection is of the same magnitude as that of NI-SIM (see below). Usually, however, the inherent sensitivity of the EC detector cannot fully be taken advantage of due to difficulties in removing the halogenated reagents and components originating from contaminated glassware, solvents, etc., prior to analysis. The column switching technique provides a means for on-line purification of samples. In this technique, the sample is injected into a pre-column after which selected fractions are cold-trapped and subsequently introduced into the analytical capillary column connected, e.g., to an EC detector. Thus, excess of the halogenated reagents need not be removed from the sample prior to injection. Several column-switching units are commercially available. The principle is illustrated in Figure 1.

DETECTION

Electron capture detection

The EC detector contains an ionization chamber where high energy electrons, usually originating from ^{63}Ni are bombarding the carrier gas molecules, usually N_2. Thermal electrons are formed and collected with an applied potential. The detector's response is due to a decrease in the resulting standing current when electrophilic sample molecules capture these electrons. The sensitivity of the EC detector to halogen-containing

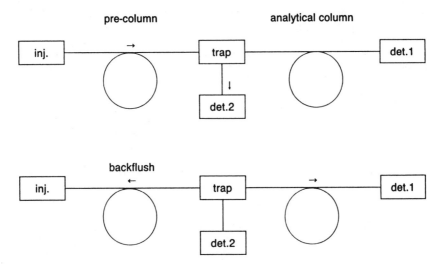

Figure 1.   Principle for the column switching technique.
The sample is separated in a pre-column after which
selected fractions are cold-trapped and introduced onto
the analytical column. The technique is suitable,
e.g., for increasing the separation efficiency by using
stationary phases of different polarity in the two
columns and for electron capture detection of
halogenated components.

compounds increases in the order F << Cl < Br < I and synergistically with
multiple substitution of the same carbon atom. The sensitivity is also
influenced by the structure of the organic compound.

Ion trap and mass selective detectors

     Small, dedicated mass spectrometers or instrumentations whose function
is based on other than "conventional" means of recording of ions, are rapidly
moving into the microbiologists' laboratories. Although being instruments
for separation and measurement of masses, they are normally considered as GC
detectors of high specificity and versatility. Some are units integrated
together with a specific GC model whereas others can be linked to practically
any GC instrumentation. They have in common a capillary gas chromatograph as
an inlet system and an integrated data system (often based upon
microcomputers). Software typically allows full extraction ion current
profile (EICP) and SIM recordings. The instrumentation and hardware are
robust, easy to handle, and well suited for routine operations. These
features add to great versatility and a favorable price. The main drawbacks
as compared with full-size mass spectrometers are limited mass range (maximum
m/z 600-800), inlet systems restricted to GC, and absence of facilities for
conventional PI/NI CI.

     The ion trap detector (ITD) (Finnegan MAT, San Jose, CA, USA) may be
described as a three dimensional quadrupole utilizing electrodes with
hyperbolic inner surfaces as shown in Figure 2. The effluent from the
capillary GC column enters the quadrupole cavity via a heated transfer line.
A filament produces electrons which are pulsed into the same cavity with the
aid of the electron gate. EI results and the variety of ions produced are
trapped in an RF-only mode of the electrodes. The ion trap is then scanned
by changing the RF-amplitude on the electrodes allowing the ions to be
consecutively ejected into the electron multiplier.

The ITD is a very compact device and differs from conventional quadrupole MS in that ionization and mass analysis occur in the same place. An interesting new option consists of possibilities for adjustable scanning sequences allowing (1) ionization and trapping of CI reagent ions, (2) CI reagent ions reaction with neutral sample, (3) selection of low mass for start of scan, and (4) scan resulting in the mass spectrum. In this way, the option can provide PI CI information at reagent precursors as low as 10^{-5} torr.

The ITD provides sensitivities in EI comparable with full size MS instrumentations. It should be borne in mind that the ITD works at a principle different from that of conventional MS instruments. A disadvantage of the ITD has been, and still is, the fact that the mass spectra produced are not always directly comparable with those of "classical" mass spectrometers.

The <u>mass selective detector</u> (MSD) (Hewlett-Packard, Palo Alto, CA, USA) is, in contrast to the ITD, a "true" quadrupole mass spectrometer where great efforts have been made to design an extremely compact set-up. The MSD analyzer operates with four molybdenum rods, 203 mm long and precisely formed to a hyperbolic shape (Figure 3). In the scanning mode, a mass range between

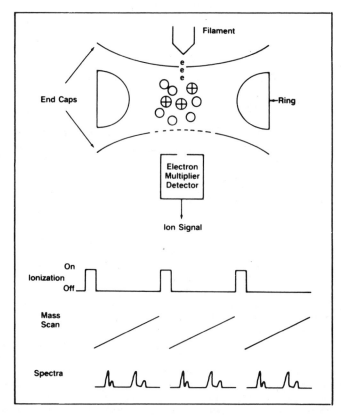

Figure 2. Operation of the ion trap detector. Reprinted with permission from Finnegan MAT, San Jose, CA, USA.

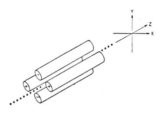

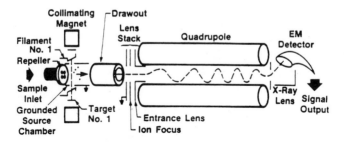

ION SOURCE PRESSURE 10 E-4 TO 10 E-6 TORR

ELECTRON BEAM 300 mA @ 70 eV

IONIZATION EFFICIENCY APPROXIMATELY 0.01 %

Figure 3.   Schematic of quadrupole mass selective detector.
Reprinted with permission from Hewlett-Packard Inc.,
Palo Alto, CA, USA.

10 and 800 atomic mass units (amu) may be utilized. The maximum scan speed obtainable matches the full size instruments (1500 amu/s) and in SIM up to 10 groups of 20 masses may be selected.

A single air-cooled turbomolecular pump backed by a prevacuum mechanical pump provides the necessary low pressure for EI in the ion source, typically 5×10^{-5} torr and 0.5 mL/min of the carrier gas. The absence of differential pumping makes CI impossible.

<u>Full size mass spectrometers with chemical ionization and positive/negative ion recording</u>

Mass spectrometers with quadrupole analyzers dominate today among low resolution MS instruments. There are several reasons for this including instrumental robustness, facile shifting between EI and CI as well as between PI and NI recording, and suitability for computerization. Most bench-top MS systems and certain singly pumped instruments still allow only for EI. Consequently, much work has been devoted to developing new derivatives that allow increased sensitivity by the formation of abundant molecular-related ions in EI spectra (see above).

CI has, however, particularly with the increased access to facilities for NI detection, continued to be the preferred ionization method in quantitative MS when utmost sensitivity and selectivity are required. An important contribution and prerequisite for NICI has been the parallel development of reproducible methods for preparing suitable halogenated derivatives, preferably such containing several fluorine atoms (see above).

The important reaction mechanism in NICI consists of capture of electrons by the sample molecules, and ion-molecule reactions between sample and reagent gas ions. The reagent gas has double functions providing means for formation of thermal electrons

$$e^- \xrightarrow{CH_4} e^-_{th} \qquad \text{(methane as reagent gas)}$$

and by contributing the excess of molecules necessary for the subsequent collisional stabilization process.

A common scheme of mechanisms for fatty acid fluoro derivatives may be formalized as follows:

$$AB + e^-_{th} \rightarrow AB^-. \quad \text{(associative resonance capture)}$$

$$AB^-. \xrightarrow{CH_4} AB^-. \quad \text{(collisional stabilization)}$$

$$AB^-. \rightarrow A^- + B \quad \text{(dissociative resonance capture)}$$

In Figure 4, a series of NICI mass spectra of different fluorinated derivatives of 3-hydroxymyristic acid has been reproduced.[15] None of the spectra show molecular radical anions. Instead, intense dissociation at the parent molecule-derivative occurs. Interestingly, in a seemingly unpredictable manner, the molecule-related fragments carry the negative charge for certain derivative types whereas for other derivatives the fluorinated derivative moiety carries the charge. For structure determination and for certainty in analyte identity in quantitative work the former derivatives are preferred.

The relative sensitivity obtained by positive and negative ion CI and EI has been studied in some detail.[23] These results show that NICI may exceed PICI considerably, and that the total ionization is comparable with that of EI obtained under standard conditions.

Implications of use of tandem mass spectrometry

Tandem mass spectrometry (MS-MS) is a very powerful analytical technique for the highly selective and sensitive determination of a wide range of chemical compound classes. The recent introduction on the market of reasonably easy-to-operate instruments with triple quadrupole geometry opens up interesting possibilities for studying lipid biomarkers in complex matrices.

The triple quadrupole mass spectrometer makes use of a quadrupole mass filter (Q_1) followed by a quadrupole collision cell (Q_2) (RF-only) and a second quadrupole mass filter (Q_3) (Figure 5). The mass analyzers Q_1 and Q_3 can be used to select both parent and daughter ions for analyzing the reaction occurring in the collision cell.

The collision cell usually consists of a small diameter, RF-only quadrupole, where the target gas (e.g., Ar or Xe) is present at a relatively high pressure, typically 10^{-3} torr. The occurring collisions are of relatively low energy (5-200 eV) leading to efficient fragmentation. The triple quadrupole mass spectrometers can be operated in several basic modes, the most important in biological work probably being that where daughter ions of a specific parent are studied.

A recent study (Carlsson et al., unpublished) comprising detection and identification of helminthosporol, a sesquiterpenoid fungal toxin isolated

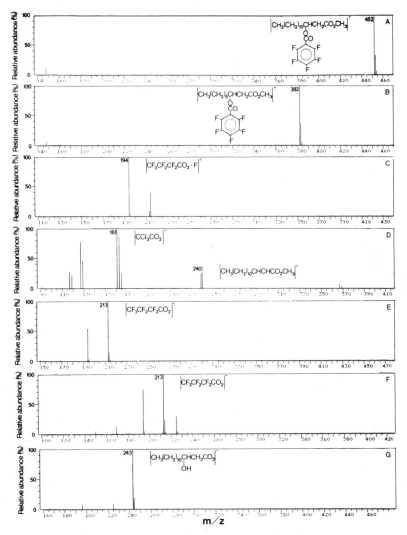

Figure 4. Negative ion-chemical ionization (methane) mass
spectra of different halogenated derivatives:
(A) 3-pentafluorobenzoyl-methyl 14:0;
(B) 3-pentafluorobenzoyl-methyl 9:0;
(C) 3-heptafluorobutyryl-methyl 14:0;
(D) 3-trichloroacetyl-methyl 14:0;
(E) 3-heptafluorobutyryl-trichloroethyl 14:0;
(F) 3-heptafluorobutyryl-pentafluorobenzyl 14:0;
(G) 3-OH-pentafluorobenzyl 14:0. Reprinted from
reference 15 with permission.

from *Cochliobolus sativus*, may illustrate the situation. Figure 6 shows the Q_1 positive ion plasmaspray mass spectrum of a chromatographic eluate using a 20 cm diol liquid chromatography column and a mixture of tetrahydrofuran:water (88:12 v/v) as solvent. The spectrum (Figure 6a) indicates the presence of a complex mixture in the fraction including the toxin biomarker with the expected molecular weight of 236 ($(M + 1)^+$ adduct ions). Adjusting the instrument for collisions between the parent adduct ions of m/z 237 and Xe in Q_2 at 1.0×10^{-3} torr resulted in the plasmaspray daughter ion spectrum (Q_3) reproduced in Figure 6b. The spectrum indicates, apart from the parent ion peak, intense fragmentation leading to ions of m/z 219 (loss of a molecule of water).

The preliminary findings involving liquid chromatography and MS-MS suggest that it should be possible to determine a wide range of lipid biomarkers in complex matrices with limited work-up and without the need for derivatization.

APPLICATIONS

Initial fouling of surfaces

Fouling of surfaces exposed to seawater results in considerable economic losses. The initial event consists of a rapid coating of the surface with a polymer that attracts bacteria. The initial bacterial film then attracts a fouling succession that eventually leads to the formation of a biofilm. Possibly, the initially attached bacteria control the complex structure of the later succession.

It was thought of interest to study the initial bacterial community and define its characteristics by the analysis of fatty acid biomarkers. Experiments were carried out by exposing Teflon strips (100 cm x 4 cm) to running seawater and extract and analyze the fatty acid profiles each day during a five-day exposure.[12] Since conventional capillary GC with FI detection of the fatty acid methyl esters was found too insensitive for detecting the fatty acids present, NICI (methane) MS determination of the corresponding PFB esters was used. A SIM fragmentogram obtained from analysis of a 1/250-th fraction of the sample of biofilm formed on the first day of exposure using five sets of combinations of different ion monitoring is shown in Figure 7. Comparison with fragmentograms of authentic bacterial fatty acid derivatives using retention time and fragment mass data as well as the uptake of dose-response curves ensured the identity and quantity of each of the individual fatty acid components.

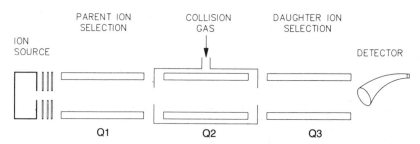

PARENT ION SELECTION   COLLISION GAS   DAUGHTER ION SELECTION

ION SOURCE   DETECTOR

Q1   Q2   Q3

Figure 5.   Schematic diagram showing the principal components of a simple triple quadrupole mass spectrometer. Reprinted with permission from VG Masslab (Altrincham, UK).

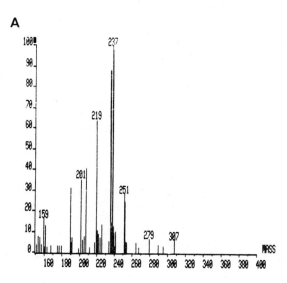

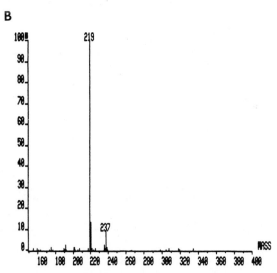

Figure 6. Positive ion plasmaspray mass spectrum of liquid
chromatography fraction containing helminthosporol,
$(M + 1)^+$ m/z 237 (A), and positive ion plasmaspray MS-
MS mass spectrum of the same fraction using Xe as
collision gas at 2 mtorr with formation of daughter
ions from $(M + 1)^+$ m/z 237 (B).

In Table 1 results from the uptake of fatty acid profiles, of typical bacterial origin, during the five-day experiment have been reproduced. The biofilm did not change its community structure significantly between the first, second and fourth and fifth day. On the third day, the strip had to be cleaned from sediments caused by storm-induced turbulence. Possibly this manipulation lowered the observed total fatty acid content.

The amounts of fatty acids present on the Teflon strips corresponded to a cell density equivalent to 2.5×10^5 bacteria the size of *Escherichia coli*/cm^2. The low detection limit for PFB esters of fatty acids by the NI-SIM technique implies possibilities to study the effect of, e.g., changes in the seawater composition and surface properties at bacterial densities approaching 100 bacteria/cm^2. No doubt highly sensitive MS determination of fatty acid biomarkers will be of great importance for the understanding of the initiation of microbe mediated biodeterioration processes such as biofouling and corrosion.

Mycobacterial infections

Tuberculostearic acid (TSA, 10-methyloctadecanoic acid), a laevorotatory branched-chain fatty acid present in several organisms belonging to *Actinomycetales* including all pathogenic mycobacterial species, has been proposed as a suitable biomarker for demonstrating mycobacteria in clinical samples. The lowest detectable amount of the methyl ester of TSA was found to be about 20 pg when monitoring at the molecular radical ion (m/z 312) using EI and 1 pg using either CI with isobutane as reagent gas or EI analyzing the t-BDMS ester derivative monitoring at m/z 355 $(M - 57)^+$. The first clinical applications included several sputum samples from patients with pulmonary tuberculosis. Both EI and CI MS-SIM detection were used.[24-26]

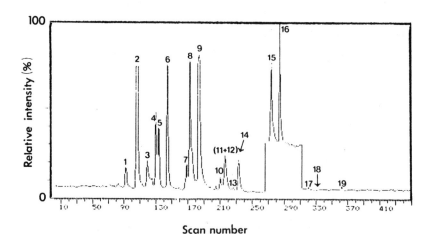

Figure 7. Mass fragmentogram using negative ion-chemical ionization (methane) of pentafluorobenzyl esters of acids in a hydrolysate of an initial marine microbial fouling community after exposure of the Teflon surface for one day. Selected ion monitoring comprising five sets of ion parameters are used. 1, Br-14:0; 2, 14:0; 3, Br-15:0; 4, *i*-15:0, 5, a-15:0; 6, 15:0; 7, 3-OH 14:0; 8, 16:1; 9, 16:0; 10, *i*-17:0; 11, a-17:0; 12, cycl 17:0; 13, unknown; 14, 17:0; 15, 18:1; 16, 18:0; 17, cycl 19:0; 18, 20:4 (internal standard); and 19, 20:0. The signals for 15:0 correspond to 1.4 pmol. Reprinted from reference 12 with permission.

Table 1. Ester-linked fatty acids of the phospholipids of the initial marine microfouling community of Teflon strips exposed to running seawater (from reference 12 with permission)

Fatty acid	m/z	pmol per total surface area (800 cm^2)				
		Day 1	Day 2	Day 3	Day 4	Day 5
Br-14:0[a]	227	177	281	104	147	186
14:0	227	537	583	499	545	552
Br-15:0[a]	241	89	63	43	52	54
i-15:0	241	374	419	230	349	430
a-15:0	241	187	209	115	174	215
15:0	241	342	369	292	342	367
3-OH 14:0	225	8	3	4	10	8
16:1	253	443	403	317	354	361
16:0	255	599	607	513	577	664
i-17:0	269	50	68	nd	52	81
a-17:0	269	55	74	63	48	101
17:0	267	110	90	73	39	68
Unknown	267	19	29	12	3	36
18:1	281	153	217	58	217	328
18:0	283	300	293	228	253	330
19:0	295	12	8	7	nd	11
20:0	311	11	15	6	7	21
Internal standard						
20:4 6	303	5	4	7	7	4
Total bacterial fatty acids		3466	3731	2564	3169	3813
Total fatty acids (pmol/cm^2)		4	5	3	4	5
Mole%						
Total saturates		52	50	60	54	51
Total unsaturates		17	17	15	18	18
Total branched		27	30	22	26	28
Total cyclopropane		4	3	3	1	2

[a] indicates the position of the branch is not known; nd not detected

Demonstration of TSA in cerebrospinal fluid from a patient with tuberculous meningitis using CI (ammonia) MS-SIM at m/z 330 (M + 18)$^+$ was also reported.[27] Recently, results from studies comprising more than 400 sputa and 100 cerebrospinal fluid specimens using EI-SIM detection of TSA demonstrated excellent sensitivity and selectivity of this method for the rapid diagnosis of mycobacterial infections.[28,29]

In subsequent studies, use of the PFB derivative of TSA in NI-SIM analyses gave a detection limit of around 0.6 pg routinely. The results from PFB-TSA detection of sputa from patients with pulmonary tuberculosis and nontuberculous pneumonia tallied with results from culturing provided that sufficient sputum (at least 0.5 mL) was available for analysis.[30] The results thus far have shown that use of NI-SIM is highly advantageous for detecting TSA in complex clinical samples (Figure 8). Recent studies indicate that two-dimensional GC with EC detection may also be useful for determining specific mycobacterial fatty acids (PFB derivatives) and alcohols (PFBO derivatives) at trace levels.[31]

Despite the reported results, it should be borne in mind that findings of TSA may indicate presence not only of mycobacteria but also of several other pathogenic bacteria, e.g., belonging to *Nocardia*, *Corynebacterium*, and *Streptomyces*. In order to avoid false-positive diagnostic results due to the limited specificity of TSA, we investigated the possibility of using mycocerosic acids as markers. These acids are present (as phthioceroldimycocerosate waxes) exclusively in a limited number of mycobacterial species including the *Mycobacterium tuberculosis* complex. A C_{32} mycocerosic acid was demonstrated in mycobacterium-positive sputum specimens after five days of incubation on Lowenstein-Jensen medium.[32] Minnikin et al.[33] reported NI-SIM detection of the mycocerosic acids as PFB esters. Lipids were extracted from cultivated cells of *Mycobacterium tuberculosis* and purified by thin-layer chromatography. The phthiocerol dimycocerosate fraction was hydrolyzed in alkali overnight and the released mycocerosic acids and phthiocerols were converted to PFB esters and pentafluorobenzylidene acetals, respectively. The PFB esters gave, as expected, mass spectra dominated by the mycocerosate anion. Further studies have shown that mycocerosic acids can be demonstrated directly in sputum specimens (see Chapter 9).

Lipopolysaccharides

Maitra et al.[10] reported use of 2- and 3-hydroxylated fatty acids, characteristic constituents of the LPS in all genuine Gram negative bacteria, as endotoxin biomarkers. The lowest detectable amount of the TMS methyl ester derivative of 3-hydroxymyristic acid was found to be approx. 200 fmol with EI-SIM analysis monitoring at the ions m/z 315.4 forming the base peak (Figure 9). Silica gel chromatography was used to separate the methyl esters of the hydroxy fatty acids from non-substituted other methyl esters present in serum. Subsequent applications reported detection of 3-hydroxymyristic acid in cerebrospinal fluid specimens from 22 out of 24 patients with meningitis caused by Gram negative bacteria.[11]

NI-SIM has shown excellent sensitivity and selectivity for the detection of hydroxy fatty acids derived from lipid A as PFBO-methyl ester derivatives.[15] This ionization method was applied in the determination of endotoxin in dust samples collected on filters in poultry confinement buildings in Sweden.[34] Two-dimensional GC with EC was used to detect halogenated derivatives of β-hydroxy fatty acids as well as diaminopimelic acid in urine from patients with infection due to Gram negative bacteria.[35] Further research is required to establish the use of GC and GC/MS techniques for the determination of endotoxins.

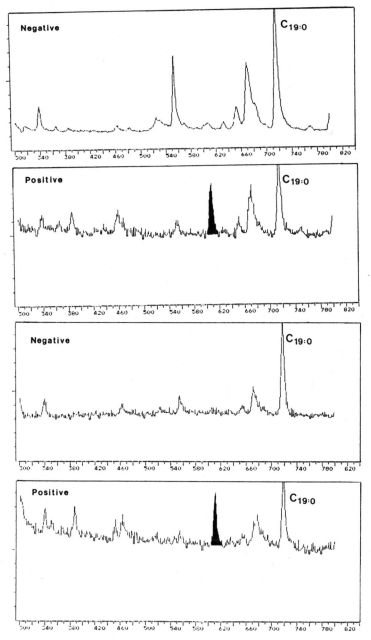

Figure 8. Representative chromatograms from analyses of sputum
samples from four patients with tuberculous (positive)
and nontuberculous (negative) pneumonia by using
selected ion monitoring detection in negative ion-
chemical ionization mode monitoring at m/z 297.3.
The darkened peaks represent pentafluorobenzyl
tuberculostearate. Reprinted from reference 30 with
permission.

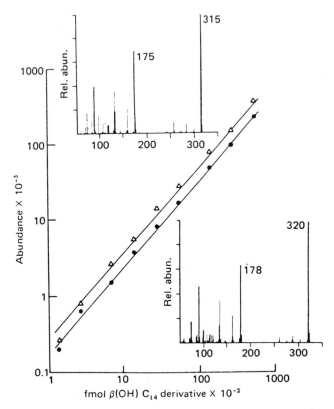

Figure 9. Dose-response curves and mass spectra of
trimethylsilylated 3-hydroxymyristic acid methyl
ester (Δ) and the corresponding pentadeuterated
derivative (•), using electron impact selected ion
monitoring detection focussing at m/z 315.4 and
320.4, respectively. Reprinted from reference 10
with permission.

Sud and Feingold[18] studied purified *Neisseria meningitidis* as well as plasma or cervical specimens spiked with these bacteria. Using EC detection of the heptafluorobutyryl-butyl ester derivative, these authors were able to detect 3-hydroxydodecanoic acid in vaginal fluid specimens from women with gonorrhoea and proposed that their method might be useful for rapid diagnosis of gonorrhoea. The detection limit of the method corresponded to about 10^5 bacteria although interfering substances made it difficult to work at maximum sensitivity. The TMS methyl ester derivative of 3-hydroxydodecanoic acid, analyzed by EI-SIM, has also been demonstrated in sera from patients with meningitis caused by *Neisseria meningitidis*, and the quantitative results obtained tallied well with those of the *Limulus* test.[36]

CONCLUSIONS

Methods for the highly sensitive and selective determination of specific microbial fatty acids have been developed. These techniques are useful both in research work and for routine screening of samples in clinical and ecological microbiology. Obvious potential applications include, for example, diagnosis of infections (not only in man but also in animals and plants), monitoring of therapy, and gain of insight in initial microbial fouling processes. Work is in progress for simplifying and developing more efficient sample preparation methods including automation. Use of liquid chromatography combined with SIM detection for the sensitive and selective determination of more complex microbial lipid structures, as well as routine use of MS-MS on complex mixtures, should open new doors to further useful applications of biomarker detection in microbiological research.

REFERENCES

1. G. Odham, L. Larsson, and P. -A. Mardh, eds., "Gas Chromatography/Mass Spectrometry Applications in Microbiology," Plenum Press, New York (1984).
2. E. Jantzen, Analysis of cellular components in bacterial classification and diagnosis, in: "Gas Chromatography/Mass Spectrometry Applications in Microbiology," G. Odham, L. Larsson, and P. -A. Mardh, eds., Plenum Press, New York (1984).
3. L. Larsson, Gas chromatography and mass spectrometry, in: "Automation in Clinical Microbiology," J. Jorgensen, ed., CRC Press Inc., Boca Raton (1987).
4. A. Tunlid, Chemical signatures in studies of bacterial communities. Highly sensitive and selective analyses by gas chromatography and mass spectrometry, Thesis, University of Lund (1986).
5. E. Jantzen and K. Bryn, Whole-cell and lipopolysaccharide fatty acids and sugars of gram-negative bacteria, in: "Chemical Methods in Bacterial Systematics," M. Goodfellow and D. Minnikin, eds., The Society of Applied Microbiology Technical Series, Academic Press, London (1985).
6. G. Bengtsson and G. Odham, A micromethod for the analysis of free amino acids by gas chromatography and its application to biological systems, Anal. Biochem. 92:426 (1979).
7. C. Asselineau and J. Asselineau, Fatty acids and complex lipids, in: "Gas Chromatography/Mass Spectrometry Applications in Microbiology," G. Odham, L. Larsson, and P. -A. Mardh, eds., Plenum Press, New York (1984).
8. G. Odham, L. Larsson, and P. -A. Mardh, Quantitative mass spectrometry and its application in microbiology, in: "GC/MS Applications in Microbiology," G. Odham, L. Larsson, P. -A. Mardh, eds., Plenum Press, New York (1984).

9. G. Phillipou, D. A. Bigham, and R. F. Seamark, Subnanogram detection of t-butyldimethylsilyl fatty acid esters by mass fragmentography, Lipids 10:714 (1975).

10. S. K. Maitra, M. C. Scholtz, T. T. Yoshikawa, and L. B. Guze, Determination of lipid A and endotoxin in serum by mass spectroscopy, Proc. Natl. Acad. Sci., USA 75:3993 (1978).

11. S. Maitra, R. Nachum, and F. Pearson, Establishment of beta-hydroxy fatty acids as chemical marker molecules for bacterial endotoxin by gas chromatography-mass spectrometry, Appl. Environ. Microbiol. 52:510 (1986).

12. G. Odham, A. Tunlid, G. Westerdahl, L. Larsson, J. Guckert, and D. C. White, Determination of microbial fatty acid profiles at femtomolar levels in human urine and the initial microfouling community by capillary gas chromatography-chemical ionization mass spectrometry with negative ion detection, J. Microbiol. Methods 3:331 (1985).

13. J. B. Brooks, Gas-liquid chromatography as an aid in rapid diagnosis by selective detection of chemical changes in body fluids, in: "The Direct Detection of Microorganisms in Clinical Samples," J. D. Coorod, J. J. Kunz, and M. J. Ferraro, eds., Academic Press, New York (1983).

14. M. I. Daneshvar, J. B. Brooks, and R. M. Winstead, Disposable reversed-phase chromatography columns for improved detection of carboxylic acids in body fluids by electron-capture gas-liquid chromatography, J. Clin. Microbiol. 25:1216 (1987).

15. A. Sonesson, L. Larsson, G. Westerdahl, and G. Odham, Determination of endotoxins by gas chromatography: Evaluation of electron capture and negative ion chemical ionization mass spectrometric detection of halogenated derivatives of beta-hydroxymyristic acid, J. Chromatogr. Biomed. Appl. 417:11 (1987).

16. J. M. Rosenfeld, O. Hammerberg, and M. C. Orvidas, Simplified methods for preparation of microbial fatty acids for analysis by gas chromatography with electron-capture detection, J. Chromatogr. Biomed. Appl. 378:9 (1986).

17. J. M. Rosenfeld, M. Mureika-Russell, and S. Yeroushalmi, Solid-supported reagents for the simultaneous extraction and derivatization of carboxylic acids from aqueous matrices. Studies on optimization of reaction conditions, J. Chromatogr. 358:137 (1986).

18. I. J. Sud and M. D. Feingold, Detection of 3-hydroxy fatty acids at picogram levels in biological specimens. A chemical method for the detection of Neisseria gonorrhoeae, J. Invest. Dermatol. 73:521 (1979).

19. G. Odham, A. Tunlid, G. Westerdahl, and P. Marden, Combined determination of poly-β-hydroxyalkanoic and cellular fatty acids in starved marine bacteria and sewage sludge by gas chromatography with flame ionization or mass spectrometric detection, Appl. Environ. Microbiol. 52:905 (1986).

20. A. Sonesson, L. Larsson, and J. Jimenez, Use of pentafluorobenzyl and pentafluoropropionyl/pentafluorobenzyl esters of bacterial fatty acids for gas chromatographic analysis with electron capture detection, J. Chromatogr. Biomed. Appl. 417:366 (1987).

21. L. Larsson, A. Sonesson, and J. Jimenez, Ultrasensitive analysis of microbial fatty acids using gas chromatography with electron capture detection, Europ. J. Clin. Microbiol. 6:729 (1987).

22. C. F. Poole and S. A. Schuette, "Contemporary practice of chromatography," Elsevier, Amsterdam (1984).

23. E. A. Bergner and R. C. Dougherty, Detection of urinary primary amines through negative chemical ionization mass spectrometry of fluresciamine derivatives, Biomed. Mass Spectrom. 8:208 (1981).

24. L. Larsson, P. -A. Mardh, and G. Odham, Detection of tuberculostearic acid in mycobacteria and nocardiae by gas chromatography and mass spectrometry using selected ion monitoring, J. Chromatogr. Biomed. Appl. 163:221 (1979).

25. L. Larsson, P. -A. Mardh, G. Odham, and G. Westerdahl, Detection of tuberculostearic acid in biological specimens by means of glass capillary gas chromatography/electron and chemical ionization mass spectrometry, utilizing selected ion monitoring, J. Chromatogr. Biomed. Appl. 182:402 (1980).

26. G. Odham, L. Larsson, and P. -A. Mardh, Demonstration of tuberculostearic acid in sputum from patients with pulmonary tuberculosis by selected ion monitoring, J. Clin. Invest. 63:813 (1979).

27. P. -A. Mardh, L. Larsson, N. Hoiby, H. C. Engbaek, and G. Odham, Tuberculostearic acid as a diagnostic marker in tuberculous meningitis, The Lancet 8320:367 (1983).

28. G. French, C. Chan, S. Cheung, R. Teoh, M. Humphries, and G. O'Mahony, Diagnosis of tuberculous meningitis by detection of tuberculostearic acid in cerebrospinal fluid, The Lancet 8551:117 (1987).

29. G. French, C. Chan, S. Cheung, and K. Oo, Diagnosis of pulmonary tuberculosis by detection of tuberculostearic acid in sputum by using gas chromatography-mass spectrometry with selected ion monitoring, J. Infect. Dis. 156:356 (1987).

30. L. Larsson, G. Odham, G. Westerdahl, and B. Olsson, Diagnosis of pulmonary tuberculosis by selected ion monitoring: Improved analysis of tuberculostearate in sputum using negative ion mass spectrometry, J. Clin. Microbiol. 25:893 (1987).

31. L. Larsson, J. Jimenez, A. Sonesson, and F. Portaels, Two-dimensional gas chromatography with electron capture detection for the sensitive determination of specific mycobacterial lipid constituents, J. Clin. Microbiol. 27: in press (1989).

32. L. Larsson, P. -A. Mardh, G. Odham, and G. Westerdahl, Use of selected ion monitoring for detection of tuberculostearic and C_{32} mycocerosic acid in mycobacteria and in five-day-incubated cultures of sputum specimens from patients with pulmonary tuberculosis, Acta Path. Microbiol. Scand. Sect. B 89:245 (1981).

33. D. E. Minnikin, R. Bolton, G. Dobson, and A. Mallet, Mass spectrometric analysis of multimethyl branched fatty acids and phthiocerols from clinically significant mycobacteria, Proc. Jap. Soc. Med. Mass Spectr. 12:23 (1987).

34. A. Sonesson, L. Larsson, A. Fox, G. Westerdahl, and G. Odham, Determination of environmental levels of peptidoglycan and lipopolysaccharide using gas chromatography with negative-ion chemical-ionization mass spectrometry utilizing bacterial amino acids and hydroxy fatty acids as biomarkers, J. Chromatogr. Biomed. Appl. 431:1 (1988).

35. A. Sonesson, L. Larsson, and J. Jimenez, Two-dimensional gas chromatography with electron capture detection used in the determination of specific peptidoglycan and lipopolysaccharide constituents of Gram-negative bacteria in infected human urine, J. Chromatogr. Biomed. Appl. 490:71 (1989).

36. K. Byrn, P. Brandtzaeg, E. Jantzen, and G. Becker, Quantification of meningococcal LPS in plasma: comparison of ion trap mass spectrometry and a chromogenic LAL test, in: International Conference on Endotoxin (Amsterdam), abstract 103 (1987).

CHAPTER 15

GAS CHROMATOGRAPHIC AND MASS SPECTROMETRIC TECHNIQUES FOR THE DIAGNOSIS OF
DISSEMINATED CANDIDIASIS

John Roboz

Department of Neoplastic Diseases
Mount Sinai School of Medicine
One Gustave Levy Place
New York, NY 10029

INTRODUCTION

Opportunistic infections are an increasingly important cause of
morbidity and mortality in the immunocompromised patient.[1,2] The
introduction into clinical practice of many new antibacterial agents of
increased potency and broader spectrum led to an increased frequency of
fungal infections over the past several decades.[3] In a series of three
studies during the period 1954–1979, major fungal infections were found to
increase from 10% to 40% among dying cancer patients.[4] Candidiasis accounted
for the majority of serious fungal infections: a four year study on 233
patients revealed that *Candida* species (including *Torulopsis glabrata*)
accounted for 74%, 61% and 92% of fungal infections in patients with acute
leukemia, lymphoma, and solid tumors, respectively.[5] In 42 autopsies on
patients with leukemia and lymphoma, severe superinfection by *Candida* and
Aspergillus species was found in 52%; most patients were not diagnosed and
hence not treated antemortem.[6] According to different estimates[7-12] up to 70%
of dying cancer patients may have fungal infections.

Systemic fungal infection also complicate renal transplantation,[13]
surgical procedures,[14] the management of high risk neonates,[15] burn
patients,[16] and recently, the treatment of drug abusers[17] and AIDS victims.[18]
The considerable current interest in the clinical[2,5,19] as well as the
molecular biological[20] aspects of candidiasis is well justified; a major new
textbook covers virtually all aspects of candidiasis and lists almost 6000
references.[21]

Candida species are normal inhabitants of mucocutaneous body surfaces
but can overgrow, invade tissue, and disseminate into deep-seated organs
(kidney, lung and liver are the main targets) in patients predisposed by
prolonged periods of neutropenia (granulocyte count < 1,000/mL) due to
aggressive cytotoxic, broad-spectrum antibiotic or corticosteroid
chemotherapy.[5,21-26] Also at risk are patients receiving parenteral
hyperalimentation, and those exposed to indwelling catheters and respiratory
manipulations. The term candidiasis (known as candidosis in Europe) is used

to denote a local or systemic infection caused by a *Candida* species (infections by *Torulopsis glabrata* are also included). "Colonization" refers to a situation when there is an active growth of the microorganism without a clinical infection in the patient. "Dissemination" means that the fungus spread from a focus (usually at a mucous surface or in the skin) to one or more deep organs, accompanied in most cases, by clinical signs of infection. The term systemic fungal infection denotes the establishment of an acute infective state of the patient. The clinical spectrum of candidiasis is very broad, ranging from mucocutaneous candidiasis, colonization of the alimentary and urinary tracts, to candidemia (presence of *Candida* cells in peripheral blood) and disseminated candidiasis with multiple organ involvement. Although several *Candida* species have been found in multiple-infected cancer patients,[27] *Candida albicans* has been the most common species in disseminated candidiasis. *Candida tropicalis* has often been associated with acute leukemia[24,28] while *Candida parapsilosis* is frequently seen in patients receiving total parenteral nutrition.[29]

The high mortality rates of disseminated candidiasis are not necessarily due to the inherent failure of chemotherapy but rather to the delay in treatments. Because of the lack of reliable diagnostic procedures, the majority of cases are not recognized early enough to permit effective treatment. The only definite method of diagnosis of disseminated candidiasis is histologic demonstration of tissue invasion by fungi, e.g., the biopsy confirmed presence of endophthalmitis or *Candida* containing macronodular skin lesions are suggestive of disseminated candidiasis. Obviously, histologic proof is not available in many cases of deep-seated organ invasion. There are no distinctive clinical manifestations; presumptive clinical diagnosis is based on fever of unclear etiology, macronodular skin lesions, endophthalmitis, and diffuse severe tenderness.[30] Patients must be evaluated individually with all available clinical data considered,[31,32] particularly when invasive gastrointestinal candidiasis is suspected because of the difficulties in differentiating from colonization.[33,34]

Due to the toxicity of systemic antifungal therapy, treatment is often undertaken with considerable reluctance in the absence of definitive diagnosis. Nevertheless, the benefits of early therapy, even on an empirical basis, have been emphasized[35-38] although there are few criteria for the evaluation of therapeutic responses to antifungal drugs.[39] The *in vitro* susceptibility testing of antifungal agents is also controversial.[40] An observation of potential importance is that systemic candidiasis is often preceded by oropharyngeal candidiasis.[41,42]

BLOOD CULTURING

The term candidemia refer to cases when *Candida* sp. are cultured, using conventional culturing techniques[30] from blood specimens with or without evidence of additional organ invasion. A study of 185 patients, treated for various diseases, in whom fungi were isolated in blood cultures, revealed disseminated fungal infection in 45% of the cases, 70% of which were caused by *C. albicans*; mortality was 58%.[43] In general, the results of blood culturing are not reliable. Blood cultures have been found positive in only 20-50% of the cases[26,44] and in about one-third of those cases the results were only available after the patient died. Patients with invasive renal candidiasis or *Candida* pneumonia diagnosed at autopsy often have a series of negative antemortem blood cultures. Alternatively, positive blood cultures are indicative of systemic treatment only if confirmed in repeated tests because catheter-related transient fungemia is frequent.[29] In addition, blood culturing is relatively slow: using the Bactec radiometric system (Johnston Laboratories, Towson, MD) reportedly takes 1.5 to 4 days while using biphasic media may require as much as 5 to 9 days.[45]

SERODIAGNOSIS

Serologic tests for the diagnosis of systemic candidiasis may be divided into three groups: detection of antibodies to *Candida*, detection of circulating candidal antigens, and quantification of circulating characteristic metabolites or cell wall constituents of the organism. The reviews of various approaches to serodiagnosis[46-48] reveal that the attempted correlations with invasive disease have resulted in a large number of false positives and false negatives and have not been proven reliable in clinical trials. More recent techniques include characterization of human sera precipitins by co-counterimmunoelectrophoresis,[49,50] several versions of latex agglutination tests for the detection of *Candida* antigens,[51-54] and testing for the presence of anti-*Candida albicans* antibodies by indirect immunofluorescence.[55] Affinity chromatography was used to isolate immunodominant candidal antigens of 47 and 60 kilodalton.[56,57] A promising approach is based on the development of monoclonal antibodies against mannan of *C. tropicalis* and the detection of the antigen by enzyme immunoassay and immunofluorescence.[58] A comparison of tests based on enzyme immunoassay (ELISA) and counterimmunoelectrophoresis for the detection of an anti-candidal antibody in 85 subjects revealed the former to be superior, however, it was concluded that the combination of serological tests may be advantageous to improve the accuracy of diagnosis.[59] Comparative tests are now carried out in several laboratories to evaluate a commercially available *Candida* detection system (CAND-TEC, Ramco Laboratories, Inc., Houston, TX). In one study of 83 specimens from 24 patients, sensitivity and specificity were 71% and 98%, respectively.[60] A review of serological techniques based on antigen detection suggests this approach to be promising although improvements in sensitivity and practicality are clearly needed.[61]

DETECTION AND IDENTIFICATION OF METABOLITES

In 1974, gas chromatographic (GC) analysis of methyl and trimethylsilyl (TMS) derivatives from hydrolyzed extracts from *Candida* cultures and serum samples from patients with confirmed candidiasis revealed the presence of several peaks considered to be potential markers. Peaks corresponding to palmitic, oleic, linoleic, and stearic acids were identified by electron ionization mass spectrometry.[62] In a follow-up paper, mannose, probably originating from mannan, was added to the list of potential markers.[63] Two additional publications suggested mannose as a marker for disseminated candidiasis and reported serum mannose concentrations determined from hydrolysates by gas chromatography.[64,65] In subsequent studies the technique based on the fatty acid profiles was suggested to be practical,[66] while a comparison of techniques based on serum mannan, mannose, and arabinitol (see later) for the diagnosis of disseminated candidiasis in an immunosuppressed rabbit model revealed only the serum mannan to be specific.[67] At least three laboratories, including the author's, were unable to confirm the "uniqueness" of the increases in serum fatty acids and mannose concentrations with respect to candidiasis, probably because of methodological problems with these techniques (private communication).

D-arabinitol, a five-carbon unbranched sugar alcohol (Figure 1), was identified by mass spectrometry as a major metabolite of several *Candida* species independently and almost simultaneously in three laboratories. An abstract by Wojnarowski et al.[68] reported increased urinary arabinitol, but no arabinitol in serum, in an experimental rabbit model; the abstract was not followed by detailed publications. Kiehn and co-workers[69] identified arabinitol as both TMS and peracetate derivatives using chemical ionization (isobutane) mass spectrometry, and determined, based on melting point measurements of the crystallized peracetate derivative, that the arabinitol metabolite of the *Candida* organisms studied was of the D form. They also developed and clinically tested a GC technique for the quantification of D-arabinitol in human serum. Similar mass spectrometric approach was used by Roboz and co-workers[70] for the identification of arabinitol in both culture and infected serum samples. They also confirmed the identity of arabinitol as the marker by determining the exact mass and corresponding molecular composition of the protonated molecular ion of the TMS derivative of the unknown using high resolution mass spectrometry. Their method developed for quantification in serum is based on selected ion monitoring. Both the GC and GC-MS techniques have been used by their developers as well as by several other laboratories for a number of applications with emphasis on clinical trials for the diagnosis and monitoring of disseminated candidiasis in patients and in experimental animal models.

METHODOLOGIES FOR THE DETERMINATION OF ARABINITOL

Gas chromatography

The original GC technique developed for the quantification of D-arabinitol in serum and urine samples from patients as well as experimental animals is based on the detection by flame ionization of the TMS derivative of arabinitol.[69,71,72] Starting with 200 μL human or 100 μL animal serum, the first step is the addition of an equal amount of water which contains 10 μg each of one or more internal standards (see next paragraph). Next, proteins are precipitated by adding two volumes (800 μL human, 400 μL animal samples) of acetone and centrifuging at 1,500 g for 5 min. After evaporating the supernatant to dryness with nitrogen in a water bath at 50°C, the residue is silylated with 100 μL of a mixture of pyridine-hexamethyldisilazane-trimethylchlorosilane (4:2:1 v/v) at 50°C for 30-40 min. An aliquot of the clear liquid is finally analyzed by GC using a 6 ft long glass column filled with 3% SE-30 (nonpolar silicone) stationary phase operated in a temperature programmed mode (140-220°C at 4°C/min) using nitrogen as the carrier gas at a flow rate of 40 mL/min. Under these conditions, the retention time of arabinitol is approximately 9 min.

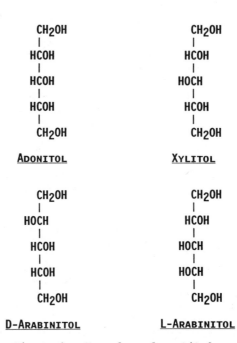

Figure 1. Formulas of pentitols.

This is a straightforward and convenient technique which, with minor modifications, has been used in several laboratories (for details see the experimental sections of the appropriate references in the sections on applications). An apparent difficulty concerns the selection of an appropriate internal standard. Meso-erythritol was originally selected as the internal standard. Because this compound is present in patients with renal failure,[72] α-methyl-D-mannopyranoside was suggested as an alternative. Subsequently, meso-erythritol was replaced by α-methylxyloside and β-methylxyloside and the use of three internal standards was suggested.[73] Although α-methyl-D-mannopyranoside has been the first choice in most subsequent GC studies, interferences in some clinical samples still presents problems in quantification.

Arabinitol in urine is determined by mixing 200 μL of urine (after appropriate dilution if warranted by high arabinitol content) with an equal volume of solution containing the internal standard, followed by evaporation to dryness and silylation.[72]

Another GC technique, developed for the simultaneous determination of arabinitol and mannose in experimental candidiasis, utilizes the per-O-acetylated aldononitrile derivatives.[65] It was subsequently shown that this technique may yield falsely high arabinitol concentrations because upon derivatization the glucose content of serum yields a pentitol peracetate byproduct with the same chromatographic retention time and chemical ionization mass spectrum as arabinitol pentaacetate.[74]

Gas chromatography-mass spectrometry

The GC-MS method developed for the quantification of arabinitol in human serum is based on the monitoring of the protonated molecular ions, obtained by chemical ionization of the TMS derivatives of arabinitol and the internal standard, from a simple extract of serum samples.[70] Sample preparation is very similar to that described above. To a 200 μL size serum sample, 100 μL of the internal standard working solution (5 μg/mL concentration) and 1 mL of acetone are added. After vortexing and centrifuging (1,200 g for 15 min at room temperature), the supernatant is evaporated with the aid of dry nitrogen in a water bath kept at 50°C. The dry residue is next silylated by adding 200 μL of a mixture of N,O-(bis)trimethylsilyltrifluoroacetamide + 1% trimethylchlorosilane and silylation grade pyridine (3:1 v/v) at 100°C for 5-10 min. Finally, a 2-4 μL size aliquot is injected into the combined GC-MS. Chromatographic separation is achieved using a 6 ft long glass column, operated isothermally at a temperature best suited to the internal standard used (see below). Isobutane is used as both the chromatographic carrier gas and the chemical ionization reagent gas at a flow rate set to optimize the ion intensity of a characteristic ion in the mass calibration compound (e.g., m/z = 414 for perfluorotributylamine). Only two masses are monitored, m/z = 513 for arabinitol and the second mass for the protonated molecular ion of the internal standard. Initially, 2-deoxy-galactitol was used as the internal standard because of its convenient chromatographic properties and because it did not have the interference problems experienced with the other compounds mentioned above. Subsequently, fully deuterated arabinitol has been used as the internal standard (protonated molecular ion at m/z = 520). This stable isotope labeled compound has identical extraction, chromatographic, and mass spectrometric properties to arabinitol, and its mass is far enough from that of normal arabinitol to preclude mass spectral interferences (contribution to the m/z = 513 peak is < 1%). In more than five years of experience, there has been no interference problem encountered with this internal standard.

Quantification with both the GC and GC-MS techniques is similar. In a conventional manner, calibration curves are obtained by adding increasing amounts of arabinitol to the serum samples (quantity of internal standard is constant) and constructing the curve using peak height or peak area ratios; from the calibration curves conventional regression analysis can be performed. Extrapolation of a calibration curve to the X axis provides the endogenous value of arabinitol in the serum used for calibration.[70] Figure 2 shows arabinitol and the internal standard peaks obtained with selected ion monitoring for a normal serum sample and samples from a patient with candidiasis before and after antifungal treatment. Quantification was accomplished by computer-generated peak area determinations. It is noted that the term total pentitols is used instead of arabinitol. This is explained in the next section.

Gas chromatographic identifications are based on retention times and it is difficult to identify small peaks among the ever present "chemical noise". Also, the background profiles of individual patients are often variable and unresolved peaks may lead to falsely high arabinitol values. In contrast, selected ion monitoring (SIM) is much more specific than GC with an FID because here particular masses, highly characteristic of the compound being analyzed, are monitored and most interferences are eliminated. Another advantage of the GC-MS technique is the possibility of using a stable isotope labeled internal standards as described above. SIM is also superior with respect to the determination of small incremental changes in arabinitol concentrations, a major requirement in marker studies where the compound of

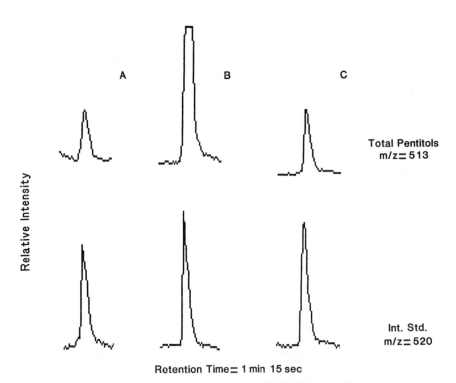

Figure 2. Selected ion monitoring of total pentitols in sera from (A) a healthy control subject, 0.7 μg/mL, (B) a cancer patient with confirmed candidiasis, 3.6 μg/mL, and (C) the same patient after successful chemotherapy, 0.8 μg/mL. Int. Std. = deuterated arabinitol.

interest is also present as an endogenous constituent. In serum spiked with arabinitol in 0.2 μg/mL increments (0.0 - 1.0 μg/mL range, four replicates at each point) it was possible to detect an increment of 0.2 μg/mL (1.35 μM) using SIM.

Determination of individual polyols

It was recognized from the beginning that the other common pentitols, adonitol and xylitol, were not separated from arabinitol using packed GC columns.[70] This was not considered to be a serious shortcoming because the endogenous serum concentrations of these two pentitols were reported to be small compared to that of arabinitol.[75] Nevertheless, "arabinitol" concentrations reported by the techniques outlined above are in fact total pentitol (arabinitol + adonitol + xylitol) concentrations (see later). The fact that increased adonitol was occasionally observed (using pentaacetate derivatization) in patients with confirmed candidiasis[70] justified the development of methodologies for the investigation of the possible role(s) of the other polyols in the diagnosis of candidiasis.

Although polyols are present in most living organisms,[75,76] relatively little is known about their biological significance[77] other than that several of them are strongly involved in the metabolism of yeasts[78] including *Candida* species.[79,80] There have been several GC[81-83] and GC-MS[84-88] techniques reported on the analysis of one or more polyols in biological fluids. Some polyols have not been detected in normal human serum by these methods, probably because of inadequate detection limits.

A GC-MS method was developed for the quantification of seven polyols in serum.[89] After adding the internal standard (2-deoxygalactitol) to serum, proteins are precipitated with alcohol and the supernatant is evaporated with dry nitrogen. The residue is acetylated with acetic anhydride-pyridine, the reaction is stopped with water, the acetates are extracted with hexane-chloroform, washed with water, dried with anhydrous Na_2SO_4 and evaporated to dryness. The residue is redissolved in hexane and an aliquot is analyzed. Either packed (3% SP-2340, Supelco Inc., Bellefonte, PA) or capillary (3% OV-1701, Scientific Glass Engineering, Austin, TX) columns can be used. The following masses are monitored: m/z = 375 for the hexitols, m/z = 303 for the pentitols, m/z = 231 for the tetritols and m/z = 317 for the internal standard. The following endogenous serum concentrations, in mean μg/mL (standard deviation) were determined in 33 normal sera:[89] erythritol 0.45 (0.14), threitol 0.20 (0.06), adonitol 0.06 (0.20), arabinitol 0.37 (0.12), xylitol 0.05 (0.02), mannitol 0.41 (0.45), galactitol 0.15 (0.11) and sorbitol 0.16 (0.11).

Another GC-MS technique[90] separates the TMS derivatives of pentitols using, conveniently, the sample preparation technique described above for packed column analysis. The separation is accomplished using a 50 m long, 0.32 mm i.d., bonded methyl silicone column (equivalent to OV-17, Quadrex Corp., New Haven, CT) operated isothermally at 195°C. Mass spectrometric conditions are the same as described above; masses monitored are m/z = 513 for the pentitols and m/z = 520 for the D7-arabinitol used as internal standard. The coefficients of variation for intraday (n = 10) and interday (n = 7 within a three months period) were arabinitol 5% and 10%, adonitol 11% and 14%, and xylitol 9% and 11%, respectively. In normals the following mean concentrations (μg/mL ± s.d.) and ranges, in parenthesis, were obtained for controls (n = 25): arabinitol 0.38 ± 0.12 (0.22-0.68), adonitol 0.07 ± 0.03 (0.02-0.12), and xylitol 0.07 ± 0.03 (0.02-0.12). The upper limit of normal was defined as the mean plus three standard deviations. Accordingly, values greater or equal to the following concentrations (μg/mL) were considered as "increased": arabinitol 0.74, adonitol 0.17, xylitol 0.17.

ARABINITOL IN THE DIAGNOSIS OF CANDIDIASIS

Normal and increased arabinitol

Some publications[75,87] reported arabinitol in normal sera as "non-detectable", others reported increased arabinitol in patients with sarcoidosis[91] and also in cancer patients and even in healthy subjects[92,93] without infection. However, the methodologies used in these investigations appear to have either inadequate sensitivity or other methodological problems and none included data on incremental sensitivity and reproducibility. The following means plus standard deviation were obtained[94,95] for arabinitol concentrations in three control populations with normal (< 1.5 mg/dL) creatinine: healthy blood donors (16-59 yr old, n = 45) 0.46 ± 0.13 μg/mL; cancer patients, treated for various malignancies but without known opportunistic infections at the time of sampling (14-64 yr old, n = 33) 0.46 ± 0.17 μg/mL; the same group as before but age 65-84 yr (n = 26) 0.52 ± 0.11 μg/mL. There was no significant difference between normal and cancer patients without candidiasis. Based on extensive studies of serum samples from normal subjects[69,70,72,95,96] the upper limit of normal serum arabinitol concentration, defined as the mean of normal plus two or three standard deviation, is in the 0.9-1.2 μg/mL range. Circulating serum arabinitol concentrations above the upper limit are considered as indicative of invasive candidiasis.

Renal Dysfunction

It has been recognized[69,70] that renal dysfunction, which is often concurrent with disseminated candidiasis, also produces increased arabinitol concentrations. This significantly reduces the diagnostic utility of the arabinitol method. Armstrong and co-workers proposed the use of arabinitol/creatinine ratios for differentiation.[71,72] Most subsequent work done using the GC technique (but not GC-MS, see later) attempted to use the arabinitol/creatinine ratio to circumvent the renal dysfunction problem.[96-99]

Subsequent studies confirmed the presence of increased arabinitol in renal dysfunction and suggested that additional problems with increased arabinitol concentrations occur in patient under dialysis.[96,97] In another study,[100] serum concentrations of erythritol, threitol, and arabinitol were found to be increased above normal in all patients (n = 7) with renal failure. Mannitol and xylitol concentrations ranged from normal to increased while adonitol, galactitol, and sorbitol were within normal range. Although a 4 h dialysis removed the polyols, except xylitol and adonitol, to about 50% of the pre-dialysis value at a linear rate, the serum concentrations of erythritol, threitol, and arabinitol were still more than double the normal in all patients at the end of the dialysis. There was no correlation between the increased polyol concentrations and the creatinine values in the pre-dialysis samples.

Clinical studies

After the extensive initial clinical studies of Kiehn and co-workers,[69,71-73] who even attempted to correlate the rate of arabinitol production of the most important *Candida* species,[101] and a smaller and inconclusive study by Eng et al.,[96] there have been three more recent clinical studies using essentially the same GC method with results ranging from negative through tentative to positive.

A patient population of 40 (all with positive *Candida* cultures) was subdivided, based on clinical and pathological findings, into groups representing superficial, possible deep and definite deep invasive candidiasis.[99] Arabinitol and arabinitol/creatinine ratios were not capable

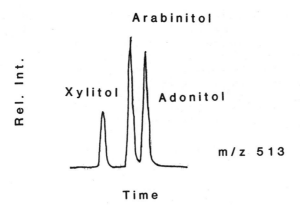

Figure 3.   Selected ion monitoring (m/z = 513) of xylitol,
            arabinitol and adonitol in a patient using a capillary
            column.[90]   This is a case where total pentitol
            determination yielded false positive because of
            increased adonitol and xylitol in the presence of
            normal arabinitol.

to distinguish between superficial and invasive candidiasis with
statistically adequate reliability. It was concluded that until further
refinements in methodology, the arabinitol technique is of very limited
clinical value for the diagnosis of invasive candidiasis.

In another study,[98] a patient population of 134 (plus 11 normals) were
subdivided into groups with *Candida* sepsis, *Candida* colonized, and bacterial
sepsis, with further subdivision according to renal status. Using
arabinitol/creatinine ratios, 69%, 36%, and 0% of patients were shown to
belong to the three categories, respectively. It was tentatively concluded
that while arabinitol alone was of limited use, the application of the
arabinitol/creatinine ratio might be of diagnostic value.

Using arabinitol values in 157 serum samples from 95 patients (plus 10
normals) Deacon has recently evaluated[97] the utility of the GC technique and
calculated a 92.5% sensitivity and 88.2% specificity. After a thoughtful
discussion of several positive and negative aspects of the technique,
particularly those concerning renal dysfunction and the potential effects of
dialysis on arabinitol, he concluded that the arabinitol technique should be
useful in a conventional hospital setting as long as adequate instrumentation
is available.

"Arabinitol" and total pentitols.

As already noted, GC techniques currently in use are based on packed
columns, thus all reported values of "arabinitol" concentrations are, in
fact, total pentitol (TP) concentrations. It has been recommended[90] to call
TP of what has been formerly known as "arabinitol", i.e., total pentitols
determined by using packed GC columns. The term "arabinitol concentration"
should be used only when the pentitols were separated by capillary GC. As
mentioned earlier, the concentration of endogenous arabinitol is
approximately 50 times larger than the concentrations of xylitol and
adonitol.[89] Figure 3 shows a case where TP determination yielded a false
positive because of both adonitol and xylitol concentrations were increased
in the presence of normal arabinitol.

A population consisting of 173 samples (73 patients) from a population
of 1250 samples (285 cancer patients) for which total pentitol data were

available (see later) was studied using capillary GC to separate individual pentitols.[90] Selection criteria included culture confirmed candidiasis, clinically suspected candidiasis with negative blood culture, and unexplained increased TP during monitoring of cancer patients undergoing immunosuppressive chemotherapy. There were 37 samples (23 patients) with culture-positive candidiasis. Increased TP (0.9-7.8 μg/mL) was found in all, yielding a sensitivity of 100%. However, in one sample the increased TP (1.0 μg/mL) resulted from normal arabinitol and increased xylitol (0.19 μg/mL), indicating a false positive by the TP method. This reduces the sensitivity to 97%. Increased TP was found in 56 samples in the culture-negative and 40 samples in the no-culture available groups. Of these, eight (14%) and seven (17%), respectively, had normal arabinitol and were, therefore, false positives. All false positives were due to the presence of larger than normal concentrations of either adonitol (0.05-4.9 μg/mL) or xylitol (0.02-0.36 μg/mL) or both. (The reason for the increase in the endogenous concentrations of these pentitols has not been determined; there is some evidence that some *Candida* species produce small amounts of adonitol). Combining the three groups studied, 12% of the samples were false positives by the TP method. The samples for this study were selected in a somewhat biased manner (see above), hence a smaller than 12% false positive rate is projected when the TP method is employed to the regular patient population.

It was concluded[90] that the technique developed for the quantification of individual pentitols can be used to eliminate potential false positives caused by increased adonitol and/or xylitol concentrations when using the TP technique. The continued use of the TP method was still recommended because of its simplicity and speed. However, when a sample with increased total pentitols and normal creatinine is encountered, it should be analyzed for individual pentitols. Conveniently, the same sample can be used for both determinations.

Evaluation of the GC-MS method for TP

The GC-MS technique for the quantification of TP has been undergoing initial clinical trials, not only as an aid in the diagnosis of invasive candidiasis but also for the prospective monitoring of immunosuppressed patients at risk of developing disseminated candidiasis with the objective of recognizing the onset of candidiasis early enough to permit more effective and less toxic treatment. This work has been carried out as part of a general approach to utilize mass spectrometric techniques for the identification and quantification of characteristic markers in bacterial as well as in fungal infection.[102]

The study population consisted of 1250 samples from 284 patients.[95] Positive autopsy and positive blood cultures were considered as diagnostic of invasive candidiasis. Negative cultures were considered as inconclusive. Frequently both positive and negative culture results were obtained for a patient with multiple cultures.

In 11 autopsy-confirmed candidiasis cases, with TP data obtained <8 days before death, the diagnostic sensitivity of the TP method was 100%. The corresponding sensitivity of the blood culturing technique (multiple samplings) was 50% with 81% of the samples being false negatives. In the culture-positive group the diagnostic sensitivity of the TP method was 98% (59/60 samples); there was one false negative. It is noted that 42% of the samples had normal creatinine, indicating no renal dysfunction. It was impossible to establish the specificity because of the inconclusive nature of the results of negative blood cultures.

The death rate (within 8 days of the last TP determination) was 74% among those with increased TP (n = 34) in the culture-positive group. The

248

combined death rate of 102 patients with increased TP in the culture-negative and no-culture available groups was 70%. The corresponding death rate for the 148 patients with normal TP was 11%. The death rates were not affected by renal status. Of the 96 patients who died with increased TP, 72% had increased TP values available more than one week before death (>2 weeks: 57%, >3 weeks: 44%). The significant difference in death rate with increased TP does not, of course, prove the presence of disseminated candidiasis, however, it does suggest that increased TP should be taken seriously.

Ninety-five patients at risk of developing candidiasis were monitored (weekly analysis) for at least one month (808 samples). In 33 patients (281 samples) where the TP (and also creatinine) remained normal throughout monitoring and where there was no confirmed candidiasis, the death rate was 12%. There were three patients (22 samples), all with confirmed and treated candidiasis, where TP went from increased to normal; there were no deaths in this group. Among six patients (33 samples) with variable TP but the last one being normal, there was no confirmed candidiasis and the death rate was 33%. There were 12 patients (66 samples) with always increased TP, 39 patients (391 samples) with "normal-to-increased" TP and five patients (37 samples) with variable TP, the last being an increased one. The death rate in these three groups was in the 80-90% range, candidiasis was confirmed in 42%, 18, and 0%, respectively.

Of the 18 patients in the "normal-to-increased" group who developed renal dysfunction, 13 (72%) had increased TP one to six weeks prior to developing renal dysfunction. The corresponding number in the "always increased" group was 17%. This indicates the importance of monitoring patients at risk because of the possibility of detecting the development of candidiasis before renal dysfunction occurs. The fact that the first increased TP often occurs weeks before death could be utilized to initiate treatment and suggests the need for systematic monitoring of patients at risk of developing disseminated visceral candidiasis.

SEPARATION OF D- and L-ARABINITOL

Although the results presented above strongly suggest that increased arabinitol in the presence of normal creatinine indicates disseminated candidiasis, even when blood cultures are negative, it is obvious that the arabinitol technique cannot be generally accepted until the problem of interference from renal dysfunction is satisfactorily resolved. The source of endogenous arabinitol in serum and urine is unknown. It may be derived, at least partially, from *Candida* species in the bowel. Studies on the metabolism of pentoses in *Candida albicans*[79,80] and the utilization of sugars by yeast[78] suggest that arabinose may be metabolized to arabinitol with the aid of NADP and a dehydrogenase. It has been assumed,[103] partially on the basis of indirect evidence,[104] that endogenous arabinitol, originating from mammalian metabolism is of the L form. The solution to the problem of differentiating arabinitol due to fungal and non-fungal origin is not the use of arabinitol/creatinine ratios but rather the separation of the D and L stereoisomers of arabinitol.

The first demonstration of the potential of using excess D-arabinitol in serum, urine, and tissues to diagnose candidiasis was based on a combined microbiological-GC technique.[103] First, total D+L arabinitol concentration was determined by a previously developed GC technique.[69] Next, an aliquot of the serum or urine sample was incubated with a special strain of *Candida tropicalis* which utilizes D-arabinitol as a sole carbon source for growth when preferred substrates are unavailable. L-arabinitol concentration was determined by a second GC analysis of the incubated sample. Due to serious experimental difficulties, the technique was abandoned.

The use of bacterial D-arabinitol dehydrogenase (ADH) to remove D-arabinitol was proposed in two techniques. In a non-chromatographic method, ADH, isolated from *Enterobacter aerogenes*, was used to oxidize D-arabinitol to D-xylulose. D-arabinitol concentration was estimated from the rate of the concomitant reduction of NAD which, in turn, was determined by following, with a spectrophotofluorimeter, the coupled reaction between NADH and resazurin.[105,106] Although reasonably good correlation was established between D-arabinitol concentrations and arabinitol/creatinine ratios,[107] this technique lacks substrate specificity because all results include endogenous D-mannitol (also oxidized in the course of the analysis) which is known to increase in renal dysfunction.[89]

A combined enzymatic-GC technique[108] utilized capillary GC to determine arabinitol (separated from xylitol and adonitol) in aliquots of the original serum sample (providing D+L concentrations) and in aliquots after treatment with ADH from *Klebsiella pneumoniae* (providing L-arabinitol concentration).33 Endogenous concentrations of D- and L-arabinitol (n = 27) were 0.22 ± 0.052 and 0.16 ± 0.055 μg/mL, respectively. Four patients with confirmed invasive candidiasis had significantly increased D-arabinitol concentrations. The technique was used to analyze serum samples from sarcoidosis patients[109] because of an earlier report[91] suggesting that this disease also results in increased arabinitol concentrations. It was concluded from the determination of D- and L-arabinitol in 42 serum samples from 33 steroid-treated and untreated sarcoidosis patients that neither sarcoidosis nor corticosteroid treatment results in increased serum D-arabinitol concentrations.[109]

Although the enzymatic-chromatographic technique is a new and novel approach, it still has several disadvantages, including the need for two analyses (before and after enzyme treatment) for each sample, a time-consuming incubation step, and interferences from antifungal drugs,[108] making it unreliable for treatment monitoring; in addition, purified ADH is not commercially available at this time.

The direct chromatographic separation of the D and L enantiomers of arabinitol in serum and urine has recently been accomplished.[110] The separation was achieved on glass capillary columns coated with perpentylated cyclodextrin[111,112] which is now commercially available (Macherey-Nagel GmbH, Duran, F.R.G.). The enantiomers of arabinitol, and also adonitol and xylitol, were analyzed as their O-trifluoroacetyl derivatives by monitoring the m/z − 519 ions (corresponding to the loss of an OCF_3CO group) obtained in chemical (isobutane) ionization. The internal standard was fully deuterated D-arabinitol, monitored at m/z = 526 (Figure 4). Serum was derivatized after lipid removal with hexane and protein removal with acetone. Urine was derivatized after ultrafiltration at a 10K exclusion limit.

The mean ± s.d. (range) D/L arabinitol ratio in normal serum (n = 21) was 1.40 ± 0.42 (0.72-2.21). The D/L ratios in candidiasis-free cancer patients (n = 10) were not significantly different statistically from those in normals. The upper limit of normal was taken as the mean plus 2 standard deviation, e.g., in serum, D/L ratios >2.24 were considered outside the normal range. Renal dysfunction without candidiasis yielded normal D/L despite high absolute arabinitol concentration (Figure 5). In a pilot study of 12 confirmed candidiasis cases, there were 10 with D/L ratios >2.24, one was a false negative and one was borderline. The endogenous D/L ratio in urine was nearly identical to that in serum despite a 60 times larger absolute arabinitol concentrations. This combined GC-MS technique offers several advantages: only one sample needs to be prepared and analyzed, there are no interferences because of the selected ion monitoring mode, deuterated arabinitol is nearly an ideal internal standard, sensitivity is high, specificity is much higher than in straight GC methods, and both serum and urine can be analyzed. An important additional practical advantage of the

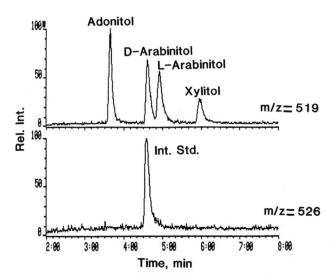

Figure 4.   Chromatographic separation of 5 ng each of adonitol, D-arabinitol, L-arabinitol and xylitol.

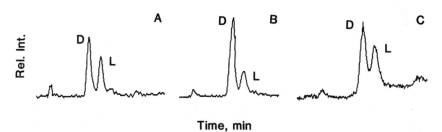

Figure 5.   Selected ion monitoring of pentitol profiles in sera from (A) a healthy control subject, D/L = 1.5; (B) confirmed disseminated candidiasis, D/L = 4.1; (C) patient on dialysis, D/L = 1.9.

proposed approach is that only D/L ratios need to be determined, thus in most cases there is no need for calibration runs for quantification.

CONCLUSIONS

If the arabinitol/polyol techniques are to be of value, they must be better than other existing methods. They must be specific, reproducible, rapid and, most importantly, must provide early indication of oncoming systemic involvement. Most of the controversies of the arabinitol technique have been based on four major methodological problems: inadequate specificity of GC, problems in quantification due to interferences with the internal standard, possible false positives due to the unpredictable presence of increased adonitol and/or xylitol and, most importantly, the inability to differentiate between increased arabinitol of fungal and non-fungal origin in renal dysfunction. Several GC-MS techniques have been developed to resolve the first three problems: selected ion monitoring for increased specificity (and also analytical sensitivity), deuterated arabinitol as internal standard, and capillary chromatographic techniques for the separation of the pentitols.

Endogenous arabinitol, both in serum and in urine, consists of a mixture of D- and L-arabinitol, at a ratio of approximately 1.5. An observed increased arabinitol concentration may be of non-fungal or fungal origin. Renal dysfunction caused by an underlying disease or renal damage caused by the side effects of antineoplastic and/or antifungal chemotherapy does result in increased arabinitol concentrations, however, the D/L ratio apparently does not change significantly. In contrast, arabinitol from the most common *Candida* species encountered in invasive candidiasis produces only the D enantiomer. The techniques described above for the determination of D/L arabinitol ratios, particularly the GC-MS method, permit differentiation of increased arabinitol of non-fungal and fungal origin in a relatively simple, specific and fast manner.

Perhaps the most important application of these methodologies will be the monitoring of patients at risk of developing disseminated candidiasis. This approach can establish individual baseline D/L ratios, and also absolute arabinitol concentrations, thereby significantly augmenting the interpretation of subsequent incremental changes. Success of this work will permit presently available antifungal therapy to become more effective and less toxic (and also less expensive) and will directly benefit cancer as well as other immunosuppressed patients with life-threatening infections.

ACKNOWLEDGEMENT

This work was supported by a grant from the Department of Health and Human Services (CA40050-03) and by the T.J. Martell Memorial Foundation for Cancer and Leukemia Research.

REFERENCES

1. J. Klastersky, ed., "Infections in Cancer Patients," Raven Press, New York (1982).
2. R. Rubin and L. Young, eds., "Critical Approach to Infection in the Compromised Host," 2nd ed., Plenum Medical Book Co., New York (1988).
3. R. Horn, B. Wong, T. Kiehn, and D. Armstrong, Fungemia in a cancer hospital: changing frequency, earlier onset, and results of therapy, Rev. Infect. Dis. 7:646 (1985).

4. D. Armstrong, Fungal infections in the compromised host, in: "Clinical Approach to Infection in the Compromised Host," R. Rubin and L. Young, eds., Plenum Medical Book Co., New York (1982).

5. A. Maksymiuk, S. Thorngprasert, R. Hopfer, M. Luna, V. Fainstein, and G. Bodey, Systemic candidiasis in cancer patients, Am. J. Med. 77(4D):20 (1984).

6. C. Singer, M. Kaplan, and D. Armstrong, Bacteremia and fungemia complicating neoplastic disease, Am. J. Med. 62:731 (1977).

7. J. Klastersky, Control of infection: A vital step for further progress in cancer chemotherapy, in: "Rec. Adv. Cancer Treatment," H. Tagnon and M. Staquet, eds., Raven Press, New York (1977).

8. S. Ketchel and V. Rodriguez, Acute infection in cancer patients, Seminars Oncol. 5:167 (1978).

9. G. Bodey, V. Rodriguez, and H. Chang, Fever and infection in leukemia patients: a study of 494 consecutive patients, Cancer 41:1610 (1978).

10. S. Cho and H. Choi, Opportunistic fungal infection among cancer patients, Am. J. Clin. Pathol. 72:617 (1979).

11. V. Rodriguez and S. Ketchel, Acute infection in patients with malignant disease, in: "Oncologic Emergencies," J. Yarbro and R. Bornstein, eds., Grune & Stratton, New York (1981).

12. E. Whimbet, T. Kiehn, P. Brannon, A. Blevins, and D. Armstrong, Bacteremia and fungemia in patients with neoplastic disease, Am. J. Med. 82:723 (1987).

13. R. Clift, Candidiasis in the transplant patient, Am. J. Med. 77(4D):34 (1984).

14. J. Salomkin, A. Flohr and R. Simmons, Candida infections in surgical patients: dose requirements and toxicity of amphotericin B, Ann. Surg. 195:177 (1982).

15. D. Weese-Mayer, D. Fondriest, R. Brouillette, and S. Shulman, Risk factors associated with candidemia in the neonatal intensive care unit: a case control study, Pediatr. Infect. Dis. J. 6:190 (1987).

16. B. McMillan, E. Law, and I. Holder, Experience with Candida infections in the burn patient, Arch. Surg. 104:509 (1972).

17. I. Bielsa, J. Miro, O. Herrer, E. Martin, X. Latorre, and J. Mascaro, Systemic candidiasis in heroin abusers, Int. J. Dermatol. 26:314 (1987).

18. Confronting AIDS, Inst. Medicine, Natl. Acad. Sciences, Natl. Academy Press, Washington, DC, pp. 24 and 316 (1986).

19. V. Fainstein and G. Bodey, eds., "Candidiasis," Raven Press, New York (1984).

20. E. Reiss, Candida albicans, in: "Molecular Immunology of Mycotic and Actinomycotic Infections," E. Reiss, ed., Elsevier, New York, p. 191 (1986).

21. F. C. Odds, "Candida and Candidosis," 2nd ed., Bailliere, London (1988).

22. F. Meunier-Carpentier, T. Kiehn, and D. Armstrong, Fungemia in immunocompromised host: changing patterns, antigenemia, high mortality, Am. J. Med. 71:363 (1981).

23. M. Degregorio, W. Lee, C. Linker, R. Jacobs, and C. Ries, Fungal infections in patients with acute leukemia, Am. J. Med. 73:543 (1982).

24. D. Warnock and M. Richardson, eds., "Fungal Infection in the Compromised Patient," John Wiley, New York (1982).

25. J. Gold, Opportunistic fungal infections in patients with neoplastic disease, Am. J. Med. 76:458 (1984).

26. M. Shepherd, R. Poulter, and P. Sullivan, Candida albicans: biology, genetics, and pathogenicity, Ann. Rev. Microbiol. 39:579 (1985).

27. R. Hopfer, V. Fainstein, M. Luna, and G. Bodey, Disseminated candidiasis caused by four different Candida species, Arch. Pathol. Lab. Med. 105:454 (1981).

28. J. Wingard, W. Merz, and R. Sarall, *Candida tropicalis*: a major pathogen in immunocompromised patients, Ann. Intern. Med. 91:539 (1979).

29. G. Bodey, Candidiasis in cancer patients, Am. J. Med. 77(4D):13 (1984).

30. J. Rippon, "Medical Mycology," 3rd ed., Saunders, Philadelphia (1988).

31. J. Edwards, ed., Severe candidal infections. Clinical perspective, immune defense mechanism, and current concepts of therapy, Ann. Intern. Med. 89:91 (1978).

32. G. Medoff and G. Kobayashi, Strategies in the treatment of systemic fungal infections, New Engl. J. Med. 302:145 (1980).

33. D. Armstrong and M. Gersten, Laboratory diagnosis of fungal and parasitic diseases in patients with neoplastic disease, in: "Infections in Cancer Patients," J. Klastersky, ed., Raven Press, New York (1982).

34. J. Trier and D. Bjorkman, Esophageal, gastric and intestinal candidiasis, Am. J. Med. 77(4D):39 (1984).

35. J. Commers and P. Pizzo, The role of empiric antifungal therapy for granulocytopenic cancer patients with persistent fever, in: "Infections in Cancer Patients," J. Klastersky, ed., Raven Press, New York (1982).

36. F. Meunier, Prevention and mycoses in immunocompromised patients," Rev. Infect. Dis. 9:408 (1987).

37. G. Medoff, Controversial areas in antifungal chemotherapy: short course and combination therapy with Amphotericin B, Rev. Infect. Dis. 9:403 (1987).

38. F. Meunier, Prevention of mycoses in immunocompromised patients, Rev. Infect. Dis. 9:408 (1987).

39. W. Dismukes, J. Bennnett, D. Drutz, J. Hraybill, J. Remington, and D. Stevens, Criteria for evaluation of therapeutic response to antifungal drugs, Rev. Infect. Dis. 2:535 (1980).

40. D. Drutz, *In vitro* antifungal susceptibility testing and measurement of levels of antifungal agents in body fluids, Rev. Infect. Dis. 9:392 (1987).

41. M. DeGregorio, W. Lee, and C. Ries, *Candida* infections in patients with acute leukemia: ineffectiveness of Nystatin prophylaxis and relationship between oropharengeal and systemic candidiasis, Cancer 50:2780 (1982).

42. R. Quintiliani, N. Owens, R. Quercia, J. Klimek, and C. Nightingale, Treatment of prevention of oropharengeal candidiasis, Am. J. Med. 77(4D):44 (1984).

43. T. Eilard, Isolation of fungi in blood cultures. A review of fungal infections in the western part of Sweden 1970-1982, Scan. J. Infect. Dis. 19:145 (1987).

44. F. Meunier-Carpentier, Significance and clinical manifestations of fungemia, in: "Infection in Cancer Patients," J. Klastersky, ed., Raven Press, New York (1982).

45. R. Hopfer, Mycology of *Candida* infections, in: "Candidiasis," G. Bodey and V. Fainstein, eds., Raven Press, New York (1985).

46. R. Penn, R. Lambert, and R. George, Invasive fungal infections: the use of serologic tests in diagnosis and management, Arch. Intern. Med. 143:1215 (1983).

47. P. Kozinn and C. Taschdjian, Laboratory diagnosis of candidiasis, in: "Candidiasis," G. Bodey and V. Fainstein, eds., Raven Press, New York, pp. 85 (1985).

48. L. Young and P. Stevens, Serodiagnosis of invasive candidiasis, in: "Antifungal Drugs", V. Georgiev, ed., Ann. New York Acad. Sci. 544:575 (1988).

49. D. Poulain and J. Pinon, Diagnosis of systemic candidiasis: development of co-counterimmunoelectrophoresis, Eur. J. Clin. Microbiol. 5:420 (1986).

50. D. Poulain, J. Fruit, L. Fournier, E. Dei-Cas, A. Vernes, Diagnosis of systemic candidiasis: application of co-counterimmunoelectrophoresis, Eur. J. Clin. Microbiol. 5:427 (1986).

51. J. Bailey, E. Sada, C. Brass, and J. Bennett, Diagnosis of systemic candidiasis by latex agglutination for serum antigen, J. Clin. Microbiol. 21:749 (1985).

52. J. Burnier, A reverse latex agglutination test for the diagnosis of systemic candidosis, J. Immunol. Methods 82:267 (1985).

53. F. Kahn and J. Jones, Latex agglutination tests for detection of *Candida* antigens in sera of patients with invasive candidiasis, J. Infect. Dis. 153:579 (1986).

54. M. Ness, Wm. Vaugham and G. Woods, *Candida* antigen latex test for detection of invasive candidiasis in immunocompromised patients, J. Inf. Dis. 159:495 (1989).

55. E. Fiss and H. Buckley, Purification of actin from Candida albicans and comparison with the *Candida* 48,000-Mr protein, Infect. Immun. 55:2324 (1987).

56. R. Matthews, J. Burnie, A. Fox, A. Baskervilles, C. Wells, S. Strachan and I. Clark, The diagnostic and therapeutic potential of monoclonal antibody to the 60-kilodalton nuclear antigen of *Candida albicans*, Serodiagn. Immunother. Inf. Dis. 3:75 (1989).

57. R. Matthews and J. Burnie, Diagnosis of systemic candidiasis by an enzyme-linked dot immunobinding assay for a circulating immunodominant 47-kilodalton antigen, J. Clin. Microbiol. 26:459 (1988).

58. E. Reiss, L. deRepentigny, R. Kuykendall, A. Carter, R. Galindo, P. Auger, S. Grann, and L. Kaufman, Monoclonal antibodies against *Candida tropicalis* mannan: antigen detection by enzyme immunoassay and immunofluorescence, J. Clin. Microbiol. 24:796 (1986).

59. J. Fisher, R. Trincher, J. Agel, T. Buxton, C. Walker, D. Johnson, R. Cormier, W. Chew, and J. Rissing, Disseminated candidiasis: a comparison of two immunologic techniques in the diagnosis, Am. J. Med. Sci. 290:135 (1985).

60. J. Fung, S. Donta, R. Tilton, *Candida* detection system (CAND-TEC) to differentiate between *Candida albicans* colonication and disease, J. Clin. Microbiol. 24:542 (1986).

61. J. Bennett, Rapid diagnosis of candidiasis and aspergillosis, Rev. Infect. Dis. 9:398 (1987).

62. G. Miller, M. Witver, A. Braude, and C. Davis, Rapid identification of *Candida albicans* septicemia in man by gas-liquid chromatography, J. Clin. Invest. 54:1235 (1974).

63. C. Davis and R. McPherson, Rapid diagnosis of septicemia and meningitis by gas-liquid chromatography, in: "Microbiology-1975", O. Schlessinger, ed., Am. Soc. Microbiol., Washington, D.C., p. 55 (1975).

64. R. Marier, E. Milligan, and Y. Fan, Elevated mannose levels detected by gas-liquid chromatography in hydrolyzates of serum from rats and humans with candidiasis, J. Clin. Microbiol. 16:123 (1982).

65. L. deRepentigny, R. Kuykendall, and E. Reiss, Simultaneous determination of arabinitol and mannose by gas-liquid chromatography in experimental candidiasis, J. Clin. Microbiol. 17:1166 (1983).

66. N. Maliwan, R. Reid, and R. Katzen, Gas-liquid chromatography for rapid diagnosis and monitoring of invasive candidal infections and candidemia, Arch. Pathol. Lab. Med. 108:108 (1984).

67. L. deRepentigny, R. Kuykendal, F. Chandler, J. R. Broderson, and E. Reiss, Comparison of serum mannan, arabinitol, and mannose in experimental disseminated candidiasis, J. Clin. Microbiol. 19:804 (1984).

68. W. Wojnarowsky, H. Jaquet, and M. Glauser, Arabinitol, a metabolite of *Candida albicans*, Experientia 35:945 (abstract), (1979).

69. T. Kiehn, E. Bernard, J. Gold, and D. Armstrong, Candidiasis: detection by gas-liquid chromatography of D-arabinitol, a fungal metabolite, in human serum, Science 206:577 (1979).

70. J. Roboz, R. Suzuki, and J. F. Holland, Quantification of arabinitol in serum by selected ion monitoring as a diagnostic technique in invasive candidiasis, J. Clin. Microbiol. 12:594 (1980).

71. B. Wong, E. Bernard, J. Gold, D. Fong, A. Silber, and D. Armstrong, Increased arabinitol levels in experimental candidiasis in rats: arabinitol appearance rates, arabinitol/creatinine ratios, and severity of infection, J. Infect. Dis. 146:346 (1982).

72. B. Wong, E. Bernard, J. Gold, D. Fong, and D. Armstrong, The arabinitol appearance rate in laboratory animals and humans: estimation from the arabinitol/creatinine ratio and relevance to the diagnosis of candidiasis, J. Infect. Dis. 146:353 (1982).

73. J. Gold, B. Wong, E. Bernard, T. Kiehn, and D. Armstrong, Serum arabinitol concentrations and arabinitol/creatinine ratios in invasive candidiasis, J. Infect. Dis. 147:504 (1983).

74. B. Wong, R. Roboz, E. Bernard, R. Suzuki, J. F. Holland, and D. Armstrong, Evaluation of aldononitrile peracetate method for measuring arabinitol in serum, J. Clin. Microbiol. 21:478 (1985).

75. C. Servo, J. Palo, and E. Pitkanen, Gas chromatographic separation and mass spectrometric identification of polyols in human cerebrospinal fluid and plasma, Acta Neurol. Scandinav. 56:104 (1977).

76. O. Touster and D. Shaw, Biochemistry of acyclic polyols, Physiol. Rev. 42:181 (1962).

77. O. Touster, The metabolism of polyols, in: "Sugars in Nutrition," H. Sipple and K. McKnutt, eds., Academic Press, New York (1974).

78. J. Barnett, The utilization of sugars by yeasts, Adv. Carbohydrate Chem. Biochem. 32:125 (1976).

79. L. Veiga, M. Bacila, and B. Horecker, Pentose metabolism, in: Candida albicans, Biochem. Biophys. Res. Commun. 2:440 (1960).

80. L. Veiga, Polyol dehydrogenases in Candida albicans. I. Reduction of D-xylose to xylitol, J. Gen. Appl. Microbiol. 14:65 (1968).

81. L. Dooms, D. Declerck, and H. Verachte, Direct gas chromatographic determination of polyalcohols in biological media, J. Chromatogr. 42:349 (1969).

82. C. Pfaffenberger, J. Szafranek, M. Horning, and E. Horning, Gas chromatographic determination of polyols and aldoses in human urine as polyacetates and aldononitrile polyacetates, Anal. Biochem. 63:501 (1975).

83. C. Pfaffenberger, J. Szafranek, and E. Horning, Gas chromatographic study of free polyols and aldoses in cataractous human lens tissue, J. Chromatogr. 126:535 (1976).

84. A. Schoots, F. Mikkers, and C. Cramers, Profiling of uremic serum by high-resolution gas chromatography-electron impact, chemical ionization mass spectrometry, J. Chromatogr. 164:1 (1979).

85. B. Petit, G. King, and K. Blau, The analysis of hexitols in biological fluids by selected ion monitoring, Biomed. Mass Spectrom. 7:309 (1980).

86. T. Marunaka, E. Matsushima, Y. Umeno, and Y. Minami, GLC-mass fragmentographic determination of mannitol and sorbitol in plasma, J. Pharm. Sci. 72:87 (1983).

87. T. Niwa, N. Yamamoto, K. Maeda, and K. Yamada, Gas chromatographic-mass spectrometric analysis of polyols in serum of uremic patients, J. Chromatogr. 277:25 (1983).

88. T. Niwa, H. Asada, K. Maeda, and K. Yamada, Profiling of organic acids and polyols in nerves of uremic and non-uremic patients, J. Chromatogr. 377:15 (1986).

89. J. Roboz, D. Kappatos, J. Greaves, and J. F. Holland, Determination of polyols in serum by selected ion monitoring, Clin. Chem. 30:1611 (1984).

90. J. Roboz, D. Kappatos, and J. F. Holland, Role of individual serum pentitol concentrations in the diagnosis of disseminated visceral candidiasis, Europ. J. Clin. Microbiol. 6:708 (1987).

91. G. Karam, A. Elliott, S. Polt, and G. Cobbs, Elevated serum D-arabinitol levels in patients with sarcoidosis, J. Clin. Microbiol. 19:26 (1984).

92. J. Corcoran, M. Friduss, A. Corcoran, S. Crawford, and H. Sommers, Problems with gas-liquid chromatography in the diagnosis of systemic candidiasis, abstract #164, ICAAC meeting (1982).

93. L. deRepentigny, R. Kuykendall, F. Chandler, R. Broderson, and E. Reiss, Comparison of serum mannan, arabinitol, and mannose in experimental disseminated candidiasis, J. Clin. Microbiol. 19:804 (1984).

94. J. Roboz, R. Suzuki, J. Greaves, M. Johnson, and J. F. Holland, Serum arabinitol levels in the diagnosis of invasive candidiasis, Cancer 23:108 (abstract 423) (1982).

95. J. Roboz, D. Kappatos, and J. F. Holland, Diagnosis and monitoring of disseminated candidiasis using serum arabinitol concentrations, Cancer Res. (abstract 889) 28:224 (1987).

96. R. Eng, H. Cheml, and M. Buse, Serum levels of arabinitol in the detection of invasive candidiasis in animals and humans, J. Infect. Dis. 143:677 (1981).

97. A. Deacon, Estimation of serum arabinitol for diagnosing invasive candidiasis, J. Clin. Pathol. 39:842 (1985).

98. C. Wells, M. Sirany, and D. Blazevic, Evaluation of serum arabinitol as a diagnostic test for candidiasis, J. Clin. Microbiol. 18:353 (1983).

99. E. Holak, J. Wu, and S. Spruance, Value of serum arabinitol for the management of *Candida* infections in clinical practice, Mycopathologia 93:99 (1986).

100. J. Roboz, D. Kappatos, and J. F. Holland, Monitoring serum polyols during hemodialysis by selected ion monitoring, presented 33rd Ann. Conf. Mass Spectr., abstract RPA 10 (1985).

101. E. Bernard, K. Christiansen, S. Tsang, T. Kiehn, and D. Armstrong, Rate of arabinitol production by pathogenic yeast species, J. Clin. Microbiol. 14:189 (1981).

102. J. Roboz, Some applications of mass spectrometry in microbiology, Rec. Dev. Mass Spectrom. 7:1 (1983).

103. E. Bernard, B. Wong, and D. Armstrong, Stereoisomeric configuration of arabinitol in serum, urine and tissues in invasive candidiasis, J. Infect. Dis. 151:711 (1985).

104. O. Touster and S. Harwell, The isolation of L-arabinitol from pentosuric urine, J. Biol. Chem. 230:1031 (1958).

105. K. Soyama and E. Ono, Enzymatic fluorimetric method for the determination of D-arabinitol in serum by initial rate analysis, Clin. Chim. Acta 149:149 (1985).

106. K. Soyama and E. Ono, Improved procedure for determining serum D-arabinitol by resazurin-coupled method, Clin. Chim. Acta 168:259 (1987).

107. K. Soyana and E. Ono, Enzymatic and gas-liquid chromatographic measurement of D-arabinitol compared, Clin. Chem. 34:432 (1988).

108. B. Wong and K. Brauer, Enantioselective measurement of fungal D-arabinitol in the sera of normal adults and patients with candidiasis, J. Clin. Microbiol. 26:1670 (1988).

109. B. Wong, R. Baughman, and K. Brauer, Levels of the *Candida* metabolite D-arabinitol in sera of steroid-treated and untreated patients with sarcoidosis, J. Clin. Microbiol. 27:1859 (1989).

110. J. Roboz, E. Nieves, and J. F. Holland, Quantification by GC/MS of separated enantiomers of arabinitol to aid the differential diagnosis of disseminated candidiasis, J. Chromatogr., in press (1989).

111. W. Konig, S. Lutz, and G. Wenz, Modified cyclodextrins-- novel, highly
      enantioselective stationary phases for gas chromatography, Angew.
      Chem. Int. Ed. Engl. 27:979 (1988).
112. W. Konig, P. Mischnick-Lubecke, B. Brassat, S. Lutz and G. Wenz,
      Improved gas chromatographic separation of enantiomeric carbohydrate
      derivatives using a new chiral stationary phase, Carbohyd. Res.
      183:11 (1988).

CHAPTER 16

USE OF LIPID BIOMARKERS IN ENVIRONMENTAL SAMPLES

Anders Tunlid[a] and David C. White

Institute for Applied Microbiology
10515 Research Drive, Suite 300
Knoxville, TN 37932-2567

INTRODUCTION

Even in a water column the classical methods of microbiology, which involve the isolation and subsequent culturing of organisms on petri plates, can lead to gross underestimations of the numbers of organisms detectable in direct counts of the same waters.[1] With sediments, soils, and biofilms, the problems with classical methods are more severe. In addition to the problems in providing a universal growth medium in the petri plate, the organisms must be quantitatively removed from the surfaces and from each other. Microscopic methods that require quantitative release of the microorganisms from the biofilm can have the problem of inconsistent removal from the surfaces.[2] Direct microscopy can sometimes be performed on thin biofilms by making estimations for organisms rendered invisible by particles or overlapping organisms in the biofilm.[3] This methodology works best when the density of organisms is low and overlapping is minimal. However, direct microscopic examinations offer a limited insight into the metabolic function or activity of the cells. Methane bacteria, for example, come in all sizes and shapes.[4] The problem is further complicated by the fact that in many environments only a tiny fraction of the organisms is active at any one level and aside from the observation of bacterial doubling time,[5] the morphology gives little evidence of the activity of the cells. The most direct method of determining the proportion of active cells in a given biofilm involves a combination of autoradiography and electron or epifluorescence microscopy. These methods require metabolic activity in the presence of the substrates and are subject to the limitations of density of organisms and thickness of the biofilm in the field of view. With the necessity for inducing metabolic activity there is a danger of inducing artificially high levels of activity with the addition of the substrates.[6]

The attachment and activity of microbes at surfaces is an extremely important feature of microbial ecology.[7] Not only do microbes attach to surfaces, but there is abundant evidence that they exist in consortia of multiple metabolic types.[8]

---

[a]Present address: Department of Ecology, Helgonavagen 5,
 S-223 62 Lund, Sweden

Analytical Microbiology Methods
Edited by A. Fox *et al.*
Plenum Press, New York, 1990

Isolation of microbes from the environment for viable counting or direct microscopic examination may offer limited insight into the interactions taking place in microbial communities. Since these consortia have much more versatile metabolic propensities than simple species, it is important in ecological studies to preserve as much as possible of the structure and metabolic interactions of the microcolonies.

THE SOLUTION-BIOMARKERS

To improve the situation intensive efforts have been put into developing alternative methods that require neither growth, with its attendant problem of microbial selection, nor removal of cells from surfaces. With the alternative methods, the community as a whole can be examined with the microstructure of its multi-species consortia preserved. The methods involve the measurement of biochemical properties of the microbial cells and their extracellular products. Those components generally distributed in cells are utilized as markers for biomass. Components restricted to subsets of the microbial communities can be utilized to define the community structure. The concept of "signatures" for subsets of the microbial community, based on the limited distribution of specific components, has been shown for many monocultures.[9] A range of biomarkers exist, from those found in all organisms to those more or less specific to groups or species of microorganisms. Because of their comparable ease of extraction from even very complex environmental samples and their great taxonomic value,[10] lipid components have so far been shown to be the most useful biomarkers for microorganisms in environmental studies (Table 1).

Biomass estimations

Phospholipids are found in the membranes of all living cells. Under the conditions expected in natural communities the bacteria contain a relatively constant proportion of their biomass as phospholipids.[11] Phospholipids are not found in storage lipids and they have a relatively rapid turnover rate in sediments so the assay of these lipids gives a measure of the "viable" cellular biomass.[12] By using appropriate conversion factors, an estimate of biomass from the content of phospholipids corresponds well to estimates based on measurements of cell counts and cell volumes as well as analysis of cell wall muramic acid, lipopolysaccharide lipid A and adenosine triphosphate (ATP) in samples from subsurface sediments.[13]

Gram negative bacteria contain unique hydroxy acids in the lipid portion (lipid A) of the lipopolysaccharide (LPS) in the cell wall.[14] For example, β-hydroxy myristic acid has been used as a biomarker for Gram negative bacteria in sediments and body fluids (see Table 1).

Ergosterol is the predominant sterol in most fungi[15] and is absent or a minor constituent in plants.[16] It has been used as a marker for fungal biomass in soil and plant material (see Table 1).

Microbial community structure

The ester-linked fatty acids in the phospholipids (PLFA) are presently the most sensitive and the most useful chemical measures of microbial biomass and community structure thus far developed.[17,18] The composition of fatty acids varies widely between different groups of organisms.[19] Bacteria unlike most other organisms do not usually synthesize polyunsaturated fatty acids. Furthermore, bacteria characteristically contain odd chain and branched fatty acids as well as cyclopropane and α- or β-hydroxy derivatives. Compositional patterns of fatty acids have been used for taxonomic and phylogenetic classification of bacteria.[10] By utilizing fatty acid patterns of bacterial

Table 1. Lipid biomarkers for biomass and community structure

Biomarker	Organism group	References
Phospholipids	All cells	87,45,13
Ergosterol	Fungi	88-91
LPS hydroxy-acids	Gram negative bacteria	43,51,53,55-56,92-94
Fatty acids in phospholipids	Sulfate reducing bacteria	57-62
	Methane-oxidizing bacteria	78,80
	Thiobacillus sp.	70,71
	Mycobacteria	95-97
	Francisella tularensis	74
	Dinoflagellates	54,98,99
Phytanyl and bi-phytanyl ether lipids	Archaebacteria	100-104 46,47
Plasmalogens	Anaerobic bacteria	12,24
Respiratory quinones	Aerobic/anaerobic bacteria	30
Sterols	Dinoflagellates	105-108

monocultures, Myron Sasser of the University of Delaware in cooperation with Hewlett Packard has been able to distinguish over 8000 strains of bacteria.[20] Isolation of bacteria from natural environments has shown that subsets of microbial communities contain specific patterns of fatty acids. For example, biomarker fatty acids have been identified in sulfate reducing, methane oxidizing, autotrophic mineral acid secreting, and pathogenic bacteria (see Table 1).

Although the analysis of PLFA cannot provide an exact description of each species or physiological type of microorganisms in a given environment, a quantitative description of the microbiota in the particular environment sampled is provided. With the techniques of statistical pattern recognition it is possible to provide a quantitative estimate of the differences between samples with PLFA analysis.

Potential problems with defining the community structure by analysis of PLFA come with the shift in fatty acid composition of some monocultures with changes in temperature and media composition, some of which were made in this laboratory.[21-23] There is as yet little published evidence for such shifts in PFLA in nature where the growth conditions that allow survival in highly competitive microbial consortia would be expected to severely restrict the survival of specific microbial strains to much narrower conditions of growth.

The community structure of the microbial consortia can be further defined by the analysis of plasmalogens and ether lipids (Table 1). The occurrence of plasmalogens in microbes is restricted to specific groups of anaerobic bacteria.[24,25] Archaebacteria (methanogens, halophiles, and thermophiles) are characterized by unique biphytanyl and di-biphytanyl ether lipids which are not found in other organisms.[26] These lipids have been utilized as biomarkers for archaebacteria in sediments, petroleum, hot-spring mats and fermenters (see Table 1).

The common bacterial respiratory quinones are ubiquinones, menaquinones, and desmethylmenaquinones. The type of quinones and length of the side chain varies with the type of bacteria, and their distribution has been utilized as taxonomic markers for bacteria.[27] The redox potential of the respiratory quinones suggests that the terminal electron acceptors of those bacteria containing ubiquinones should be of high potential as compared with those of bacteria containing naphthoquinones. Aerobes contain benzoquinones, some, but not all, anaerobes contain naphthoquinones.[28,29] Recently, Hedrick and White[30] showed that analyses of respiratory quinones can be utilized as sensitive biomarkers of aerobic versus anaerobic bacterial metabolism in environmental samples.

Nutritional status

Chemical methods can also be utilized to indicate the nutritional status of organisms. Many microorganisms accumulate lipid storage material under specific environmental conditions.[31] The nutritional status of microorganisms can be analyzed by measuring the proportions of these polymers relative to the cellular biomass. The nutritional status of microeukaryotes (algae, fungi and protozoa) has been monitored by measuring the ratio of triacyl glycerols to the cellular biomass.[32] Certain bacteria form the unique reserve polymer poly-β-hydroxybutyrate (PHB) under conditions in which the organisms can accumulate carbon, but have insufficient amounts of total nutrients to allow growth and cell division.[33,34]

Starvation induces the formation of minicells in some bacteria.[35] There is also a loss of cell components including membrane lipids but there is a marked increase in the portion of cyclopropane and monoenoic PLFA with the double bond in the trans configuration.[36] These changes can be utilized as biomarkers of the nutritional status of these bacteria.

Metabolic activity

The analyses described above all involve the isolation of components from microbial consortia. Since each of these components is isolated, the incorporation of labeled isotopes from precursors can be utilized to provide rates of synthesis or turnover in properly designed experiments. For example, incorporation of $H_3{}^{32}PO_4$ into phospholipids has been used to estimate the activity of the total microbiota, and incorporation of ^{35}S sulfate into sulpholipids has been used as a measure of the activity of microeukaryotes.[37,38] The ratio of the synthesis of PLFA (cellular growth) to that of PHB (carbon accumulation) has been shown to reflect the growth conditions in bacterial habitats.[6]

Mass spectrometry analyses allow the use of stable isotope-labelled precursors to study the rate of synthesis of biological markers. Stable isotope-labelled precursors are superior to radioactive isotopes in that the former have higher specific activities, in many cases approaching 100%. This enables the use of marker precursors at natural concentrations, which would help to avoid the distortions induced by the addition of high concentrations of substrates required for many radioactive precursors. A highly sensitive and selective method has been developed to measure ^{13}C enrichment in bacterial fatty acids using mass spectrometry.[39]

THE METHOD

Generally, chemical analyses of lipid biomarkers involve the extraction of the sample with a suitable combination of organic solvents, followed by isolation and separation by various chromatographic techniques. In the past ten years this laboratory has developed a battery of methods for the analysis of lipid biomarkers in environmental samples.[9,40] These analyses are based on an efficient one-phase chloroform:methanol:water extraction of the sample (Figure 1).[12,41,42] The extracted lipids are then fractionated into three different classes on a silicic acid column: neutral lipids, containing triglycerides, sterols, and quinones; glycolipids with, e.g., PHB; and polar lipids that contain phospholipids, plasmalogens and ether lipids (Figure 1). Covalently bound lipids, such as lipid A in LPS, can be recovered from the residue of the Bligh and Dyer extraction.[43]

After extraction and isolation of the different lipid classes, the compounds are converted to suitable derivatives and separated by capillary gas chromatography (GC) (sterols, PHB-monomers, PLFA, aldehydes from plasmalogens and hydroxy acids from lipid A)[11,17,18,34,44,45] or high performance liquid chromatography (HPLC) (quinones and ether lipids).[30,46,47] Environmental samples contain a very complex mixture of lipids and care must be taken to correctly identify the biomarkers. For example, analysis of PLFA in marine sediments often yields 100-150 different fatty acids. The direct chemical verification of the lipids is, in most cases, performed with mass spectrometry (MS).[48] Special techniques are needed in some cases, for example, to determine the configuration and localization of double bonds in unsaturated fatty acids.[49]

The sensitivity for analyzing signatures with these techniques is at the picomolar level. For PLFA, this corresponds to the content in 5×10^6 bacterial cells like *Escherichia coli*.[9] However, a detailed examination of microbial communities can often require the analysis of components from a much smaller number of organisms. Introduction of more sensitive and selective detector systems such as selected ion monitoring (SIM) MS have made it possible to examine microbial constituents at sensitivities several orders of magnitude higher than more conventional GC techniques.[50] With this technique using chemical ionization (CI) and negative ion detection, microbial fatty acids can be determined at femtomolar levels, permitting detection of about 600 bacteria the size of *E. coli*.[51,52]

Nomenclature. Fatty acids are designated as total number of carbon atoms: number of double bond closest to the aliphatic (ω) end of the molecule indicated with the geometry "c" for *cis* and "t" for *trans*. The prefixes "i," "a," and "br" refer to *iso-, anteiso-*, and methyl-branching of unconfirmed position, respectively. Other methyl-branching is indicated as position from the carboxylic acid end, i.e., $C_{10\ Me\ 10:0}$. Cyclopropyl fatty acids are designated as "cy" with the ring position in parenthesis relative to the aliphatic end.

ENVIRONMENTAL STUDIES

The use of biomarkers for *in situ* studies of microbial communities has been validated in a number of studies by (1) inducing shifts in communities by altering the microenvironment, (2) by isolating specific organisms or groups of organisms for biomarkers and then detecting the markers after induction, (3) by detecting shifts in community nutritional status with alteration in the environment, (4) by detecting specific organisms and their activity, and (5) by consequences of specific predation by grazers. These studies have been discussed in detail by White.[40] This section discusses the application of lipid biomarker analysis to a number of environmental studies.

Microbial communities in sediments

Groups of fatty acids have been identified as valid biomarkers for microorganisms in marine sediments. Perry et al.[53] proposed that *iso-* and *anteiso-* branched fatty acids, 10-methyl palmitic acid ($C_{10 \, Me \, 16:0}$), cyclopropyl $C_{17:0}$ and $C_{19:0}$ acids, *cis* vaccenic acid ($C_{18:1\omega7c}$), the $C_{15:1}$, $C_{17:1\omega6}$ and $C_{17:1\omega8}$ isomers, and the branched monoenoic fatty acids $C_{i-15:1\omega8}$ and $C_{i-17:1\omega8}$ can be utilized as bacterial markers in sediment samples. Gillan and Hogg[54] defined nine subgroups (chemotypes) of bacteria in sediments based on fatty acid profiles. Gram negative bacteria in sediments can be detected using β-hydroxy acids as biomarkers.[53,55,56]

However, only in a few cases have detailed experiments been performed to study the relationship between specific bacterial biomarkers and the activity and dynamics of the corresponding microbiota. The sulfate reducing bacteria contain PLFA patterns which can be utilized to identify the lactate-utilizing *Desulfovibrio*, the acetate-utilizing *Desulfobacter*, and the propionate-utilizing *Desulfobulbos*.[57-62] These biomarkers allow the differentiation between sulfate reducing bacteria utilizing lactate, propionate, or those using acetate or higher fatty acids. Detailed analysis of sulfate reducing bacteria by N. Dowling of this laboratory strongly suggests that the majority of sulfate reducing bacteria found in marine sediments and in waters used in secondary recovery of oil are the acetate-utilizing strains. Sediments from a Scottish loch, when amended with lactate or propionate, show increases in the biomarkers for *Desulfovibrio* or *Desulfobulbos* respectively, that parallel increases in specific substrate induced sulfate reducing activity (Parkes, Dowling, and White, unpublished data). The biomarkers for sulfate reducing bacteria have also been utilized to study the interactions between consortia of glutamate fermenting and sulfate reducing bacteria in a continuous flow system inoculated with estuarine sediments.[63]

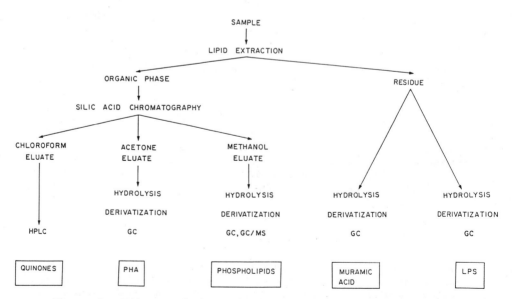

Figure 1. Diagram of the analytical scheme for the biochemical analysis of microorganisms in environmental samples.

Biofilms and fouling

The fouling of surfaces exposed to seawater generates enormous economic problems. The initial event in the fouling process involves an immediate coating of the surface with biopolymers that attract bacteria that binds to the surface.[7] The structure of the initial attached community of microorganisms was analyzed by SIM with chemical ionization and negative ion detection.[51] Examination of the PLFA profile indicated several properties of this community. The prominent fatty acids were characteristic of bacteria and the absence of polyenoic fatty acids indicated that microeukaryotes were lacking in the community.

The initial attached bacterial community attracts a succession of microorganisms leading to the formation of a complex biofilm containing fungi, protozoa, algae as well as bacteria.[64] The properties of the surface influence the development and structure of the biofilm. The microbiota colonizing the surface of non-degradable polyvinylchloride needles contain a significantly lower biomass and activity than the community found on biodegradable pine needles.[65] The microtopography of the surface also markedly affects the biomass and structure of the biofilm.[66]

Corrosion and biodeterioration

There is increasing evidence that corrosion of metals exposed to seawater is facilitated by the presence of microbes and their products at the surface.[67] Microbes of different physiological types when acting in consortia appear to be more destructive than monocultures. The sulfate reducing bacteria are of particular importance in microbially facilitated corrosion and it has been proposed that these obligate anaerobic bacteria can function in an apparently aerobic environment by occupying anaerobic microniches in biofilms formed by the utilization of oxygen by facultative anaerobic consortia. Analysis of PLFA markers has shown that sulfate reducing bacteria can form biofilms with the heterotrophic bacteria *Vibrio natrigiens* on stainless steel coupons exposed to aerobic seawater that facilitates corrosion.[68,69]

The acid-forming *Thiobacillus* form a remarkable collection of unique PLFA: methoxy-, cyclopropyl-, hydroxy-monounsaturated, hydroxy-cyclopropyl, and monounsaturated components with the double bond in unusual positions.[70] The signature PLFA of these organisms were readily detected in a microcosm designed to measure the degradation of concrete exposed to biologically generated acid and in concrete samples from sewers that suffered structural failure.[71]

Microorganisms on plant roots

Plant roots are important sites for microbial activity in soil and the root surface represents an area where microbial biomass is often substantially increased when compared with the soil itself. The root microorganisms influence the growth and development of the plant by competing for nutrients and by taking place in symbiotic or pathogenic interactions.

The dynamics of bacteria associated with the rape plant *Brassica napus* have been examined by analyzing PLFA and PHB.[72] Bacteria isolated from soil surrounding rape plant roots showed profiles of fatty acids that were distinctively different from those of sterile roots. The prominent PLFA of the bacteria were short and branched saturated, cyclopropane and monoenoic fatty acids. The bacteria were added to sterile rape roots. After three weeks of growth bacteria associated with the roots showed active growth but no formation of PHB whereas organisms recovered from the soil away from the roots showed less growth and the accumulation of high amounts of PHB.

Fatty acid analysis has also been utilized to characterize bacteria that suppress *Rhizoctonia* damping-off on cucumber roots grown in bark compost media.[73] Cucumber roots grown in a naturally suppressive medium had higher proportion of $C_{18:1\omega7c}$ and $C_{i-17:1\omega8}$ but lower proportion of several methyl branched fatty acids compared to roots grown in a conducive medium. These results suggest major differences in bacterial community composition between suppressive and conducive systems. Fatty acid analysis also demonstrated that colonization of the cucumber roots by a biocontrol agent (*Flavobacterium balustinum*) induced major changes in the composition of the rhizosphere bacterial community.

Detection of pathogens

The pathogenic bacterium *Francisella tularensis* has been shown to contain unusual long-chained monoenoic PLFA that can be used as biomarkers.[74] Organisms with this pattern have been isolated from patients with serological evidence of infection. The biomarker PLFA pattern has been detected in human and animal tissues with serological, clinical and cultural evidence for the infection as well as in soils and vaccines that have been shown to contain the organism.

Detection of the covalently bound hydroxy acids of the lipid A of LPS in Gram negative bacteria has proved to be a valuable assay in the definition of Gram negative infection. With this assay it is possible to detect Gram negative bacterial infection in mammalian tissue or secretions.[51]

Degradation of organic pollutants

Groundwater is becoming increasingly important as a resource of clean fresh water for industrial and domestic use. At the same time, more of this vital resource is found to be contaminated with potentially toxic wastes. Analysis of lipid biomarkers in subsurface samples, collected with careful attention to avoid contamination from surface soil, revealed the presence of a specific microflora.[75,76] The microbiota was sparse compared with that present in surface soils. PLFA analysis showed absence of long chained polyenoic fatty acids typical of microeukaryotes and high proportions of fatty acids typical of bacteria. The bacteria of the uncontaminated subsurface aquifer sediments showed nutritional stress as evidenced by high levels of PHB and extracellular polymers. Contamination of the sediments increased the microbial biomass, decreased the rate of PHB formation and shifted the community to a more Gram negative bacterial consortium.

Trichloroethylene (TCE) is a major contaminant of the subsurface ground water aquifers and methane oxidizing bacteria have been shown to be associated with a consortium that can degrade it.[77] Analysis for PLFA biomarkers in type I and type II methane-oxidizing bacteria has shown the presence of unique monoenoic fatty acids.[78] Type I methane-oxidizers have predominantly C_{16} fatty acids with unusual double bond positions: $C_{16:1\omega8c}$, $C_{16:1\omega8t}$, $C_{16:1\omega5c}$ and $C_{16:1\omega5t}$. Type II methane-oxidizers have 18 carbon monoenoic fatty acids with similar double bond positions: $C_{18:1\omega8c}$, $C_{18:1\omega8t}$, $C_{18:1\omega7t}$ and $C_{18:1\omega6c}$. Type II methanogens also produce PHB.[79] Both the total microbial biomass and signature components typical of type II methane-oxidizing bacteria as well as the level of PHB increased dramatically in soil exposed to natural gas.[80] The soil columns enriched in microbial biomass and specific type II methane-oxidizing bacteria showed that methane disappearance was correlated with rapid biodegradation of TCE.[80] Preliminary evidence indicates similar increases in signature lipids in the subsurface sediments recovered from zones where the TCE concentration increases the microbial metabolic activity (Phelps, Davis, Fliermans, and White, unpublished data).

Monitoring of fermenters

The rate limiting step in fermentations is the degradation of polymers.[81] A second consortia of microorganisms converts the carbohydrates and amino acids released from biopolymers into organic acids, alcohols, hydrogen and carbon dioxide. These are the anaerobic fermenters and some of the organisms contain plasmalogen phospholipids that are limited to this physiological group of anaerobes in the microbial world.[24] Other groups of anaerobic fermenters contain phosphosphingolipids with unusual sphingosine bases. These were detected in *Bacteriodes*.[82]

The microbial biomass, community structure and nutritional status of microbes in thermophilic methane producing digesters have been analyzed with lipid biomarkers.[83] Analysis of PLFA showed a simple PLFA pattern of saturated and unsaturated normal fatty acids. The loss of polyenoic fatty acids was utilized to monitor degradation of the plant biomass feedstock. After inoculation and incubation the PLFA showed a markedly different pattern with increases in PLFA typical of microbes from anaerobic digesters. The continuous addition of volatile fatty acids such as propionate and butyrate, or the terminal electron acceptors nitrate or sulfate induced markedly shifts in the community structure as indicated in the PLFA analysis.[83]

In experiments in cooperation with P. D. Brooks, C. A. Mancuso (of this laboratory) and D. P. Chynoweth (University of Florida) at the Disney World water hyacinth digester, the eubacterial community structure was markedly different in samples recovered from different parts of the digester. Comparing the feedstock chopped hyacinth material, the stored feedstock, the bottom material, the top of the bottom material, the bottom of the floating mat, and the top of the floating mat in the digester, there was a progressive decrease in the proportions of polyenoic and monoenoic PLFA. This decrease correlated with an increase in the proportion of branched and cyclic as well as saturated PLFA. The proportion of methanogenic archaebacteria measured as the di- and tetraphytanylglycerol ether phospholipids increased from the feedstock, to traces in the stored and partially fermented feedstock with more in the material at the bottom of the digester to the highest levels at the top of the digester.[84]

Changes in microbial pigments during digestion in the Disney World fermenter has been shown by J. J. Olie of this laboratory. Chlorophyll a and b, phaephytin a, phaephorbide as well as several bacterial carotenoids increased between the feedstock hyacinths and the digester samples. Possibly the most interesting finding of these analyses was the detection of large amounts of bacteriochlorophyll in the digester samples. None was detected in the feedstock. The role of these photosynthetic organisms that are known to utilize hydrogen in the dark to fix nitrogen is currently under investigation.

CONCLUSION AND PROSPECTS

Chemical measures for the biomass, community structure and nutritional status of microbial consortia based on analysis of lipid biomarkers provide a quantitative definition useful for studies of interactions between microorganisms and their environment. Areas that will benefit greatly from the application of these methods are the role of microbes in the facilitation of corrosion, studies of interaction between plant roots and microorganisms, detection of pathogens in the environment, the role of microorganisms in the degradation of pollutants and monitoring of fermenting processes.

Recently developed chemical analytical techniques will make it possible to determine new classes of biomarkers, to improve the detection limit and to

assay complex biological microenvironments. One approach to increase the
sensitivity and speed of the analysis is to utilize the extraordinary
separation efficiency of tandem mass spectrometry (MS-MS).[85] In combination
with extended mass range, soft ionization techniques and new inlet systems,
it is now possible to analyze large and complex lipid molecules.[86]
Techniques for desorption of ions directly from a surface also have great
potential in studies of microbial consortia. In combination with
microprobes, desorption techniques can be used to analyze the distribution of
specific molecules on micrometer scale. Such methods can be of great
significance in studies of the critical initial phases of biodeterioration
processes, such as biofouling and corrosion and in studies of interactions
between pathogenic microorganisms and plants or animals.

ACKNOWLEDGEMENTS

     This work would have been impossible without the dedicated work of the
colleagues who formed this laboratory. This work has been supported by
grants N00014-82-C-0404 and N00014-83-K-0056 from the Office of Naval
Research; OCE-80-19757, DPP-82-13796 and INT-83-12117 from the National
Science Foundation; NAG2-149 from the Advanced Life Support Office, National
Aeronautics and Space Administration; contracts CR-80-9994 and CR-81-2504
from the Robert S. Kerr Environmental Research Laboratory of the U. S.
Environmental Protection Agency; and AX-681901 from the E. I. Dupont de
Nemours and Co., Atomic Energy Division, Savannah River Laboratory, Aiken,
SC; and the generous gift of the Hewlett Packard HP-1000 RTE-6/VM data system
for the HP5996A GC/MS system. Grants from the Swedish Natural Science
Research Council to A. Tunlid are also acknowledged.

REFERENCES

1. H. W. Jannasch and G. E. Jones, Bacterial populations in seawater as
    determined by different methods of enumeration, Limnol. Oceanogr.
    4:128 (1959).
2. D. J. W. Moriarty, Biogeochemistry of ancient and modern sediments, in:
    "Problems in the Measurement of Bacterial Biomass in Sandy
    Sediments," Australian Academy of Science, Canberra (1980).
3. D. E. Caldwell and J. J. Germida, Evaluation of difference imagery for
    visualizing and quantitating microbial growth, Canad. J. Microbiol.
    31:35 (1984).
4. J. G. Zeikus, The biology of methanogenic bacteria, Bacteriol. Rev.
    41:514 (1977).
5. A. Hagstrom, U. Larsson, P. Horstedt, and S. Normark, Frequency of
    dividing cells, a new approach to the determination of bacterial
    growth rates in aquatic environments, Appl. Environ. Microbiol.
    37:805 (1979).
6. R. H. Findlay, P. C. Pollard, D. J. W. Moriarty, and D. C. White,
    Quantitative determination of microbial activity and community
    nutritional status in estuarine sediments: evidence for a
    disturbance artifact, Canad. J. Microbiol. 31:493 (1985).
7. K. C. Marshall, "Interfaces in Microbial Ecology," Harvard University
    Press, Cambridge (1976).
8. J. W. Costerton, T. J. Marrie, and K. -J. Cheng, Phenomena of bacterial
    adhesion, in: "Bacterial Adhesion Mechanisms and Physiological
    Significance," C. D. Savage and M. Fletcher, eds., Plenum Press, New
    York (1985).
9. D. C. White, Analysis of microorganisms in terms of quantity and
    activity in natural environments, in: "Microbes in Their Natural
    Environments," Society for General Microbiology Symposium, 34:37
    (1983).

10. M. P. Lechevalier, Lipids in bacterial taxonomy-- a taxonomist's view, Crit. Rev. Microbiol. 7:109 (1977).

11. D. C. White, R. J. Bobbie, J. S. Herron, J. D. King, and S. J. Morrison, Biochemical measurements of microbial mass and activity from environmental samples, in: "Native Aquatic Bacteria: Enumeration, Activity and Ecology," ASTM STP 695, American Soc. for Testing and Materials (1979).

12. D. C. White, W. M. Davis, J. S. Nickels, J. D. King, and R. J. Bobbie, Determination of the sedimentary microbial biomass by extractible lipid phosphate, Oecologia 40:51 (1979).

13. D. L. Balkwill, F. R. Leach, F. T. Wilson, J. F. McNabb, and D. C. White, Equivalence of microbial biomass measures based on membrane lipid and cell wall components, adenosine triphosphate, and direct counts in subsurface aquifer sediments, Microb. Ecol. 16:73 (1988).

14. O. Luderitz, M. A. Freudenberg, C. Galanos, V. Lehmann, E. Th. Rietschel, and D. H. Shaw, Lipopolysaccharides of Gram negative bacteria, in: "Current Topics in Membranes and Transport," vol. 17, S. Razin and S. Rottem, eds., Academic Press, New York (1982).

15. J. D. Weete, Sterols of the fungi: distribution and biosynthesis, Phytochemistry 12:1843 (1973).

16. W. R. Nes, The biochemistry of plant sterols, Adv. Lipid Res. 15:233 (1977).

17. R. J. Bobbie and D. C. White, Characterization of benthic microbial community structure by high resolution gas chromatography of fatty acid methyl esters, Appl. Environ. Microbiol. 39:1212 (1980).

18. J. B. Guckert, C. B. Antworth, P. D. Nichols, and D. C. White, Phospholipid, ester-linked fatty acid profiles as reproducible assays for changes in prokaryotic community structure of estuarine sediments, FEMS Microbiol. Ecology 31:147 (1985).

19. J. L. Harwood and N. J. Russell, "Lipids in Plants and Microbes," George Allen and Unwin, London (1984).

20. M. Sasser, Identification of bacteria by fatty acid composition, Am. Soc. Microbiol. Meet. March 3-7, 1985.

21. G. H. Joyce, R. K. Hammond, and D. C. White, Changes in membrane lipid composition in exponentially growing Staphylococcus aureus during shift from 37 to 25°C, J. Bacteriol. 104:323 (1970).

22. F. F. Frerman and D. C. White, Membrane lipid changes during formation of a functional electron transport system in Staphylococcus aureus, J. Bacteriol. 94:1854 (1967).

23. P. H. Ray, D. C. White, and T. D. Brock, Effect of growth temperatures on the lipid composition of Thermus aquaticus, J. Bacteriol. 108:227 (1971).

24. H. Goldfine and P. O. Hagen, Bacterial plasmalogens, in: "Ether Lipids: Chemistry and Biology," E. Snyder, ed., Academic Press, New York (1972).

25. Y. Kamio, S. Kanegasaki, and H. Takanshi, Occurrence of plasmalogens in anaerobic bacteria, J. Gen. Appl. Microbiol. 15:439 (1969).

26. M. DeRosa, A. Gambacorta, and A. Gliozzi, Structure, biosynthesis, and physical properties of archaebacterial lipids, Microbiol. Rev. 50:70 (1986).

27. M. D. Collins and D. Jones, Distribution of isoprenoid quinone structural types in bacteria and their taxonomic implications, Microbiol. Rev. 45:316 (1981).

28. C. R. Whistance, J. F. Dillon, and D. R. Threlfall, The nature, intergeneric distribution and biosynthesis of isoprenoid quinones and phenols in gram-negative bacteria, Biochem. J. 111:461 (1969).

29. R. Hollander, G. Wolf, and W. Mannheim, Lipoquinones of some bacteria and mycoplasmas, with considerations on their functional significance, Antonie van Leeuwenhoek J. Microbiol. 43:177 (1977).

30. D. B. Hedrick and D. C. White, Microbial respiratory quinones in the environment I. A sensitive liquid chromatographic method, J. Microbiol. Methods 5:243 (1986).

31. E. A. Dawes and P. J. Senior, The role and regulation of energy reserve polymers in micro-organisms, Adv. Microb. Physiol. 20:135 (1973).

32. M. J. Gehron and D. C. White, Quantitative determination of the nutritional status of detrital microbiota and grazing fauna by triglyceride glycerol analysis, J. Exp. Mar. Biol. Ecol. 64:145 (1982).

33. J. S. Nickels, J. D. King, and D. C. White, Poly-β-hydroxybutyrate metabolism as a measure of unbalanced growth of the estuarine detrital microbiota, Appl. Environ. Microbiol. 37:459 (1979).

34. R. H. Findlay and D. C. White, Polymeric β-hydroxy alkanoates from environmental samples and Bacillus megaterium, Appl. Environ. Microbiol. 45:71 (1983).

35. R. Y. Morita, Starvation-survival of heterotrophs in the marine environment, Adv. Microb. Ecol. 6:171 (1982).

36. J. B. Guckert, M. A. Hood, and D. C. White, Phospholipid, ester-linked fatty acid profile changes during nutrient deprivation of Vibrio cholerae: increases in the trans-cis ratio and proportions of cyclopropyl fatty acids, Appl. Environ. Microbiol. 52:794 (1986).

37. D. C. White, R. J. Bobbie, S. J. Morrison, D. K. Oosterhof, C. W. Taylor, and D. A. Meeter, Determination of microbial activity of estuarine detritus by relative rates of lipid biosynthesis, Limnol. Oceanogr. 22:1089 (1977).

38. D. J. W. Moriarty, D. C. White, and T. J. Wassenberg, A convenient method for measuring rates of phospholipid synthesis in seawater and sediments: its relevance to the determination of bacterial productivity and the disturbance artifacts introduced by measurements, J. Microbiol. Methods 3:321 (1985).

39. A. Tunlid, H. Ek, G. Westerdahl, and G. Odham, Determination of ^{13}C-enrichment in bacterial fatty acids using chemical ionization mass spectrometry with negative ion detection, J. Microbiol. Methods 7:77 (1987).

40. D. C. White, Validation of quantitative analysis for microbial biomass, community structure, and metabolic activity, Arch. Hydrobiol. Beih. Ergeben Limnol. 31:1 (1988).

41. E. G. Bligh and W. J. Dyer, A rapid method of total lipid extraction and purification, Can. J. Biochem. Physiol. 37:911 (1959).

42. J. D. King, D. C. White, and C. W. Taylor, Use of lipid composition and metabolism to examine structure and activity of estuarine detrital microflora, Appl. Environ. Microbiol. 33:1177 (1977).

43. J. H. Parker, G. A. Smith, H. L. Fredrickson, J. R. Vestal, and D. C. White, Sensitive assay, based on hydroxy-fatty acids from lipopolysaccharide lipid A for gram negative bacteria in sediments, Appl. Environ. Microbiol. 44:1170 (1982).

44. D. C. White, R. J. Bobbie, J. S. Nickels, S. D. Fazio, and W. M. Davis, Nonselective biochemical methods for the determination of fungal mass and community structure in estuarine detrital microflora, Botanica Marina 23:239 (1980).

45. M. J. Gehron and D. C. White, Sensitive assay of phospholipid glycerol in environmental samples, J. Microbiol. Methods 1:23 (1983).

46. R. F. Martz, D. L. Sebacher, and D. C. White, Biomass measurement of methane-forming bacteria in environmental samples, J. Microbiol. Methods 1:53 (1983).

47. C. A. Mancuso, P. D. Nichols, and D. C. White, A method for the separation and characterization of Archaebacterial signature ether lipids, J. Lipid Res. 27:49 (1986).

48. C. Asselineau and J. Asselineau, Fatty acids and complex lipids, in: "Gas Chromatography/Mass Spectrometry Applications in Microbiology," G. Odham, L. Larsson, and P. -A. Mardh, eds., Plenum Press, New York (1984).

49. P. D. Nichols, J. B. Guckert, and D. C. White, Determination of monunsaturated double bond position and geometry for microbial monocultures and complex consortia by capillary GC-MS of their dimethyl disulphide adducts, J. Microbiol. Methods 5:49 (1986).

50. A. Tunlid and G. Odham, Ultrasensitive analysis of bacterial signatures by gas chromatography/mass spectrometry, in: "Perspectives in Microbial Ecology," Proc. of the Fourth International Symposium on Microbial Ecology, Ljubljana, Yugoslavia, (1986).

51. G. Odham, A. Tunlid, G. Westerdahl, L. Larsson, J. B. Guckert, and D. C. White, Determination of microbial fatty acid profiles at femtomolar levels in human urine and the initial marine microfouling community by capillary gas chromatography-chemical ionization mass spectrometry with negative ion detection, J. Microbiol. Methods 3:331 (1985).

52. G. Odham, A. Tunlid, G. Westerdahl, P. Marden, Combined determination of poly-β-hydroxyalkanoic and cellular fatty acids in starved marine bacteria and sewage sludge using gas chromatography with flame ionization or mass spectrometry detection, Appl. Environ. Microbiol. 52:905 (1986).

53. G. J. Perry, J. K. Volkman, and R. B. Johns, Fatty acids of bacterial origin in contemporary marine sediments, Geochem. Cosmochim. Acta 43:1715 (1979).

54. F. T. Gillan and R. W. Hogg, A method for the estimation of bacterial biomass and community structure in mangrove-associated sediments, J. Microbiol. Methods 2:275 (1984).

55. P. A. Cranwell, The stereochemistry of 2- and 3-hydroxy acids in a recent lacustrine sediment, Geochem. Cosmochim. Acta 45:547 (1981).

56. H. Gossens, W. I. C. Rijpstra, R. R. Duren, J. W. deLeeuw, and P. A. Schenck, Bacterial contribution to sedimentary organic matter; a comparative study of lipid moieties in bacteria and recent sediments, Org. Geochem. 10:683 (1986).

57. J. J. Boon, J. W. deLeeuw, G. J. v.d. Hoek, and J. H. Vosjan, Significance and taxonomic value of iso and anteiso monoenoic fatty acids and branched β-hydroxy acids in Desulfovibrio desulfuricans, J. Bacteriol. 129:1183 (1977).

58. R. J. Parkes and J. Taylor, The relationship between fatty acid distributions and bacterial respiratory types in contemporary mariane sediments, Estuarine Coastal Mar. Sci. 16:173 (1983).

59. J. Taylor and R. J. Parkes, The cellular fatty acids of the sulfate-reducing bacteria, Desulfobacter sp., Desulfobulbus sp., and Desulfovibrio desulfuricans, J. Gen. Microbiol. 129:3303 (1983).

60. A. Edlund, P. D. Nichols, R. Roffey, and D. C. White, Extractable and lipopolysaccharide fatty acid and hydroxy acid profiles from Desulfovibrio species, J. Lipid Res. 26:982 (1985).

61. R. J. Parkes and A. G. Calder, The cellular fatty acids of three strains of Desulfobulbus, a propionate-utilizing sulfate-reducing bacterium, FEMS Microbiol. Ecol. 31:361 (1985).

62. N. J. E. Dowling, F. Widdel, and D. C. White, Analysis of phospholipid ester-linked fatty acid biomarkers of acetate-oxidizing sulfate reducers and other sulfide forming bacteria, J. Gen. Microbiol. 132:1815 (1986).

63. J. E. Dowling, P. D. Nichols, and D. C. White, Phospholipid fatty acid and infra-red spectroscopic analysis of a sulphate-reducing consortium, FEMS Microbiol. Ecol. 53:325 (1989).

64. J. S. Nickels, R. J. Bobbie, D. F. Lott, R. F. Martz, P. H. Benson, and D. C. White, Effect of manual brush cleaning on the biomass and community structure of the microfouling film formed on aluminum and titanium surfaces exposed to rapidly flowing seawater, Appl. Environ. Microbiol. 41:1442 (1981).

65. R. J. Bobbie, S. J. Morrison, and D. C. White, Effects of substrate biodegradability on the mass and activity of the associated estuarine microbiota, Appl. Environ. Microbiol. 35:179 (1978).

66. J. S. Nickels, R. J. Bobbie, R. F. Martz, G. A. Smith, D. C. White, and N. L. Richards, Effect of silicate grain shape, structure and location on the biomass and community structure of colonizing marine microbiota, Appl. Environ. Microbiol. 41:1262-1268 (1981).

67. N. J. E. Dowling, J. Guezennec, and D. C. White, Methods for insight into mechanisms of microbially influenced metal corrosion, in: "Biodeterioration," D. R. Hougton, R. N. Smith, and H. O. U. Eggins, eds., Elsevier Applied Science Publ., London (1988).

68. N. J. E. Dowling, J. Guezennec, and D. C. White, Facilitation of corrosion of stainless steel exposed to aerobic seawater by microbial biofilms containing both facultative and absolute anaerobes, in: "Microbial Problems in the Offshore Oil Industry," E. C. Hill, J. L. Sherman, and R. J. Watkinson, eds., John Wiley, Chichester (1986).

69. N. J. E. Dowling, J. Guezennec, M. L. Limoine, A. Tunlid, and D. C. White, Corrosion analysis of carbon steels affected by aerobic and anaerobic bacteria in mono and coculture using AC impedance and DC techniques, Corrosion 44:869 (1988).

70. B. D. Kerger, P. D. Nichols, C. A. Antworth, W. Sand, E. Bock, J. C. Cox, T. A. Langworthy, and D. C. White, Signature fatty acids in the polar lipids of acid producing *Thiobacilli*: methoxy, cyclopropyl, α-hydroxy-cyclopropyl and branched and normal monoenoic fatty acids, FEMS Microbiol. Ecology 38:67 (1986).

71. B. D. Kerger, P. D. Nichols, W. Sand, E. Bock, and D. C. White, Association of acid producing *Thiobacilli* with degradation of concrete: analysis by "signature" fatty acids from the polar lipids and lipopolysaccharide, J. Industrial Microbiol. 2:63 (1986).

72. A. Tunlid, B. H. Baird, M. B. Trexler, S. Olsson, R. H. Findlay, G. Odham, and D. C. White, Determination of phospholipid ester-linked fatty acids and poly-β-hydroxybutyrate for the estimation of bacterial biomass and activity in the rhizosphere of the rape plant *Brassica napus* (L), Canad. J. Microbiol. 31:1113 (1986).

73. A. Tunlid, H. A. J. Hoitink, C. Low, and D. C. White, Characterization of bacteria that suppress *Rhizoctonia* damping-off in bark compost media by analysis of fatty acid biomarkers, Appl. Environ. Microbiol. 55:1368 (1989).

74. P. D. Nichols, W. R. Mayberry, C. P. Antworth, and D. C. White, Determination of monounsaturated double bond position and geometry in the cellular fatty acids of the pathogenic bacterium *Francisella tularensis*, J. Clin. Microbiol. 21:738 (1984).

75. D. C. White, G. S. Smith, M. J. Gehron, J. H. Parker, R. H. Findlay, R. F. Martz, and H. L. Fredrickson, The ground water aquifer microbiota: biomass, community structure and nutritional status, Developments in Industrial Microbiol. 24:201 (1983).

76. G. A. Smith, J. S. Nickels, B. D. Kerger, J. D. Davis, S. P. Collins, J. T. Wilson, J. F. McNabb, and D. C. White, Quantitative characterization of microbial biomass and community structure in subsurface material: A prokaryotic consortium responsive to organic contamination, Canad. J. Microbiol. 32:104 (1986).

77. J. T. Wilson and B. H. Wilson, Biotransformation of trichlorethylene in soil, Appl. Environ. Microbiol. 49:242 (1985).

78. P. D. Nichols and D. C. White, Accumulation of poly-β-hydroxybutyrate in a methane-enriched halogenated hydrocarbon-degrading soil column, Proceed. Fourth Intl. Symp. Interaction between Sediments and Water, Melbourne, Australia (1987).

79. P. D. Nichols, G. A. Smith, C. P. Antworth, R. S. Hanson, and D. C. White, Phospholipid and lipopolysaccharide normal and hydroxy fatty acids as potential signatures for the methane-oxidizing bacteria, FEMS Microbiol. Ecology 31:327 (1985).

80. P. D. Nichols, J. M. Henson, C. P. Antworth, J. Parsons, J. T. Wilson, and D. C. White, Detection of a microbial consortium including type II methanotrophs by use of phospholipid fatty acids in an aerobic halogenated hydrocarbon-degrading soil column enriched with natural gas, Environ. Toxicol. Chem. 6:8997 (1987).

81. N. R. Wolin, The rumen fermentation: a model for microbial interactions in anaerobic ecosystems, Adv. Microbiol. Ecol. 3:49 (1979).

82. B. Rizza, A. N. Tucker, and D. C. White, Lipids of *Bacteroides melanogenicus*, J. Bacteriol. 101:84 (1970).

83. J. M. Henson, G. A. Smith, and D. C. White, Examination of thermophilic methane-producing digesters by analysis of bacterial lipids, Appl. Environ. Microbiol. 50:1428 (1985).

84. A. T. Mikell, T. J. Phelps, and D. C. White, Phospholipids to monitor microbial ecology in anaerobic digesters, in: "Methane from Biomass. A Systems Approach," W. Smith, ed., Elsevier, New York (1986).

85. J. V. Johnson and R. A. Yost, Tandem mass spectrometry for trace analysis, Anal. Chem. 57:758A (1985).

86. K. L. Busch and R. G. Cooks, Mass spectrometry of large, fragile, and involatile molecules, Science 218:247 (1982).

87. D. C. White, R. J. Bobbie, J. D. King, J. S. Nickels, and P. Amoe, Lipid analysis of sediments for microbial biomass and community structure, in: "Methodology for Biomass Determinations and Microbial Activities in Sediments," ASTM STP 673, American Society for Testing and Materials, pp. 87 (1979).

88. M. Seitz, D. B. Sauer, R. Burroughs, H. E. Mohr, and J. D. Hubbard, Ergosterol as a measure of fungal growth, Phytopathology 69:1202 (1979).

89. S. E. Matcham, B. R. Jordan, D. A. Wood, Estimation of fungal biomass in a solid substrate by three independent methods, Appl. Microbiol. Biotechnol. 21:108 (1985).

90. W. D. Grant and A. W. West, Measurements of ergosterol, diaminopimelic acid and glucosamine in soil: evaluation as indicators of microbial biomass, J. Microbiol. Methods 6:47 (1986).

91. W. F. Osswald, W. Holl, and E. F. Elstner, Ergosterol as a biochemical indicator of fungal infection in spruce and fir needles from different sources, Z. Naturforsch. 41c:542 (1986).

92. S. K. Maitra, M. C. Schotz, T. T. Yoshikawa, and L. B. Guze, Determination of lipid A and endotoxin in serum by mass spectroscopy, Proc. Natl. Acad. Sci. USA 75:3993 (1978).

93. N. Saddler and A. C. Wardlaw, Extraction, distribution and biodegradation of bacterial lipopolysaccharides in estuarine sediments, Antonie van Leeuwenhoek J. Microbiol. 46:27 (1980).

94. K. Kawamura and R. Ishiwatari, Tightly bound β-hydroxy acids in recent sediment, Nature 297:144 (1982).

95. L. Larsson, P. -A. Mardh, G. Odham, and G. Westerdahl, Detection of tuberculostearic acid in biological specimens by means of glass capillary gas chromatography--electron and chemical ionization mass spectrometry, utilizing selected ion monitoring, J. Chromatogr. 182:402 (1980).

96. G. Odham, A. Tunlid, L. Larsson, and P. -A. Mardh, Mass spectrometric determination of selected microbial constituents using fused silica and chiral glass capillary gas chromatography, Chromatographia 16:83 (1982).

97. L. Larsson, G. Odham, G. Westerdahl, and B. Olsson, Diagnosis of pulmonary tuberculosis by selected ion monitoring: improved analysis of tuberculostearate in sputum using negative-ion mass spectrometry, J. Clin. Microbiol. 25:893 (1987).

98. J. K. Volkman and R. B. Johns, The geochemical significance of positioned isomers of unsaturated acids from an intertidal zone sediment, Nature 267:693 (1977).

99. H. L. Frederickson, T. E. Cappenberg, and J. W. De Leeuu, Polar lipid ester-linked fatty acid composition of Lake Vechten seston: an ecological application of lipid analysis, FEMS Microbiology Ecology 38:381 (1986).

100. W. Michaelis and P. Albrecht, Molecular fossils of archaebacteria in kerogen, Naturwiss. 66:420 (1979).

101. S. C. Brassell, A. M. K. Wardroper, I. D. Thomson, J. R. Maxwell, and G. Eglinton, Specific acyclic isoprenoids as biological markers of methanogenic bacteria in marine sediments, Nature 290:693 (1981).

102. B. Chappe, W. Michaelis, P. Albrecht, and G. Ourisson, Fossil evidence for a novel series of archaebacterial lipids, Naturwiss. 66:522 (1982).

103. D. M. Ward, S. C. Brassell, and G. Eglinton, Archaebacterial lipids in hot-spring microbial mats, Nature 318:656 (1985).

104. G. G. Pauly and E. S. Van Vleet, Acyclic archaebacterial ether lipids in swamp sediments, Geochim. Cosmochim. Acta 50:1117 (1986).

105. J. J. Boon, W. I. C. Rijpstra, F. de Lange, and J. W. De Leeuu, Black sea sterol-- a molecular fossil for dinoflagellate blooms, Nature 277:125 (1979).

106. S. C. Brassell and G. Eglinton, Biogeochemical significance of a novel sedimentary C_{27}-stanol, Nature 290:579 (1981).

107. J. W. De Leeuu, W. I. C. Rijpstra, and P. A. Schenk, Free, esterified and residual bound sterols in Black Sea Unit I sediments, Geochim. Cosmochim. Acta 47:455 (1983).

108. N. Robinson, G. Eglinton, S. C. Brassel, and P. A. Cranwell, Dinoflagellate origin for sedimentary 4α-methyl steroids and $5\alpha(H)$-stanols, Nature 308:439 (1984).

INDEX

Derivatization, types (continued)
  t-butyldimethylsilyl ethers
      (TBDMS), 141, 143, 220
  butyl heptafluorobutyryl (BHFB),
      90-98
  isobutyl heptafluorobutyryl, 90
  methyl esters (*see* FAME)
  O-methyl oxime acetates, 77, 79
  pentafluorobenzyl (PFB), 141,
      166-170, 222, 226-228, 230
  per-O-acetylated aldononitrile,
      243
  trifluoroacetyl, 77, 79, 250
  trimethyl silyl, 77, 79, 242
Dissemination, 240
Deuterium labelling, 81, 91, 111,
      234, 243, 250

Electrophoresis, 112
Endospores, 9
Endotoxin, 2, 8, 149-158
  in dust samples, 232
Environmental microbiology, 259-268
      (*see also* Biofilms;
      Surfaces; and Sediments)
  hazardous waste, 2
  Corrosion, 2, 228, 230, 265 (*see
      also* Biofilms; Surfaces,
      fouling of)
Eubacteria, 4, 90

FAME (*see* Chemical markers, fatty
      acids, methyl esters)
Fatty acids, (*see* Chemical markers,
      fatty acids)
Fermentation, 267
Fungi (yeasts)
  Aspergillus, 239
  Candida
    arabinitol in diagnosis, 246-249
    blood culturing, 240
    Candida albicans, 4, 12, 94-96,
        189-191, 240
    Candida parapsilosis, 240
    Candida tropicalis, 240
    candidemia, 240
    diagnosis of candidiasis, 239-
        252
    metabolites, detection and
        identification, 241
    serodiagnosis, 241
      enzyme immunoassay detection,
          241
  Cochliobolus sativus, 228, 229
  Cryptococcus neoformans, 94, 95
  measurement of background amino
      acid racemization, 94
  sterols in, 260, 261
  Torulopsis glabrata, 239, 240

Gas chromatography (GC), 19-34, 59
  bleed, 22
  carrier gas, 19, 20, 22
  capillary columns, 20, 22
  bonded phase (crosslinked), 20, 21
  borosilicate glass, 21
  fused silica, 20, 21, 46, 61, 82
  short, 202
  column switching, 222, 223 (*see
      also* Gas chromatography,
      two-dimensional)
  detector, 19, 20, 31
    electron-capture, 33, 35, 91,
        144, 146, 219-222
    flame ionization (FID), 32, 33
    Fourier-transform spectrometry
        (FTIR), 34
    linear dynamic range, 32
    microwave induced plasma (MIPD),
        34
    minimum detectable quantity
        (MDQ), 32 (*see also* Limit of
        detection)
    nitrogen-phosphorus (NPD), 32,
        34
  selectivity, 32
  sensitivity, 32
  injection (*see* sample
      introduction)
  mobile phase, 20
  oven, 19
  packed column, 20
  sample introduction, 19, 22
    cold on-column, 25
    headspace (*see* Headspace gas
        chromatography)
    programmed temperature, 25
    solvent effect, 23
    split, 22, 24
    splitless, 23, 24
  stationary phase, 19, 20, 27, 30
    Carbowax, 29, 31
    chiral, 31, 90, 91, 96-97, 250
    Chirasil-val, 29, 31
    cyanopropyl silicone, 29
    methyl silicone, 28, 29, 31
    phenyl silicone, 29, 31
    polyethylene glycol, 31
    polysiloxane, 28
    silicone, 28
    trifluoropropyl silicone, 28, 29
  Two-dimensional, 232
Gene probe, 3
Glycocalyx, 9 (*see also* Capsule)
Gonorrhoea, 235 (*see also* Bacteria,
      Neisseria)
Gram positive, 4-9, 73, 90
Gram negative, 4-9, 73, 90, 149, 232,
      260, 266
Gram stain, 6, 9, 55, 56

Gram type, 6 (*see also* Pyrolysis, differentiation of Gram types)
Group polysaccharide, 4, 192

Headspace gas chromatography, 25, 101-120
  alcohols, 104
  amines, 104, 106, 114
  anaerobes, 26, 103, 125-135
  Clausius-Clapeyron law, 101-103
  dynamic analysis, 103
  fatty acids, 103, 104, 125-135
  gas sampling valve, 26, 103, 106
  GC procedures, 106-108
  hydrocarbons
  hydrogen detection, 103
  mass spectrometry, combined with, 101
  metabolic products of clostridia, 101-120, 125-135
  methane detection, 103, 105, 108
  sulfur compounds, 104, 105, 108, 113-119
  Raoult's law, 101-103
  static analysis, 103

Internal standard, 82, 93, 243, 250

Labelling,
  isotopic, 262 (*see also* Deuterium labelling)
Library search, 66, 181
Limit of detection, 77
Limulus lysate assay, 12, 235
Linkage analysis, (*see* Structural analysis)
Lipids, 259-268 (*see also* Chemical markers, fatty acids)
Lipopolysaccharide (LPS), 4, 9, 73, 149, 260, 261
  Lipid A, 2, 9, 75, 149-158, 260
  Chromobacterium violaceum, 153, 154
  Enterobacteriaceae, 152
  Escherichia coli, 152, 153, 157
  Neisseria gonorrhoeae, 152
  Proteus mirabilis, 155
  Salmonella typhimurium, 152
  Synthetic, 155
  Xanthomonas sinensis, 153, 154
  O-antigen, 7, 9, 73
Lipoteichoic acid (LTA), 6, 7, 73

Mammalian tissues and fluids, 75-78, 95-97
Mass spectrometry (MS), 34-47
  benchtop, 35, 46
  carrier gas flows in, 37
  detectors, 20, 45
    electron multiplier, 45

Mass spectrometry, detectors (continued)
    Faraday-cup collector, 45
    photographic, 45
  double-focusing, 43
  Fourier-transform, 44, 151
  ion cyclotron resonance, 44
  ion source, 20, 37
  ion trap, 44, 201-213, 223
  ionization, 37
    chemical (CI), 20, 37, 39, 195, 196, 225, 226, 263
    negative, 40, 91, 92, 144, 146, 219, 221, 225-228, 232, 233, 263
    positive, 38, 225
    collision activation dissociation (CAD), 163-177
    desorption, 41, 46
    electron impact (EI), 20, 37, 39, 83, 164, 195, 196
    fast atom bombardment (FAB), 42, 150, 151
    field, 40, 43, 150
    laser desorption (LD), 42, 150, 151, 153
    plasma desorption (PD), 42, 150
    secondary ion, 42, 150-151
    soft, 46, 150, 202
  magnetic sector, 43
  liquid chromatography, 228, 229
  mass analyzed kinetic energy spectrometry (MIKE), 164
  mass analyzer, 20
  mass selective detector, 224 (*see also* quadrupole MS)
  mass separation, 43
  pumps,
    diffusion, 36
    rotary, 36
    turbomolecular, 36
  quadrupole MS, 36, 43, 44, 224, 2255
  sample inlet, 37
  selected ion monitoring (SIM), 46, 75, 78, 84, 94, 114, 219, 228, 232-235, 244, 263
  tandem (MS-MS), 45, 226
    of fatty acids, 165-167, 172
  time-of-flight (TOF), 43, 1501
  triple quadrupole, 226, 228
  vacuum system, 36
Mass spectrum, 20, 35, 38
Metabolic products, 4, 10, 101
Microbial identification system for fatty acids, 64-67, 201, 260, 261
Microscopy, 259
Morphology, 3, 53, 55
Mucopeptide, 6 (*see also* Peptidoglycan)

Murein, 6 (*see also* Peptidoglycan)

Nuclear magnetic resonance (NMR), 141-143
Nutritional status of organisms, 262

Opportunistic infections, 239
Outer membrane, 6, 7, 149

Pattern recognition, (*see* Statistical analysis; Similarity index)
Peptidoglycan, 4, 6, 7, 11, 13, 73, 89, 90, 185-191
Plant roots, microorganisms on, 265
Porins, 8
Prokaryote, 6
Proteins, 89
Pyrolysis, 10, 12, 27, 179-198, 201-213
  N-acetylglucosamine, 186
  N-acetylmuramic acid, 186
  acetamide, 180, 185-191
  chemical markers generated by, 179-181, 185-196
  chitin, 189
  culture conditions, 181-182
  Curie-point, 27, 28, 201-213
  dianhydroglucitol from glucitol in group B streptococci, 180, 192-195
  differentiation of Gram types, 185-191
  DNA and RNA, 186, 187
  furfuryl alcohol, 180, 185-191
  diglycerides, 203
  GC columns in, 184-185
  mass spectrometry (Py-MS), 180, 201
  muramyl dipeptide, 186
  peptidoglycan, 186
  propionamide, 185, 186
  resistively heated, 27, 28, 182-184
  variability, 181-185

Racemization, 40, 91, 94
Renal dysfunction, associated with candidiasis, 246

Sarcina, 55
Sediments, microbial communities in, 264
Sequence analysis (*see* Structural analysis)
Similarity index, 66
Spirochaeta, 55
Stationary phases, 27 (*see also* Gas chromatography, stationary phases)

Statistical analysis, 66, 126, 128, 181, 196, 212, 261
Structural analysis, 2
  of lipid A, 150
  of mycolic acids, 167
Sugars (*see* Chemical markers, sugar)
Surfaces,
  fouling of, 228, 230, 259, 265
    (*see also* Corrosion)

Teichoic acid, 6, 7, 8, 74
Teichuronic acid, 6
Trace detection, 2, 75, 95
Tuberculosis, (*see* Mycobacterium)

Water pollution, 266

Yeasts (*see* Fungi)